James Bicheno Francis

Lowell Hydraulic Experiments

James Bicheno Francis

Lowell Hydraulic Experiments

ISBN/EAN: 9783744743556

Printed in Europe, USA, Canada, Australia, Japan

Cover: Foto ©berggeist007 / pixelio.de

More available books at **www.hansebooks.com**

PREFACE TO THE SECOND EDITION.

Since the first edition of this work appeared, in 1855, the manufacturing corporations at Lowell, lessees of the water-power furnished by the Merrimack River at that point, have surrendered their leases and taken others containing new provisions for the purpose of more fully protecting all parties in the enjoyment of their respective rights; this has rendered necessary a new and elaborate series of experiments for the purpose of perfecting the method of gauging the flow of water in open channels by the use of loaded tubes. Some experiments had been made on this subject at Lowell before the publication of the first edition, the principal results of which were given; the later experiments are, however, so much more complete, and have been made under circumstances so much more favorable, that it has been found necessary to rewrite, entirely, the chapter on that subject.

The general use at Lowell of the *Diffuser*, an apparatus for utilizing the power usually lost in turbines, from the water leaving them with a considerable velocity, has created much interest in Venturi's tube, the action in which involves the same principles as the Diffuser. Experiments on Venturi's tube had been previously made only when discharging into the air; it appeared highly probable that greater results might be obtained if the tube was submerged, so as to discharge under water. Experiments made under these circumstances, and detailed at length in this edition, indicate a considerably greater flow than had been previously obtained.

The author takes this opportunity of acknowledging his obligations to Mr. Uriah A. Boyden of Boston, for useful suggestions during the last twenty-five years, on almost every subject discussed in this volume. Also to Mr. John Newell, now of Detroit, Michigan, to whom he is much indebted for assistance in the execution and reduction of some of the most important series of experiments, and to whose fidelity the precision attained in the results is in no small degree due. Also to Mr. Joseph P. Frizell, now of Davenport, Iowa, to whom he is indebted for assistance in some points involving the higher mathematics.

Lowell, Mass., *March*, 1868.

TABLE OF CONTENTS.

INTRODUCTION.

PART I.

EXPERIMENTS ON HYDRAULIC MOTORS.

PART II.

EXPERIMENTS ON THE FLOW OF WATER OVER WEIRS, AND IN SHORT RECTANGULAR CANALS.

A Method of Gauging the Flow of Water in Open Canals of uniform rectangular Section, and of Short Length.

Experiments on the Flow of Water through Submerged Orifices and Diverging Tubes.

TABLES.

PART I.

PART II.

INTRODUCTION.

The northern regions of the United States of North America, probably possess a greater amount of water-power than any other part of the world of equal extent, and the active and inventive genius of the American people, combined with the very high price of labor, has had a powerful influence in bringing this power into use. Nevertheless, the water-power is so vast, compared with the population, that only a small portion of it has, up to this time, been applied to the purposes of man. It was estimated, not long since, that the total useful effect derived from water-power in France, was about 20,000 horse-power. An amount of power far exceeding this, is already derived from the Merrimack River and its branches, in Massachusetts and New Hampshire. What must be the amount of the population and wealth of the Northern States, when the other rivers that water them are equally improved?

One of the earliest and most successful efforts to bring into use, in a systematic manner, one of the larger water-powers, was made at Lowell in Massachusetts; where, in 1821, a number of farms situated near Pawtucket Falls on the Merrimack River, were purchased by several capitalists of Boston, who obtained a charter from the State of Massachusetts under the name of *The Merrimack Manufacturing Company*. In 1826, the property was transferred to the *Proprietors of the Locks and Canals on Merrimack River*, a corporation chartered in 1792 for the purpose of improving the navigation of the Merrimack River. Previously to the transfer, the Merrimack Manufacturing Company had erected a dam of about 950 feet in length, at the head of Pawtucket Falls, and had also enlarged the Pawtucket Canal, which was originally constructed, previously to the year 1800, by the Proprietors of the Locks and Canals on Merrimack River, for the purposes of navigation. Subsequently to the enlargement, however, this canal has been used both for purposes of navigation, and to supply water to the wheels of numerous manufacturing establishments.

The dam at the head of Pawtucket Falls, in the ordinary state of the river, deadens the current of the river for about 18 miles, forming, in low water, a reservoir of about 1120 acres; this extensive reservoir is of great value in very low stages of the river, as it affords space for the accumulation of the flow of the river during the night, when the manufactories are not in operation. This accumulation is subsequently drawn off, together with the natural flow of the river, during the usual working hours.

The total fall of the Merrimack River at Pawtucket Falls, in ordinary low water, is about 35 feet, of which about 2 feet is lost in consequence of the descent in the canals, leaving a net fall of about 33 feet. About $\frac{1}{4}$ of the water is used on the entire fall, and the remainder is used twice over, on falls of about 14 and 19 feet respectively. The water-power has been granted by the Proprietors of the Locks and Canals on Merrimack River, in definite quantities called *Mill Powers*, which are equivalent to a gross power of a little less than 100 horse-power each. Grants have been made to eleven manufacturing companies, who have an aggregate capital, somewhat exceeding thirteen millions of dollars. Thus, to the Merrimack Manufacturing Company, there have been granted $21\frac{2}{3}$ mill powers, each of which consists of the right to draw, for 15 hours per day, 25 cubic feet of water per second on the entire fall. Up to this time, there have been granted at Lowell $139\frac{11}{30}$ mill powers, or a total quantity of water equal to 3595.933 cubic feet per second. A large portion of this water is used on turbines of a very superior description, and nearly all the remainder, on breast wheels of good construction, a portion of which, however, do not use quite the whole of the fall on which they are placed. We may, however, assume that, upon an average, a useful effect is derived, equal to $\frac{2}{3}$ of the total power of the water expended. Calling the fall 33 feet, and the weight of a cubic foot of water 62.33 pounds, we shall have for the effective power derived from the water-power granted by the Proprietors of the Locks and Canals on Merrimack River at Lowell.

$$\frac{3595.933 \times 62.33 \times 33 \times \frac{2}{3}}{550} = 8965.4 \text{ horse-power.}$$

In consequence of the success attending the improvement of the water-power at Lowell, several other extensive water-powers in New England have been brought into use in a similar manner. Some of these undertakings have been quite successful, whilst with others, as yet only partially developed, the success has not been so decided.

The great abundance of water-power in this country has had a strong tendency to encourage its extravagant use; the machines used in the manufactories are usually great consumers of power; the ability of a machine to turn off the greatest quantity of work with the least manual labor, and in the *least time*, has been the point mainly considered; and whether it required a greater or less amount of power, has been a secondary consideration.

The engineering operations connected with the water-power at Lowell, have frequently demanded more definite information on certain points in hydraulics, than was to be found in any of the publications relating to that science; and hence has arisen the necessity, from time to time, of making special experiments to supply the required information. Whenever such emergencies have arisen, the officers who have the general care of the interests of the several corporations, with a liberality founded on enlarged views of the true interests of the bodies they represent, have always been willing to defray such expenses as were necessary, in order that the experiments might be made in a satisfactory manner.

The experiments recorded in the following pages, are a selection from those made by the author, in the discharge of his duty, as the Engineer of the Corporations at Lowell. They may be divided into two classes, namely, *First*, those on hydraulic motors, and, *second*, those on the flow of water over weirs, and in short rectangular canals. Combined with the description of the experiments, there are also given some other investigations, which may appear somewhat out of place, but which, from their utility or novelty, will be found interesting to many persons who have cultivated the science of hydraulics.

The unit of length adopted in this work, is the English foot according to a brass standard measure made by Cary of London, now in the possession of the Lowell Machine Shop.

HYDRAULIC EXPERIMENTS.

PART I.

EXPERIMENTS ON HYDRAULIC MOTORS.

EXPERIMENTS UPON THE TREMONT TURBINE.

1. Until within a few years, the water-wheels in use in the principal manufacturing establishments in New England, were what are there generally called *breast wheels*, sometimes known also by the name of *pitch back wheels*. They are the same in principle as the *overshot-wheel*, the useful effect being produced, almost entirely, by the simple weight of the water in the buckets, and differing only from the *overshot-wheel* in this, that the water is not carried entirely over the top of the wheel, but is let into the buckets near the top, but on the opposite side from that adopted for the overshot-wheel. An apron, fitting as closely as practicable to the wheel, is used to prevent the water leaving the buckets, until it reaches very nearly the bottom of the wheel.

In Lowell, these wheels have been constructed principally of wood, many of them of very large dimensions. Those in the mills of the Merrimack Manufacturing Company, for instance, are thirty feet in diameter, with buckets twelve feet long. Four of the mills belonging to this company, have two such wheels in each of them.

Until the year 1844, the breast wheel, as above described, was considered here the most perfect wheel that could be used. Much prejudice existed here, as elsewhere, against the reaction wheels; a great number of which had, however, been used throughout the country, in the smaller mills, and with great advantage; for, although they usually gave a very small effect in proportion to the quantity of water expended, their cheapness, the small space required for them, their greater velocity, being less impeded by backwater, and not requiring expensive wheelpits of masonry, were very important considerations; and in a country where water power is so much more abundant than capital, the economy of money was generally of greater importance than the saving of water.

A vast amount of ingenuity has been expended by intelligent millwrights, on these wheels; and it was said, several years since, that not less than three hundred patents relating to them, had been granted by the United States Government. They continue, perhaps as much as ever, to be the subject of almost innumerable modifications. Within a few years, there has been a manifest improvement in them, and there are now several varieties in use, in which the wheels themselves are of simple forms, and of single pieces of cast-iron, giving a useful effect approaching sixty per cent. of the power expended.

2. The attention of American engineers was directed to the improved reaction water-wheels in use in France and other countries in Europe, by several articles in the Journal of the Franklin Institute; and in the year 1843, there appeared in that journal, from the pen of Mr. Ellwood Morris, an eminent engineer of Pennsylvania, a translation of a French work, entitled, *Experiments on water-wheels having a vertical axis, called turbines, by Arthur Morin, Captain of Artillery, etc. etc.* In the same journal, Mr. Morris also published an account of a series of experiments, by himself, on two turbines constructed from his own designs, and then operating in the neighborhood of Philadelphia.

The experiments on one of these wheels, indicate a useful effect of seventy-five per cent. of the power expended, a result as good as that claimed for the practical effect of the best overshot-wheels, which had, heretofore, in this country, been considered unapproachable, in their economical use of water.

3. In the year 1844, Uriah A. Boyden, Esq., an eminent hydraulic engineer of Massachusetts, designed a turbine of about seventy-five horse-power, for the Picking House of the Appleton Company's cotton-mills, at Lowell, in Massachusetts; in which wheel, Mr. Boyden introduced several improvements, of great value.

The performance of the Appleton Company's turbine, was carefully ascertained by Mr. Boyden, and its effective power, exclusive of that required to carry the wheel itself, a pair of bevel gears, and the horizontal shaft carrying the friction pulley of a Prony dynamometer, was found to be seventy-eight per cent. of the power expended.

4. In the year 1846, Mr. Boyden superintended the construction of three turbines of about one hundred and ninety horse-power each, for the same company. By the terms of the contract, Mr. Boyden's compensation depended upon the performance of the turbines, and it was stipulated that two of them should be tested. The contract also contained the following clause, "and if the mean power derived from

these turbines be seventy-eight per cent. of the power of water expended, the Appleton Company to pay me twelve hundred dollars for my services, and patent rights for the apparatus for these mills; and if the power derived be greater than seventy-eight per cent., the Appleton Company to pay me, in addition to the twelve hundred dollars, at the rate of four hundred dollars for every one per cent. of power, obtained above seventy-eight per cent." In accordance with the contract, two of the turbines were tested, a very perfect apparatus being designed by Mr. Boyden for the purpose, consisting, essentially, of a Prony dynamometer to measure the useful effects, and a weir to gauge the quantity of water expended.

5. A great improvement in the mode of conducting hydraulic experiments was here adopted, in making each set of observations continuous, the time of each observation being noted; thus, the observer who noted the height of the water above the wheel, recorded regularly, say every thirty seconds, the time and the height; and so with the other observers, the recorded times furnishing the means of afterwards identifying simultaneous observations.

6. The observations were put into the hands of the author, for computation, who found that the mean maximum effective power of the two turbines tested, was eighty-eight per cent. of the power of the water expended.

According to the terms of the contract, this made the compensation for engineering services, and patent rights for these three wheels, amount to fifty-two hundred dollars, which sum was paid by the Appleton Company without objection.

7. These turbines have now been in operation about eight years, and their performance has been, in every respect, entirely satisfactory. The iron-work for these wheels was constructed by Messrs. Gay and Silver, at their machine shop at North Chelmsford, near Lowell; the workmanship was of the finest description, and of a delicacy and accuracy altogether unprecedented in constructions of this class.

8. These wheels, of course, contained Mr. Boyden's latest improvements, and it was evidently for his pecuniary interest that the wheels should be as perfect as possible, without much regard to cost. The principal points in which one of them differs from the constructions of Fourneyron, are as follows.

9. *The wooden flume, conducting the water immediately to the turbine, is in the form of an inverted truncated cone, the water being introduced into the upper part of the cone, on one side of the axis of the cone (which coincides with the axis of the turbine) in such a manner, that the water, as it descends in the cone, has a gradually increasing velocity, and a spiral motion; the horizontal component of the spiral motion being in the direction of the motion of the wheel.* This horizontal motion is derived from the necessary velocity with which the water enters the truncated cone; and the arrangement is such that, if perfectly proportioned, there would be no loss of power between the nearly still water in the principal

penstock and the guides or leading curves near the wheel, except from the friction of the water against the walls of the passages. It is not to be supposed that the construction is so perfect as to avoid all loss, except from friction; but there is, without doubt, a distinct advantage in this arrangement over that which had been usually adopted, and where no attempt had been made to avoid sudden changes of direction and velocity.

10. *The guides, or leading curves, are not perpendicular, but a little inclined backwards from the direction of the motion of the wheel, so that the water, descending with a spiral motion, meets only the edges of the guides.* This leaning of the guides has also another valuable effect; when the regulating gate is raised only a small part of the height of the wheel, the guides do not completely fulfil their office of directing the water, the water entering the wheel more nearly in the direction of the radius, than when the gate is fully raised; by leaning the guides, it will be seen that the ends of the guides, near the wheel, are inclined, the bottom part standing further forward, and operating more efficiently in directing the water, when the gate is partially raised, than if the guides were perpendicular.

11. In Fourneyron's constructions, a garniture is attached to the regulating gate, and moves with it, for the purpose of diminishing the contraction; this, considered apart from the mechanical difficulties, is probably the best arrangement; to be perfect, however, theoretically, this garniture should be of different forms for different heights of gate; but this is evidently impracticable.

In the Appleton Turbine, the garniture is attached to the guides, the gate (at least the lower part of it) being a simple thin cylinder. By this arrangement, the gate meets with much less obstruction to its motion than in the old arrangement, unless the parts are so loosely fitted as to be objectionable; and it is believed that the coefficient of effect, for a partial gate, is proportionally as good as under the old arrangement.

12. *On the outside of the wheel is fitted an apparatus named, by Mr. Boyden, the Diffuser.* The object of this extremely interesting invention, is to render useful a part of the power otherwise entirely lost, in consequence of the water leaving the wheel with a considerable velocity. It consists, essentially, of two stationary rings or discs, placed concentrically with the wheel, having an interior diameter a very little larger than the exterior diameter of the wheel; and an exterior diameter equal to about twice that of the wheel; the height between the discs, at their interior circumference, is a very little greater than that of the orifices in the exterior circumference of the wheel, and at the exterior circumference of the discs, the height between them is about twice as great as at the interior circumference; the form of the surfaces connecting the interior and exterior circumferences of the discs, is gently rounded, the first elements of the curves, near the interior circumferences, being nearly horizontal. There is con-

sequently, included between the two surfaces, an aperture gradually enlarging from the exterior circumference of the wheel, to the exterior circumference of the diffuser. When the regulating gate is raised to its full height, the section, through which the water passes, will be increased by insensible degrees, in the proportion of one to four, and if the velocity is uniform in all parts of the diffuser at the same distance from the wheel, the velocity of the water will be diminished in the same proportion; or its velocity on leaving the diffuser, will be one fourth of that at its entrance. By the doctrine of living forces, the power of the water in passing through the diffuser must, therefore, be diminished to one sixteenth of the power at its entrance. It is essential to the proper action of the diffuser, that it should be entirely under water; and the power rendered useful by it, is expended in diminishing the pressure against the water issuing from the exterior orifices of the wheel; and the effect produced, is the same as if the available fall under which the turbine is acting, is increased a certain amount. It appears probable that a diffuser of different proportions from those above indicated, would operate with some advantage without being submerged. It is nearly always inconvenient to place the wheel entirely below low-water-mark; up to this time, however, all that have been fitted up with a diffuser, have been so placed; and, indeed, to obtain the full effect of a fall of water, it appears essential, even when a diffuser is not used, that the wheel should be placed below the lowest level to which the water falls in the wheelpit, when the wheel is in operation.

The action of the diffuser depends upon similar principles to that of diverging conical tubes, which, when of certain proportions, it is well known, increase the discharge; the author has not met with any experiments on tubes of this form, discharging under water, although, there is good reason to believe, that tubes of greater length and divergency would operate more effectively under water, than when discharging freely in the air; and that results might be obtained, that are now deemed impossible by most engineers.

Experiments on the same turbine, with and without a diffuser, show a gain in *the coefficient of effect*, due to the latter, of about three per cent. By the principles of living forces, and assuming that the motion of the water is free from irregularity, the gain should be about five per cent. The difference is due, in part at least, to the unstable equilibrium of water, flowing through expanding apertures; this must interfere with the uniformity of the velocities of the fluid streams, at equal distances from the wheel.

13. *Suspending the wheel from the top of the vertical shaft, instead of running it on a step at the bottom.* This had been previously attempted, but not with such success as to warrant its general adoption. It has been accomplished with complete success by Mr. Boyden, whose mode is, to cut the upper part of the shaft into a series of necks,

and to rest the projecting parts upon corresponding parts of a box. A proper fit is secured by lining the box, which is of cast-iron, with babbitt metal, a soft metallic composition consisting, principally, of tin; the cast-iron box is made with suitable projections and recesses to support and retain the soft metal, which is melted and poured into it, the shaft being at the same time in its proper position in the box. It will readily be seen that a great amount of bearing surface can be easily obtained by this mode, and also, what is of equal importance, it may be near the axis; the lining metal, being soft, yields a little if any part of the bearing should receive a great excess of weight. The cast-iron box is suspended on gimbals, similar to those usually adopted for mariners' compasses and chronometers, which arrangement permits the box to oscillate freely in all directions, horizontally, and prevents, in a great measure, all danger of breaking the shaft at the necks, in consequence of imperfections in the workmanship, or in the adjustments. Several years' experience has shown, that this arrangement, carefully constructed, is all that can be desired; and that a bearing thus constructed, is as durable, and can be as readily oiled, and taken care of, as any of the ordinary bearings in a manufactory.

14. The buckets are secured to the crowns of the wheel in a novel, and much more perfect manner, than had been previously used; the crowns are first turned to the required form, and made smooth; by ingenious machinery devised for the purpose, grooves are cut with great accuracy in the crowns, of the exact curvature of the buckets; mortices are cut through the crowns, in several places in each groove; the buckets, or floats, are made with corresponding tenons, which project through the crowns, and are riveted on the bottom of the lower crown, and on the top of the upper crown; this construction gives the requisite strength and firmness, with buckets of much thinner iron than was necessary under any of the old arrangements; it also leaves the passages through the wheel entirely free from injurious obstructions.

15. Mr. Boyden has also designed a large number of turbines for different manufacturing establishments in New England, many of them under contracts similar to that with the Appleton Company, and has accumulated a vast number of valuable experiments and observations upon them, which, it is to be hoped, he will find time to prepare for publication; as such opportunities but rarely occur to engineers so able to profit by them.

16. In the year 1849, the Manufacturing Companies at Lowell purchased of Mr. Boyden, the right to use all his improvements relating to turbines and other hydraulic motors. Since that time it has devolved upon the author, as the chief engineer of these companies, to design and superintend the construction of such turbines as might be wanted for their manufactories, and to aid him in this important undertaking, Mr. Boyden has communicated to him copies of many of his designs for turbines, together

with the results of experiments upon a portion of them; he has communicated, however, but little theoretical information, and the author has been guided, principally, by a comparison of the most successful designs, and such light as he could obtain from writers on this intricate subject.

17. The first designs, prepared by the author, after the arrangement with Mr. Boyden was entered into, were for four turbines of essentially the same dimensions; namely, two for the Suffolk Manufacturing Company, and two for the Tremont Mills, for the purpose of furnishing power for the cotton-mills of these companies at Lowell. These turbines were constructed at the Lowell Machine Shop, and were completed in January, 1851.

For the purpose, principally, of estimating the success of these turbines, one of them was fitted up with a complete apparatus for measuring its power, and gauging the quantity of water discharged; the gauging apparatus was afterwards used to make the experiments on the discharge of water over weirs of different proportions, for the purpose of determining, practically, some of the relations required to be known, in order to compute the flow of water through such apertures.

DESCRIPTION OF THE TURBINE.

18. The water is conducted from the principal feeder to the mills at Lowell, called the Northern Canal, by an arched canal, or penstock, about ninety feet in length. The forebay, inside the wheel-house, is constructed of masonry, and has a general width of twenty feet, and a depth of water of fourteen feet; the channels through which the water passes, are so capacious, that the loss of fall in passing from the Northern Canal to the forebay, is scarcely sensible. During the experiments, however, the head of the penstock was partially closed by gates, so that there was a sensible fall at that time.

The entrance of the arched canal is protected by a coarse rack, or grating, for the purpose of preventing large floating substances from entering the forebay; each turbine is also separately guarded by a fine rack, placed in the forebay, which prevents the entrance into the turbine of all floating substances that might be injurious. Both racks are made of large extent, to avoid sensible loss of head to the water in passing through them.

The extreme rigor of the New England winter renders it necessary to afford to water-wheels of all descriptions, complete protection from the cold. The result is, that less interruption from frost is experienced, than in many milder climates. The wheel-house, in which these turbines are placed, is a substantial brick building, well warmed in the winter by steam.

After passing the turbines, the water is conducted by an arched canal, or raceway, about nine hundred feet in length, to the lower level of the Western Canal, which serves as a feeder to the Mills of the Lawrence Manufacturing Company.

19. Plate I. is a vertical section through the centre of the turbine, and the axis of the supply pipe.

Plate II. is a plan of the turbine, and wheelpit.

Plate III., Figure 1, is a plan of nearly one fourth part of the disc and wheel. Figure 2 is a plan of the whole wheel, the guides, and garniture. Figure 3 is a vertical section through both crowns of the wheel.

The same letters indicate the same parts, in all these three plates.

20. *A, the forebay,* in which the level of the water is nearly the same as in the Northern Canal; it is represented at the usual working height.

21. *B, the surface of the water in the wheelpit,* represented at the lowest height at which the turbine is intended to operate.

22. *C, the masonry of the wheelpit.* The faces towards the wheel, are of granite ashlar work, in blocks containing, generally, from ten to forty cubic feet. The backing is of hard mica slate. The capping course, shown particularly on Plate II., is neatly dressed on its upper surface. The whole is compactly laid in hydraulic cement.

23. *D, the floor of the wheelpit.* This floor sustains the weight of part of the supply pipe, and of part of the water in it, and all the rest of the apparatus, excepting the wheel itself and the vertical shaft, which are supported by beams and braces, directly from the side walls of the wheelpit. It was necessary that the floor should have sufficient stiffness to resist the great upward pressure which takes place when the wheelpit is kept dry by pumps, in order to permit repairs to be made. The walls of the wheelpit are built upon the floor;— there was, consequently, no danger of the whole floor being pressed upwards, but the great width of the pit, (twenty-four feet,) would allow the floor to yield in the centre, unless it had great stiffness.

To meet these requirements, three cast-iron beams are placed across the pit, the ends extending about a foot under the walls, on each side; on these are laid thick planks which are firmly secured to the cast-iron beams, by bolts. To protect the thick planking from being worn out by the constant action of the water, they are covered with a flooring of one inch boards, which can be easily renewed when necessary.

24. *E, the wrought iron supply pipe.* This is constructed of plate iron, ⅜ inch thick, riveted together in a similar manner to steam boilers. The horizontal part is nine feet in diameter, the curved part gradually diminishes in diameter, to its junction with the upper curb. The upper end of the supply pipe is terminated by a cast-iron ring *F*, turned smooth on the face, to receive the wooden head gate. The supply pipe is also furnished with the man hole and ventilating pipe *G*, and the leak box *H*. The use of

the latter is, to catch the leakage of the head gate, whenever it is closed for repairs of the wheel; at such times, the leakage is carried off into the raceway, below the wheelpit, by a six inch pipe, furnished with a valve which can be opened and shut at pleasure.

25. *I, the cast-iron curbs.* These conduct the water from the wrought iron supply pipe, to the disc *K*. The curbs are made in four parts, for the convenience of the founder. The surfaces at which they are joined, are turned true in a lathe, packed with red lead, and bolted together with bolts one and a half inches diameter, placed about six inches apart. The general thickness of the iron is one and a quarter inches. The flanges are two inches thick. The upper curb has a projection cast on it, to receive the disc pipe. The lower curb is finished on all sides; the outside, to permit the regulating gate to be moved up and down easily; the inside, to present a smooth surface to the water, and to match accurately with the garniture *L*.

The curbs are supported from the wheelpit floor by four columns, two of which are shown at *N N*, plate I., resting on the cast-iron beam *O*; this is placed on the floor, for the purpose of distributing the weight. The centres of the columns are thirteen inches from the outside circumference of the wheel. The beams *N'* rest immediately upon the columns, and the curb upon the beams, the latter projecting over the columns far enough for that purpose. The beams *N'* also act as braces from the wheelpit wall to the curb, and are strongly bolted at each end.

26. *K, the disc.* This is of cast-iron, one and a half inches thick, and is turned smooth on the upper surface, and also on its circumference. It is suspended from the upper curb, by means of the disc pipes *M M*. The disc carries on its upper surface thirty-three guides, or leading curves, for the purpose of giving the water, entering the wheel, proper directions. They are made of Russian plate iron, one tenth of an inch in thickness, secured to the disc by tenons, passing through corresponding mortices, cut through the disc, and are riveted on the under-side. The upper corners of the guides, near the wheel, are connected by the garniture *L*, which is intended to diminish the contraction of the streams entering the wheel, when the regulating gate is fully raised. The garniture is composed of thirty-three pieces of cast-iron, or one to fill each space between the guides; these pieces of cast-iron are, necessarily, of irregular form; for a top view of them see *L*, plate III., figure 2. They are also shown in section at plate I. They are carefully fitted to fill the spaces between the guides; above the top of the guides, the adjoining pieces are in contact; they are strongly riveted to the guides, and to each other. After they were all fitted and riveted, the disc was put in a lathe, and the top, the periphery, and a part of the inside of the garniture, were turned off, so that it would fit accu-

rately, but easily, to the corresponding part of the lower curb. The disc is not fastened to the lower curb, but is retained in its place, horizontally, by the latter.

27. *M M, the disc pipe.* The disc is fastened to the bottom of the disc pipe by fifteen tap screws, one and a quarter inches in diameter. As there is a vertical pressure on the disc, due to the pressure of the whole head, on its horizontal area, the disc pipe and its fastenings require to be very strong. The pipe is eight and a half inches diameter, inside, or one and a half inches larger than the shaft passing through it, and is one and a quarter inches thick. The upper flange is furnished with adjusting screws, by which the weight is supported upon the upper curb, and which afford the means of adjusting the height of the disc. The escape of water between the upper curb and the upper flange of the disc pipe, is prevented by a band of leather on the outside, which is retained in its place by the wrought iron ring *P*. This ring is made in two segments. The top of the disc pipe, just below the upper flange, has two projections, or wings, which fit into corresponding recesses in the top of the curb; these are to prevent the disc from rotating in the opposite direction to the wheel, to which there is a powerful tendency, arising from the reaction of the water issuing from the guides.

28. *R R, the regulating gate.* This is represented on the section, at plate I., as fully raised, and in this position the wheel would be giving its full power. The gate is of cast-iron, the cylindrical part is one inch thick, the upper part of the cylinder is stiffened by a rib, to which are attached three brackets, one of which is shown at *S*, plate I., and the two others at *S S*, plate II. To these brackets are attached wrought iron rods, by which the gate is raised and lowered. The brackets are attached to the gate at equal distances, and therefore the rods support equal parts of its weight. To one of the rods is attached the rack *V*. The other two rods are attached, by means of links, to the levers *T T*, plate II. The other ends of these levers carry geared arch heads, into which, and into the rack *V*, work three pinions, *W*, of equal pitch and size, fastened to the same shaft. As the fulcrums of the levers *T T*, plate II., are exactly in the middle, between the pitch lines of the arch heads and the points to which the rods are attached, it will be seen, that by the revolution of the pinion shaft, the gate must be moved up or down, equally on all sides. The shaft on which the pinions are fastened, is driven by the worm wheel *X*, plates I. and II.; this is driven by the worm *a*, either by the governor *Y*, or the hand wheel *Z*. The shaft on which the worm *a* is fastened, is furnished with movable couplings, which, when the speed gate is at any intermediate points between its highest and lowest positions, are retained in place by spiral springs; in either of the extreme positions, the couplings are separated by means of a lever, moved by pins in the rack *V*; by this means both the

regulator and hand wheel are prevented from moving the gate in one direction, when the gate has attained either extreme position. If, however, the regulator or hand wheel should be moved in the opposite direction, the couplings would catch, and the gate would be moved. The weight of the gate is counterbalanced by weights attached to the levers $T\,T'$, and by the intervention of a lever to the rack V; by this arrangement, both the governor and hand wheel are required to operate, with only the force necessary to overcome the friction of the apparatus.

29. *b b The wheel.* This consists of a central plate of cast-iron, and of two crowns, *c c*, of the same material, to which the buckets are attached. The central plate and the crowns are turned accurately in a lathe, for the purpose of balancing them, and also to diminish, as much as possible, the resistances in moving rapidly through the water. The lower crown is fastened to the central plate, as shown at figures 1 and 3, plate III. These figures also show, at *c c*, the form of the crowns; the upper and lower crowns are precisely alike; they are nine and a half inches wide. At the inner edge, and at the circumference, the thickness is 0.625 inches, and at 5.5 inches from the inner edge, where they have the greatest thickness, they are one inch thick.

By reference to figure 1, plate III., it will be seen that the buckets do not extend to the circumference of the crowns. In the direction of the radius, the ends of the buckets are 0.25 inches from the circumference. This is for the purpose of permitting the wheel to be handled with less danger of injuring the ends of the floats; as these are filed down to an edge, they would be very likely to be damaged during the construction of the wheel, if they were not guarded by the slight projection of the crowns. This construction also enables the grooves in the crowns to afford more perfect support to the ends of the buckets, and also permits a tenon to be nearly at the extremity of the bucket.

The buckets are forty-four in number, and are of the form represented on plate III., figure 1. They are made of plate iron of excellent quality, imported from Russia for the purpose, they are $\frac{9}{64}$ of an inch in thickness, and are secured to the crowns in the following manner.

The crowns having been first turned to the required form, grooves are cut in them of the exact form of the buckets, to the depth *e e*, figure 3, plate III.; this depth is 0.1 inches at the edges and 0.5 inches near the middle. These grooves are cut in a machine contrived for the purpose, in which the cutting tool is guided by a cam. Three mortices for each bucket are then cut through each crown; corresponding tenons are left on the buckets; the latter are bent to the required form, by means of a pair of dies, prepared for the purpose, the plate iron having been first moderately heated. The tenons of all the buckets are then entered into the mortices in both crowns, the latter are then drawn together, by means of a number of screws applied to different parts of the circumference, and when the edges of the buckets are drawn into the bottom of the grooves, the tenons are riveted on the opposite sides. This construction gives great stability to the buckets, and permits the use of very thin iron.

30. *d d The vertical shaft.* This is of wrought iron, and is accurately turned in every part.

The diameters are as follows: —

Below the hub of the wheel, ...	7	inches.
In the hub of the wheel, ...	$7\frac{1}{4}$	"
Between the top of the hub and the lower bearing,	7	"
Between the bottom of the lower bearing and the hub of the bevel gear, ...	8	"
In the hub of the bevel gear, ...	$8\frac{1}{4}$	"
From the top of the hub of the bevel gear to the suspension box, ...	8	"

By reference to plate I., it will be seen that the shaft does not run upon a step at the bottom, but upon a series of collars, resting upon corresponding projections in the suspension box *e'*. The part of the shaft on which the collars are placed, is made separate from the main shaft, and is joined to it at *f*, by means of a socket in the top of the main shaft, which receives a corresponding part of the collar piece. The collars are made of cast steel; they are separately screwed on, and keyed to a wrought iron spindle.

31. *e' The suspension box.* This is made in two parts, to admit of its being taken off, and put on the shaft; it is lined with babbit metal, a soft composition consisting principally of tin. It is found that bearings thus lined will carry from fifty to a hundred pounds to the square inch, with every appearance of durability.

32. *f' f', The upper and lower bearings.* These are of cast-iron, lined with babbit metal; they are retained in position, horizontally, by means of adjusting screws; vertically, their weight is sufficient. The parts of the shaft inside the hubs of the wheel and the bevel gear, are made slightly tapering, about $\frac{1}{16}$ of an inch in diameter in the length of the hubs; the hubs are bored out with the same taper, but a very little smaller in diameter; they are then drawn on by a powerful screw purchase, and in this manner are made to fit very tight. To prevent danger of bursting the hubs, they are before being drawn on or bored out, strongly hooped with wrought iron hoops, driven on hot.

33. The suspension box *e'* (art. 31,) rests upon the gimbal *g*, plates I. and II. The gimbal itself is supported on the frame *h h*, by adjusting screws, which give the means

of raising and lowering the suspension box, and, with it, the vertical shaft and wheel. It will be perceived, by the arrangement of the bearings above and below the bevel gear, that no lateral strain can be thrown upon the suspension box. The construction of the shaft will evidently not admit, with safety, of lateral strain at the suspension box, and it is accordingly so arranged that this box is free to oscillate horizontally in any direction, a small quantity, in case any irregularity in the form of the shaft should require it.

The lower end of the shaft is fitted with a cast steel pin *i*, plate I. This is retained in its place by the step, which is made in three parts, and lined with casehardened wrought iron. The step is furnished with adjusting screws, by means of which the shaft can be moved horizontally in any direction, a small distance.

The weight of the wheel, upright shaft, and bevel gear, is supported by means of the suspension box *c*,' on the frame *k*, which rests upon the long beams *m*, reaching across the wheelpit, and supported at the ends by the masonry, and also at intermediate points by the braces *n n*.

From economical considerations *the diffuser*, described at art. 12, was omitted at the Tremont Turbine; a large majority of the turbines in use at Lowell, however, are fitted up with that apparatus.

34. The following are some of the dimensions of the turbine, carefully taken after the parts were finished:—

Height between the upper and lower crowns, at the outer extremities of the buckets, a mean of 44 measurements, ...	0.9314	feet.
Height between the upper and lower crowns, at the inner extremities of the buckets, a mean of 44 measurements, ...	0.9368	"
Height between the crowns, at a point 5.5 inches from the inner edges of the crowns, (designed to be 0.75 inches less than at the inner edges,)	0.8743	"
Shortest distance between the outer extremities of the buckets and the next adjacent buckets, a mean of 132 measurements,	0.18757	"
Shortest horizontal distance between two adjacent guides, taken at' the top of the circumferential part of the disc, a mean of 33 measurements, ...	0.1960	"
Do. do. at the bottom of the garniture, ...	0.2117	"
Do. do. half-way up between the disc and the garniture, . .	0.2044	"
The shortest distance between the guides, by a mean of the whole 99 measurements, ...	0.20403	"
Height from the top of the circumferential part of the disc to the bottom of the garniture, a mean of 33 measurements, . .	0.97090	"

35. The following are some of the most important dimensions of the apparatus; they are taken from the original designs, which were very closely followed in the construction.

Diameter of the exterior circumference of the crowns of the wheel,			8.333	feet.
"	"	outer extremities of the buckets, ...	8.292	"
"	"	interior edges of the crowns, and inner edges of the buckets, ...	6.750	"
"	"	outside of the cylindrical part of the regulating gate,	6.729	"
"	"	inside of the cylindrical part of the regulating gate,	6.562	"
"	"	of the outside of the lower curb, taken below the flange, ...	6.542	"
"	"	inside of the lower curb, taken at the top, . . .	6.333	"
"	"	inside of the lower curb, taken at the top of the guides, ...	6.167	"
"	"	lower part of the disc, ...	6.729	"

DESCRIPTION OF THE APPARATUS USED IN THE EXPERIMENTS ON THE TREMONT TURBINE.

36. The details of this apparatus are represented on plate IV.

The useful effect was measured with a Prony dynamometer, represented in sectional elevation at figure 1, and in plan at figure 2.

37. *The friction pulley* A is of cast-iron 5.5 feet in diameter, two feet wide on the face, and three inches thick. It is attached to the vertical shaft by the spider B, the hub of which occupies the place on the shaft intended for the bevel gear.

The friction pulley has, cast on its interior circumference, six lugs, C C, corresponding to the six arms of the spider. The bolt holes in the ends of the arms are slightly elongated in the direction of the radius, for the purpose of allowing the friction pulley to expand a little as it becomes heated, without throwing much strain upon the spider. When the spider and friction pulley are at the same temperature, the ends of the arms are in contact with the friction pulley. The friction pulley was made of great thickness for two reasons. When the pulley is heated, the arms cease to be in contact with the interior circumference of the pulley, consequently they would not prevent the pressure of the brake from altering the form of the pulley. This renders great stiffness necessary in the pulley itself. Again, it is found that a heavy friction pulley insures more regularity in the motion, operating, in fact, as a fly-wheel, in equalizing small irregularities.

38. *The brakes E and F* are of maple wood; the two parts are drawn together by the wrought iron bolts *G G*, which are two inches square.

39. *The bell crank F'* carries at one end the scale *I*, and at the other the piston of the hydraulic regulator *K*; this end carries also the pointer *L*, which indicates the level of the horizontal arm. The vertical arm is connected with the brake *F*, by the link *M*, figure 3.

40. *The hydraulic regulator K*, figures 1, 2, and 5, is a very important addition to the Prony dynamometer, first suggested to the author by Mr. Boyden in 1844. Its office is to control and modify the violent shocks and irregularities, which usually occur in the action of this valuable instrument, and are the cause of some uncertainty in its indications.

The hydraulic regulator used in these experiments, consisted of the cast-iron cylinder *K*, about 1.5 feet in diameter, with a bottom of plank, which was strongly bolted to the capping stone of the wheelpit, as represented in figure 1. In this cylinder, moves the piston *N*, formed of plate iron 0.5 inches thick, which is connected with the horizontal arm of the bell crank by the piston rod *O*. The circumference of the piston is rounded off, and its diameter is about $\frac{1}{10}$ inch less than the diameter of the interior of the cylinder. The action of the hydraulic regulator is as follows. The cylinder should be nearly filled with water, or other heavy inelastic fluid. In case of any irregularity in the force of the wheel, or in the friction of the brake, the tendency will be, either to raise or lower the weight; in either case the weight cannot move, except with a corresponding movement of the piston. In consequence of the inelasticity of the fluid, the piston can move only by the displacement of a portion of the fluid, which must evidently pass between the edge of the piston and the cylinder, and the area of this space being very small, compared to the area of the piston, the motion of the latter must be slow; giving time to alter the tension of the brake screws before the piston has moved far. It is plain that this arrangement must arrest all violent shocks, but, however violent and irregular they may be, it is evident that, if the mean force of them is greater in one direction than in the other, the piston must move in the direction of the preponderating force, the resistance to a slow movement being very slight. A small portion of the useful effect of the turbine must be expended in this instrument; probably less, however, than in the rude shocks the brake would be subject to without its use.

41. For the purpose of ascertaining the velocity of the wheel, a counter was attached to the top of the vertical shaft, so arranged that a bell was struck at the end of every fifty revolutions of the wheel.

42. To lubricate the friction pulley, and at the same time to keep it cool, water was let on to its surface in four jets, two of which are shown in figure 2, plate IV.

These jets were supplied from a large cistern, in the attic of the neighboring cotton-mill, kept full, during the working hours of the mill, by force-pumps. The quantity of water discharged by the four jets was, by a mean of two trials, 0.0288 cubic feet per second.

In many of the experiments with heavy weights, and consequently slow velocities, oil was used to lubricate the brake, the water, during the experiment, being shut off. It is found that, with a small quantity of oil, the friction between the brake and the pulley, is much greater than when the usual quantity of water is applied; consequently, the requisite tension of the brake screws is much less with the oil, as a lubricator, than with water. This may not be the whole cause of the phenomenon, but, whatever it may be, the ease of regulating in slow velocities is incomparably greater with oil as a lubricator, than with water applied in a quantity sufficient to keep the pulley cool. The oil was allowed to flow on in two fine continuous streams;—it did not, however, prevent the pulley from becoming heated sufficiently to decompose the oil, after running some time, which was distinctly indicated by the smoke and peculiar odor. When these indications became very apparent, the experiment was stopped, and water let on by the jets, until the pulley was cooled. As the pulley became heated, the brake screws required to be gradually slackened.

In the experiments, in table II., the lubricating fluid was as follows.

In the first twenty-six experiments, water alone was used.

In the four experiments numbered from 27 to 30, three gallons of linseed oil were used.

In all the experiments requiring a lubricator, and numbered from 31 to 48, inclusive, linseed oil was used.

In experiments 49 and 50, resin oil was used.

In experiments numbered from 51 to 60, inclusive, water alone was used.

In experiment 61, resin oil was used.

In experiment 62, resin oil and a small stream of water were used;—in the latter part of the experiment, a good deal of steam was generated by the heat of the friction pulley.

In experiment 63, resin oil alone was used.

In experiments numbered from 66 to 72, inclusive, water alone was used.

In experiments numbered from 73 to 79, inclusive, resin oil and a small stream of water were used.

In experiments numbered from 81 to 84, inclusive, water alone was used.

In experiments 85 and 86, resin oil and a small stream of water were used.

In experiment 87, resin oil alone was used.

In experiments 90 and 91, water alone was used.

In experiment 92, resin oil and a small stream of water were used.

43. A special apparatus was provided to indicate the direction in which the water left the wheel. For this purpose the vane *P*, figures 1, 6, and 7, plate IV., was placed near the circumference of the wheel, and was keyed on to the vertical shaft *Q*, which turned freely on a step resting on the wheelpit floor. The upper end of the shaft carried the hand *R*, figures 1 and 4, and directly under the hand was placed the graduated semicircle *S*, divided into 180°. When the vane was parallel to a tangent to the circumference of the wheel, drawn through the point nearest to the axis of the vane, and the vane was in the direction of the motion of the wheel, the hand pointed at 0°, and, consequently, when the vane was in the direction of the radius of the wheel, the hand pointed at 90°. To prevent sudden vibrations of the vane, a modification of the hydraulic regulator was attached to the lower part of the vane shaft. This apparatus is represented in detail by figures 6 and 8.

44. The quantity of water discharged by the wheel was gauged at a weir erected for the purpose at the mouth of the wheelpit. It is represented on plate V.

Figure 1 is a plan, and figure 2 a section, showing the relative positions of the turbine *A*, the grating *B*, the gauge box *C*, and the two divisions or bays of the weir, *D*, and *E*.

As the water issued from the orifices of the turbine with considerable force, particularly when the velocity of the wheel was much quicker or slower than that corresponding to the maximum coefficient of effect, there were often such violent commotions in the wheelpit, that, unless some mode was adopted to diminish them before the water reached the weir, or even the place where the depths on the weir were measured, it would have been impossible to make a satisfactory gauge of the water. For this purpose the grating *B*, figures 1 and 2, was placed across the wheelpit. This grating presented numerous apertures, nearly uniformly distributed over its entire area, through which the water must pass. In the experiments with a full gate, the fall from the upper to the lower side of the grating was generally from three to four inches. The combined effect of this fall and of the numerous small apertures, was, to obliterate almost entirely the whirls and commotions of the water above the grating. About 4.5 feet in length of the grating between *F* and *G*, figure 1, was so nearly closed, that but little water passed through that part of the grating;—this made it very quiet in the vicinity of the gauge box *C*.

Figure 3, plate V., is an elevation of the weir. The two bays *D* and *E* were of nearly equal length,—the crest of the weir was almost exactly horizontal, and the extreme variation did not exceed 0.01 inch. The crest of the weir was of cast-iron, planed on the upper edge *H*, figures 2 and 4, and also on the upstream face, to a point 1.125 inches below the top;—below this, at *I*, figure 4, there was a small bevel, also planed, the slope of which, on an average, was $\frac{3}{16}$ inch in a height of $\frac{5}{8}$ inch; — the remainder of the casting was unplaned. The crest of the weir *H* was $\frac{3}{8}$ inch thick, and was horizontal. The upstream edge, at *H*, was a sharp corner. The ends of the weir *K*, figures 1, 2, and 3, were of wood, and of the same form as the crest *H*, except that there was no bevelled part corresponding to *I*, figure 4. The crest of the weir *H* was about 6.5 feet above the floor of the wheelpit. The ends of the weir *K* projected from the walls of the wheelpit, and also from the central pier, a mean distance of 1.235 feet. The length of the bay *D*, was 8.489 feet, and of the bay *E*, 8.491 feet, making the total length of the weir 16.98 feet.

45. The depth of the water on the weir was taken in the gauge box *C*, figures 1 and 2, plate V., by means of the hook gauge *L*, which is represented in detail by figures 9, 10, and 11, plate IV.

The hook gauge is the invention of Mr. Boyden,* and is an instrument of inestimable value in hydraulic experiments. All other known methods of measuring the heights of the surface of still water, are seriously incommoded by the effects of capillary attraction; this instrument, on the contrary, owes its extraordinary precision to that phenomenon. At figure 10, plate IV., the point of the hook *A*, is represented as coinciding with the surface of the water. If the point of the hook should be a very little above the surface, the water in the immediate vicinity of the hook, would, by capillary attraction, be elevated with it, causing a distortion in the reflection of the light from the surface of the water. The most convenient method of observing with this instrument, according to the experience of the author, is, first, to lower the point of the hook, by means of the screw, to a little distance below the surface; — then to raise it again slowly by the same means, until the distortion of the reflection begins to show itself, — then to make a slight movement of the screw in the opposite direction, so as just to cause the distortion to disappear; the point will then be almost exactly at the level of the surface.

With no particular arrangements for directing light on the surface, differences in height of 0.001 feet are very distinct quantities; but by special arrangements for light and vision, differences of 0.0001 feet might be easily appreciated.

As this instrument cannot be efficiently used in a current, it was placed in the box *C*, in which the communication with the exterior was maintained by the hole *M*,

* In *Versuche über den ausfluss des wassers durch schieber, hähne, klappen und ventile*, by *Julius Weisbach*, *Leipzig*, 1842, page 1, is described an instrument for observing heights of water, having a slight resemblance to the hook gauge; it was however used by Boyden in a more perfect form, several years previous to the publication of that work.

when, by partially obstructing this communication, the extent of the oscillations could be diminished at will.

For the most perfect observations, it is essential that the surface of the water should be at rest. If, however, it should oscillate a little, a good mean may be obtained by adjusting the point of the hook to a height at which it will be visible above the surface of the water only half the time.

The movable rod to which the hook was attached, was of copper, and graduated to hundredths of feet, but, by means of the vernier, thousandths were measured, and in some cases ten thousandths were estimated. In later, and more perfect forms of this instrument, the point of the hook is immediately under the graduation.

46. The heights of the water in the forebay, and in the wheelpit, were taken by means of gauges, placed in the gauge boxes *p* and *q*, plate II. These boxes were similar to the box *C*, plate V., in which the hook gauge was placed. Both gauges were graduated to feet and hundredths, and both had the same zero point, viz., the level of the crest of the weir, so that the difference in the readings at the two gauges, gives, at once, the fall acting upon the wheel; and the difference between the depths of the water on the weir, as observed at the hook gauge, and the reading at the gauge *q*, gives the fall at the grating.

In consequence of want of space in plate II., the gauge box *p* is not represented in its true position, — it was actually in front of the head gate, and about six feet distant.

47. The heights of the regulating gate were taken at the rack *V*, plate I. The weights used for measuring the useful effect, were pieces of pig-iron of various sizes, each of which had been distinctly marked with its weight by Mr. O. A. Richardson, the official sealer of weights and measures for the City of Lowell.

MODE OF CONDUCTING THE EXPERIMENTS.

48. A separate observer was appointed to note each class of data; the time of each observation was also noted, which gave the means of identifying simultaneous observations. To accomplish this, each observer was furnished with a watch having a second hand; — the watch by which the speed of the wheel was observed, was taken as the standard; all the others were frequently compared with it, and when the variations exceeded ten or fifteen seconds, they were either adjusted to the standard, or the difference noted.

This mode of observing must, evidently, lead to more precise results than that in which a single observer, however skilful, undertakes to note all the phenomena, or even several of them. By the method adopted, a regular record is made of the state of things at very short intervals, furnishing the data for a mean result for any required period, and also the means of detecting, in most cases, the causes of apparent discrepancies. It also relieves the experimenter from the distraction of having numerous exact observations to make in a very short time, and leaves him much more at liberty to exercise a vigilant watch over the general course of the experiment.

49. As it may be useful to experimenters, not accustomed to this mode of observing, and, at the same time, afford the reader some means of judging of the accuracy of the results obtained in these experiments, the following extracts are given from the original note-books. The extracts include the data observed for experiment numbered 30 in table II. This experiment is selected, simply, because it gave the maximum coefficient of effect.

WEIGHT IN THE SCALE.

Extract from the note-book of the author, who superintended the experiments.

	1,498 lbs. 10½ oz.
4^h, 43′, added	26 " 0¼ "
Weight for the next experiment, . .	1,524 lbs. 10¾ oz.

SPEED OF THE WHEEL.

Extract from the note-book of Mr. Charles Leonard.

Times at which the bell struck.			Differences.	Times at which the bell struck.			Differences.	Times at which the bell struck.			Differences.
H.	min.	sec.	Seconds.	H.	min.	sec.	Seconds.	H.	min.	sec.	Seconds.
4.	55.	58.00		5.	0.	52.00	59.00	5.	4.	47.00	59.00
	56.	56.50	58.50		1.	50.75	58.75		5.	45.50	58.50
	57.	55.25	58.75		2.	49.50	58.75		6.	44.25	58.75
	58.	54.25	59.00		3.	48.00	58.50		7.	43.00	58.75
	59.	53.00	58.75								

NOTE. The bell struck once in every fifty revolutions of the wheel.

ELEVATION OF THE POINTER ON THE BELL CRANK.

Time.			Height of pointer, in feet.	Time.			Height of pointer, in feet.	Time.			Height of pointer, in feet.
H.	min.	sec.		H.	min.	sec.		H.	min.	sec.	
4.	55.	0.	0.19	4.	59.	30.	0.20	5.	4.	0.	0.17
		30.	0.13	5.	0.	0.	0.18			30.	0.18
	56.		0.13			30.	0.19		5.		0.24
		30.	0.14		1.		0.21			30.	0.18
	57.		0.15			30.	0.17		6.		0.19
		30.	0.19		2.		0.20			30.	0.19
	58.		0.20			30.	0.19		7.		0.16
		30.	0.19		3.		0.19			30.	0.14
	59.		0.21			30.	0.19				

NOTE. The extremity of the pointer was 6.5 feet from the fulcrum of the bell crank. When the horizontal arms of the bell crank were level, the height of the pointer was 0.20 feet.

HEIGHT OF THE WATER ABOVE THE WHEEL.

Taken in the forebay by Mr. John Newell.

Time.			Height, in feet.	Time.			Height, in feet.	Time.			Height, in feet.
H.	min.	sec.		H.	min.	sec.		H.	min.	sec.	
4.	55.	0.	15.100	4.	59.	30.	15.110	5.	4.	0.	15.120
		30.	15.100	5.	0.		15.115			30.	15.120
	56.		15.100			30.	15.120		5.		15.120
		30.	15.100		1.		15.120			30.	15.115
	57.		15.110			30.	15.110		6.		15.115
		30.	15.115		2.		15.105			30.	15.140
	58.		15.110			30.	15.100		7.		15.110
		30.	15.100		3.		15.115			30.	15.110
	59.		15.105			30.	15.125				

NOTE. The top of the weir is the zero point of the gauge in the forebay.

HEIGHT OF THE WATER AFTER PASSING THE WHEEL.

Taken in the wheelpit by Mr. Lloyd Hixon.

Time.			Height, in feet.	Time.			Height, in feet.	Time.			Height, in feet.
H.	min.	sec.		H.	min.	sec.		H.	min.	sec.	
4.	56.	0.	2.20	5.	0.	0.	2.21	5.	4.	0.	2.22
		30.	2.21			30.	2.21			30.	2.21
	57.		2.21		1.		2.21		5.		2.21
		30.	2.21			30.	2.21			30.	2.21
	58.		2.21		2.		2.21		6.		2.21
		30.	2.21			30.	2.21			30.	2.20
	59.		2.20		3.		2.20		7.		2.22
		30.	2.21			30.	2.20			30.	2.20

NOTE. The top of the weir is the zero point of the gauge in the wheelpit.

HEIGHTS OF THE WATER ABOVE THE WEIR BY THE HOOK GAUGE.

Observed by Mr. Daniel Haeffely.

Time.			Height, in feet.	Time.			Height, in feet.	Time.			Height, in feet.
H.	min.	sec.		H.	min.	sec.		H.	min.	sec.	
4.	57.	5.	1.8710	5.	1.	10.	1.8690	5.	4.	35.	1.8730
	58.	15.	1.8710		1.	45.	1.8700		5.	50.	1.8725
	58.	50.	1.8720		2.	15.	1.8720		6.	25.	1.8725
	59.	20.	1.8730		2.	50.	1.8720		6.	55.	1.8725
	59.	50.	1.8715		3.	15.	1.8715		7.	20.	1.8720
5.	0.	15.	1.8715		3.	40.	1.8715		7.	45.	1.8715
	0.	45.	1.8705		4.	5.	1.8730				

NOTE. The zero of the hook gauge was 0.002 feet *below* the top of the weir.

DIRECTION OF THE WATER LEAVING THE WHEEL.

Observed at the vane index by Mr. John C. Woodward.

Time.			Direction.		Time.			Direction.		Time.			Direction.	
H.	min.	sec.	deg.	min.	H.	min.	sec.	deg.	min.	H.	min.	sec.	deg.	min.
4.	57.	0.	59.	0.	5.	1.	0.	57.	0.	5.	5.	0.	58.	0.
		30.	57.	0.			30.	59.	30.			30.	59.	30.
	58.		59.	0.		2.		58.	0.		6.		59.	30.
		30.	58.	0.			30.	57.	0.			30.	57.	0.
	59.		58.	0.		3.		60.	0.		7.		59.	0.
		30.	58.	30.			30.	58.	0.			30.	57.	30.
5.	0.		57.	0.		4.		59.	0.		8.		59.	0.
		30.	57.	30.			30.	56.	0.					

NOTE. When the vane pointed in the direction of the radius of the wheel, the reading of the index was 90°. 0° was in the direction of the motion of the wheel.

50. Previously to the commencement of the experiments, the apparatus for measuring the useful effect was carefully adjusted. The bell crank was balanced when there were no weights in the scale. For this purpose the link *M*, figure 3, plate IV., was removed, and the chamber of the hydraulic regulator filled with water;—weights were then applied to the top of the bell crank, near the end to which the hydraulic regulator was attached, until the whole was in equilibrium;—the final adjustment was made, by placing a weight of about two pounds at the extremity of one of the horizontal arms of the bell crank,—the arm was retained horizontally until a signal was given, when it was left at liberty to descend, and the time occupied in descending a certain distance was noted;—the weight was then removed to the extremity of the other arm, and the same process repeated. The balance weights were altered until the times of descent were equal. To overcome, as much as possible, the friction of the fulcrum, the pin forming it was lubricated with sperm oil, and, during the descent, the head of the pin was struck lightly and rapidly with a small hammer.

After the bell crank was satisfactorily balanced, the link *M* was reattached, and the brake adjusted, by means of the screw which formed the connection between the link and the brake. It was adjusted so that a line upon the brake was perpendicular to the axis of the link, when the horizontal arm of the bell crank was horizontal. The length of the brake was then measured upon this line.

The length of the brake as thus measured was found to be . .	9.745	feet.
The effective length of the vertical arm of the bell crank was	4.500	"
And the effective length of the horizontal arm to which the scale was hung, was	5.000	"
Consequently, the effective length of the brake was $\frac{9.745 \times 5}{4.5} =$	10.827778	"

51. The gauges in the forebay, and in the wheelpit, were carefully adjusted by levelling from the top of the weir. This was repeated by different persons, so as to remove all chance of error.

52. The hook gauge was compared with the weir, by a different method. When the regulating gate of the turbine was shut down as tight as possible, it was still found that a quantity of water leaked into the wheelpit, exceeding, a little, the quantity that leaked out of the wheelpit, so that a small quantity continued to run over the weir. The principal leak into the wheelpit was between the regulating gate and the lower curb, the leather packing not being perfectly adjusted. The hook gauge was firmly attached to a post, placed in the wheelpit for that purpose, and at a height known to be nearly correct. The regulating gate was closed, and after the water had arrived at a uniform state, the height of the water at the hook gauge was noted, and, at the same time, the depths of the water on the weir were measured directly with a graduated rule. To perform this accurately, a board, about four inches long, was held by an assistant on the crest of the weir, at the place where it was intended to measure the depth; — the author then applied the rule, previously well dried, vertically, on the top of the weir, in front of the board. On first immersing the rule, the water in contact with it did not stand at the true level of the surface, but formed a little hollow around the rule; it immediately commenced rising, however, and after a few moments came to a level, which was indicated by the reflection of a light from the surface, a lamp being held by an assistant, in a proper position, for that purpose.

The depths on the weir, taken in the manner just described, February 20th, 1851, were as follows.

	Depths on the westerly bay of the weir. Inches.	Depths on the easterly bay of the weir. Inches.
	0.37	0.36
	0.36	0.36
	0.37	0.36
	0.37	0.36
Means . .	0.3675	. . . . 0.36
Or in feet .	0.0306	. . . . 0.0300

While the heights given in the preceding table were being measured, the depth by the hook gauge was constantly 0.0318 feet; consequently, by this comparison, the zero of the hook gauge was 0.0012 feet below the mean height of the top of the weir, in the westerly bay, and 0.0018 feet below the mean height in the easterly bay, or 0.0015 feet below the mean height in both bays. A similar comparison was made February 22d, 1851, when the zero of the hook gauge was found to be 0.0024 feet below the mean height of the weir. The mean of the two comparisons, or 0.0020 was adopted as the correction to be subtracted from the reading of the hook gauge, to give the mean depth upon the weir.

53. During the experiments, the levels of the water in the upper and lower canals, were maintained nearly uniform. The height of the lower canal, at the place where the water, passing the weir, fell into it, varied a little, depending upon the quantity of water discharged by the wheel. It was highest when the wheel was running with the regulating gate fully raised, and the brake removed; under these circumstances the surface of the water was from 0.3 feet to 0.4 feet below the top of the weir. In the other experiments with the regulating gate fully raised, the fall from the top of the weir to the surface of the water in the lower canal, was from 0.4 feet to 0.6 feet. The brackets *N* and the planks *O*, figure 2, plate V., were not put on until after the turbine experiments were concluded, so that the water passing the weir, met with no obstruction until it struck the water in the lower canal.

It will be seen by the experiments on the weir, (art. 127,) that the obstruction, caused by the planks, was scarcely appreciable, which renders it certain that the effect of the lower canal, in obstructing the flow over the weir, must have been entirely inappreciable.

DESCRIPTION OF TABLE II., CONTAINING THE EXPERIMENTS UPON THE TURBINE AT THE TREMONT MILLS.

54. The data obtained by direct observation, and the results deduced from them by calculation, are arranged together, for convenience of reference, in table II.

The columns numbered 1, 2, and 3, require no further explanations than are contained in the several headings.

55. COLUMN 4. *Height of the regulating gate.* The three first experiments were made under circumstances identical in every thing, except that the height of the regulating gate was varied a little, for the purpose of ascertaining the height giving the maximum coefficient of effect. The mean height between the crowns of the wheel, at the inner edges of the buckets, was 0.9368 feet, or 11.2416 inches; the curvature of the disc and garniture, however, rendered it necessary to raise the gate rather more than this, in order to present the most favorable aperture. By a comparison of the first three experiments, it appears that the most favorable result was obtained, with the gate raised to a height of 11.50 inches, or a little less; the succeeding experiments, numbered from 4 to 50, inclusive, were made with the regulating gate raised to the full height, or to 11.50 inches, nearly. A comparison of the first three experiments will show that there could be no appreciable difference in the results, that could be attributed to the small differences in the heights of the gate, in the experiments numbered from 4 to 50, inclusive, and they are accordingly all classed together, as experiments with a full gate, the small difference in the heights being accidental.

The experiments numbered from 51 to 64, inclusive, were made with the gate raised 8.55 inches, or three-fourths of the full height, nearly. Those numbered from 65 to 76, inclusive, were made with the gate at very nearly half of the full height. Those numbered from 77 to 79, inclusive, were made with the gate at about seven eighths of the full height. Those numbered from 80 to 89, inclusive, were made with the gate at about one fourth of its full height. And the last three experiments were made with the gate raised one inch.

56. COLUMN 5. *Time.* The times entered under the heads *beginning*, and *ending of the experiments*, are taken from the notes of the "speed of the wheel," and indicate the times at which the bell, attached to the counter, was struck, which, by a comparison of the various note-books, appeared, by the regularity of the observations, to be the most suitable for the commencement and termination of the experiment.

57. Column 6. *Duration of the experiment*, is obtained by taking the differences of the times of the beginning, and ending of the experiment, as given in column 5.

58. Column 7. *Total number of revolutions of the wheel during the experiment.* This is obtained from the note-book of the "speed of the wheel," by counting the number of observations of the times at which the bell was struck; this number, less one, multiplied by 50, which is the number of revolutions of the wheel to each stroke of the bell, gives the number of revolutions during the experiment.

59. Column 8. *Number of revolutions of the wheel per second*, is obtained by dividing the total number of revolutions of the wheel, by the duration of the experiment.

60. Column 9. *The weight in the scale*, requires no explanation.

61. Column 10. *Useful effect, or the friction of the brake, in pounds avoirdupois, raised one foot per second.* This is obtained by multiplying together the weight in the scale, and the velocity that the point of application of the weight, tends to take. Or, in other words, the product of the weight into the velocity that the weight would actually take, if, for an infinitely short time, the brake, and the apparatus connecting it with the weight, were rigidly connected with the friction pulley.

The effective length of the brake, including the leverage due to the different lengths of the arms of the bell crank, was 10.827778 feet (art. 50). The circumference of a circle of this radius is 68.0329 feet. This circumference is a constant for all the experiments in which any useful effect was produced, and column 10 was obtained by the product of this constant, the weight, and the number of revolutions of the wheel per second. The computation was performed by logarithms, and if the results given in the tables should be verified by actual multiplications, minute differences would, no doubt, be detected.

62. Columns 11 and 12. *Heights of the water in the forebay and in the wheelpit.* These heights are all referred to the top of the weir, consequently, the differences give the fall acting upon the wheel.

63. Column 13. *Total fall acting upon the wheel.* These are the differences referred to in the last sentence. In experiments 27 and 28, observations were taken in the ventilating pipe *G*, plate I., for the purpose of estimating the loss of fall to this part of the supply pipe; — it was not convenient, however, to measure these heights with complete accuracy. In experiment 27, the height of the water in the ventilating pipe was 0.106 feet below the level in the forebay; — in experiment 28 the difference was found to be 0.102 feet; — in experiment 30, which gave the maximum coefficient of effect, the quantity of water discharged by the wheel, was a little less than in either experiment 27, or 28. We may, therefore,

conclude, that when the regulating gate was fully raised, and the wheel running with the velocity giving the maximum coefficient of effect, the fall acting upon the wheel being 12.903 feet, the loss of fall from the forebay to the ventilating pipe, was very nearly 0.10 feet.

64. COLUMN 14. *Depth of water on the weir.* The depths on the weir were observed with the hook gauge, described at art. 45.

65. COLUMN 15. *Quantity of water passing the weir.* These quantities have been calculated by the formula

$$Q = 3.33\,(l - 0.1\,n\,h)\,h^{\frac{3}{2}},$$

in which

Q = Quantity, in cubic feet per second.
l = The total length of the weir, in feet.
n = The number of end contractions in the weir.
h = The depth on the weir, in feet.

It is unnecessary here to discuss the reasons that have induced the author to adopt this formula, so different from any that has been used heretofore, as the subject is fully considered in another part of this work.

A small quantity of water entered the wheelpit without passing through the wheel; there was also a small quantity that leaked out by passing through the floor of the wheelpit; the latter quantity, when the depth on the weir was 0.496 feet, was estimated at 0.0409 cubic feet per second; see art. 130. As these quantities were very minute, and tended to compensate each other, they have been neglected, and the quantity computed as passing the weir is taken for the quantity discharged by the wheel.

66. COLUMN 16. *Total power of the water.* This column is obtained by multiplying together the total fall acting upon the wheel, the quantity of water passing the weir per second, and the weight of a cubic foot of water. The temperature of the water was constantly at 32° Fahrenheit, it was nearly pure, and the weight of a cubic foot was taken at 62.375 pounds avoirdupois.

The water of the Merrimack River is always remarkably free from impurities, held in solution, flowing, as it does, from, and through a primitive formation, covered with a sterile soil. In midwinter, at which season these experiments were made, it is more than ordinarily pure, as at that season the surface of the country is usually covered with snow, and the soil frozen to a considerable depth; the river itself, wherever it flows with a moderate current, is frozen over, so that heavy carriages can often pass with safety, and at the time when these experiments were made, the river for about eighteen miles before it reached the turbine, was covered with a solid coating of ice, with scarcely an opening in the whole distance. When the river is thus frozen, the water flows along under the ice, entirely free from floating particles of ice, even in the most severe weather.

As the author had frequently felt the want of a table of the absolute weights of a cubic foot of water at different temperatures, he, several years since, computed the following table.

In the *Encyclopedia Britannica*, seventh edition, vol. 21, page 846, is given the following extract from the British act of Parliament, establishing the standards for weights and measures.

"Provided always, and be it enacted, that in all cases of dispute respecting the correctness of any measure of capacity, arising in a place where recourse cannot conveniently be had to any of the aforesaid verified copies or models of the standard measures of capacity, it shall and may be lawful, to and for, any justice of the peace, or magistrate, having jurisdiction in such place, to ascertain the content of such measure of capacity by direct reference to the weight of pure or rain water which such measure is capable of containing; ten pounds avoirdupois weight of such water, at the temperature of 62° by Fahrenheit's thermometer, being the standard gallon ascertained by this act, the same being in bulk equal to 277.276, 1822 (1823, 277.274) cubic inches, and so in proportion," etc. 277.274 cubic inches was taken, as it appeared to be the latest determination.

In the first volume of the *Traité de Chimie*, by *J. J. Berzelius*, second French edition, Paris, 1846, there is given a table of the specific gravities of pure water, at different temperatures of the centigrade scale, deduced from Haellstroem's experiments.

From these two authorities were derived the data for the following table.

TABLE I.

WEIGHT OF A CUBIC FOOT OF PURE WATER AT DIFFERENT TEMPERATURES.

Temperature, in degrees of Fahrenheit's thermometer.	Weight in air, of a cubic foot of pure water. Pounds avoirdupois.	Temperature, in degrees of Fahrenheit's thermometer.	Weight in air, of a cubic foot of pure water. Pounds avoirdupois.	Temperature, in degrees of Fahrenheit's thermometer.	Weight in air, of a cubic foot of pure water. Pounds avoirdupois.
32	62.375	50	62.368	69	62.278
33	62.377	51	62.365	70	62.272
34	62.378	52	62.363	71	62.264
35	62.379	53	62.359	72	62.257
36	62.380	54	62.356	73	62.249
37	62.381	55	62.352	74	62.242
38	62.381	56	62.349	75	62.234
39 (max.)	62.382	57	62.345	76	62.225
39.38	62.382	58	62.340	77	62.217
40	62.382	59	62.336	78	62.208
41	62.381	60	62.331	79	62.199
42	62.381	61	62.326	80	62.190
43	62.380	62	62.321	81	62.181
44	62.379	63	62.316	82	62.172
45	62.378	64	62.310	83	62.162
46	62.376	65	62.304	84	62.152
47	62.375	66	62.298	85	62.142
48	62.373	67	62.292	86	62.132
49	62.371	68	62.285		

67. COLUMN 17. *Ratio of the useful effect to the power expended.* This column is obtained by dividing the numbers in column 10 by those in column 16.

68. COLUMN 18. *Velocity due to the fall acting upon the wheel.* The numbers in this column have been calculated by the formula

$$V = \sqrt{2gh}.$$

V = the velocity in feet per second.

g = the velocity acquired by a body at the end of the first second of its fall in a vacuum.

h = the fall acting upon the wheel; this is given in column 13.

The value of g has been calculated by the formula given in the second edition of the *Traité D'Hydraulique*, by *D'Aubuisson*, page 5, viz.: —

$$g = 9^{m}.8051\,(1 - 0.00284 \cos. 2\,l)\,\left(1 - \frac{2e}{r}\right);$$

l being the latitude of the place; e, its elevation above the level of the sea; and r, the radius of the terrestrial spheroid, at the level of the sea, and at the place;

$$\{\, r = 6366407^{m}\,(1 + 0.00164 \cos. 2\,l)\,\}.$$

The latitude of Lowell, as given in the American Almanac, is 42°, 38′, 46″, and the height above the sea is known to be about 25 metres. With these data, the above formula gives, in feet,

$$g = 32.1618.$$

69. COLUMN 19. *Velocity of the interior circumference of the wheel.* The diameter of the circle inscribing the inner edges of the buckets, is 6.75 feet; see art. 35. Consequently the interior circumference of the wheel is 21.20575 feet. The product of this number into the number of revolutions per second, given in column 8, gives the numbers in column 19.

70. COLUMN 20. *Ratio of the velocity of the interior circumference of the wheel, to the velocity due to the fall acting on the wheel.* This column is obtained by dividing the numbers in column 19 by the corresponding numbers in column 18. This column indicates the relative velocities of the wheel, in the different experiments, eliminated from the effects of the variations in the fall acting upon the wheel.

71. COLUMN 21. *Quantity of water which passed the wheel, reduced to a uniform fall of thirteen feet.* The numbers in this column are obtained from those in column 15, in the following manner.

Let H = the observed fall acting upon the wheel.
Q = the observed quantity.
Q' = the quantity that would have passed the wheel, if the fall had been thirteen feet, instead of H, all other circumstances being the same.

As the quantity of water discharged by the wheel, all other things being equal, will vary as the square root of the fall acting upon the wheel, we have

$$\sqrt{H} : Q :: \sqrt{13} : Q',$$

therefore

$$Q' = Q\sqrt{\frac{13}{H}}.$$

The quantities given in column 21, have been calculated by this formula.

72. COLUMN 22. *Ratio of the reduced quantity in column 21, to the reduced quantity in experiment* 30. The numbers given in this column indicate the relative quantities discharged by the wheel in the different experiments, eliminated from the effects due to the variations in the fall acting upon the wheel; the reduced quantity in experiment 30 is taken as unity, that experiment giving the maximum coefficient of effect. It will be seen by a comparison of columns 20 and 22, that the quantity discharged by the wheel, when the gate is fully raised, diminishes regularly with the velocity. The quantity discharged is a minimum in experiment 42,

in which the wheel had the least velocity. In experiments 43 and 44, however, in which the wheel was prevented from revolving, by screwing up the brake, the quantity discharged was considerably above the minimum. Whether this is due to an accidental position of the buckets relative to the guides, presenting apertures more favorable to the discharge than the average of all positions, or whether it is due to some more general cause, the author is not aware.

73. COLUMN 23. *Direction of the water leaving the wheel, as indicated by the vane.* The angles given in this column show the position of the vane, relative to a line passing through the axis of the vane, and parallel to a tangent drawn through the point in the circumference of the wheel, nearest to the axis of the vane, zero being in the direction of the motion of the wheel. The apparatus with which these angles were taken, is described at art. 43. In the experiments made when the gate was fully raised, or nearly so, the vane operated satisfactorily; but as the height of the gate was diminished, the indications of the vane became more uncertain. The vane was of nearly the same height as the orifices in the exterior circumference of the wheel; this was very suitable for the experiments with the gate fully raised, but in the experiments with the gate partially raised, a portion of the height of the vane was exposed to irregular currents, which probably interfered seriously with its operation. The observations made with the vane in the experiments in which the gate was partially raised, are much less to be relied on than those made when the gate was fully raised, the value of the indications being, in some degree, proportioned to the height of the gate.

74. COLUMN 24. *Mean elevation of the pointer on the bell crank.* The numbers in this column indicate the mean positions of the bell crank, during the experiments, in reference to a gauge placed 6.5 feet from the fulcrum of the bell crank. It will be seen by the table that the mean positions differ but little from the horizontal; the pointer was however generally a little below, which indicates that the weight was generally lifted a little too high.

The play of the brake was confined between two fixed stops, placed so that when the pointer stood at 0.20 feet below the horizontal, the brake struck,—and it struck the other stop when the pointer was at 0.21 feet above the horizontal. The brake was not allowed, however, to touch either of the stops in any of the experiments reported, in which it was undertaken to regulate the friction of the brake; the fact that it did touch was deemed a sufficient reason to reject the experiment. Little inconvenience, however, was experienced from this cause, as the hydraulic regulator afforded very perfect control over the brake.

TABLE

EXPERIMENTS UPON THE TURBINE AT THE

1	2	3		4	5						6	7	8	9	10
No. of the experiment.	DATE 1851.	Temperature of the atmosphere in degrees of Fahrenheit's thermometer.		Height of the regulating gate, in inches.	TIME.						Duration of the experiment, in seconds.	Total number of revolutions of the wheel during the experiment.	Number of revolutions of the wheel per second.	Weight in the scale, in pounds avoirdupois.	Useful effect, or the friction of the brake, in pounds avoirdupois, raised one foot per second.
		External air in the shade.	In the wheelpit.		Beginning of the experiment.			Ending of the experiment.							
					H.	min.	sec.	H.	min.	sec.					
1	February 17, P. M.	31.00	41.00	11.50	2	19	52.00	2	28	15.50	503.50	450	0.89374	1443.34	87760.8
2	" " "	29.00	36.75	11.60	2	37	8.00	2	47	24.75	616.75	550	0.89177	1443.34	87567.2
3	" " "	30.25	36.25	11.45	2	56	26.00	3	6	41.50	615.50	550	0.89358	1443.34	87745.0
4	" " "	29.00	35.25	11.49	3	23	24.00	3	29	8.50	344.50	550	1.59651	307.03	33348.3
5	" " "	29.50	35.50	"	3	29	8.50	3	37	18.00	489.50	750	1.53218	411.48	42892.0
6	" " "	29.75	35.25	"	3	37	18.00	3	44	42.75	444.75	650	1.46149	519.77	51680.6
7	" " "	29.50	35.50	"	3	45	18.00	3	52	32.00	434.00	600	1.38249	638.36	60040.8
8	" " "	29.25	35.50	"	4	4	35.00	4	9	40.50	305.50	400	1.30933	750.42	66845.5
9	" " "	29.25	35.50	"	4	10	19.75	4	15	41.00	321.25	400	1.24514	854.87	72416.3
10	" " "	29.00	35.50	"	4	15	41.00	4	24	7.50	506.50	600	1.18460	957.35	77154.6
11	" " "	29.00	35.75	"	4	24	51.00	4	33	44.25	533.25	600	1.12518	1057.49	80949.8
12	" " "	28.50	35.00	"	5	1	10.50	5	10	31.00	560.50	1000	1.78412	0.	0.
13	February 18, A. M.		35.75	"	9	14	5.50	9	22	58.00	532.50	950	1.78404	0.	0.
14	" " "	33.75	36.25	"	9	42	32.00	9	51	7.25	515.25	550	1.06744	1156.27	83969.8
15	" " "	34.25	36.75	"	9	51	7.25	10	0	4.50	537.25	550	1.02373	1229.41	85625.3
16	" " "	34.00	36.50	"	10	12	27.00	10	19	57.25	450.25	450	0.99945	1269.42	86314.4
17	" " "	33.75	36.50	"	10	20	48.00	10	29	23.50	515.50	500	0.96993	1319.22	87051.8
18	" " "	34.00	36.50	"	10	41	55.00	10	48	58.25	423.25	400	0.94507	1359.23	87392.7
19	" " "	34.75	36.75	"	10	49	52.00	10	59	48.00	596.00	550	0.92282	1397.12	87714.1
20	" " "	36.00	36.00	"	11	16	14.50	11	25	23.25	548.75	500	0.91116	1416.70	87819.8
21	" " "	36.50	36.50	"	11	25	23.25	11	35	33.00	609.75	550	0.90201	1433.43	87964.3
22	" " "	36.50	36.75	"	11	45	12.00	11	59	8.00	836.00	750	0.89713	1443.06	88076.2
23	February 18, P. M.	41.50	39.25	"	2	23	56.50	2	33	18.00	561.50	1000	1.78094	0.	0.
24	" " "	39.75	38.75	"	2	41	30.50	2	50	51.00	560.50	500	0.89206	1454.24	88257.1
25	" " "	39.00	40.00	"	2	50	51.00	3	4	2.00	791.00	700	0.88496	1464.80	88189.9
26	" " "	38.75	38.00	"	3	22	7.50	3	26	53.25	285.75	250	0.87489	1474.37	87756.6
27	" " "	38.75	38.00	"	3	27	54.00	3	42	6.25	852.25	750	0.88002	1474.37	88271.4
28	" " "	38.50	36.25	"	3	58	40.25	4	11	4.75	744.50	650	0.87307	1485.63	88242.7
29	" " "	38.50	36.25	"	4	28	54.50	4	40	27.00	692.50	600	0.86643	1498.66	88339.3
30	" " "	37.25	36.50	"	4	55	58.00	5	7	43.00	705.00	600	0.85106	1524.67	88278.9
31	February 20, A. M.			11.48	9	16	16.25	9	25	35.00	558.75	1000	1.78971	0.	0.
32	" " "	33.50	35.00	"	9	47	40.50	9	53	39.25	358.75	300	0.83624	1552.44	88320.8
33	" " "	33.75	35.00	"	10	8	35.00	10	18	49.75	614.75	500	0.81334	1597.08	88372.5
34	" " "	36.75	35.50	"	10	37	30.50	10	48	8.25	637.75	500	0.78401	1648.87	87947.8
35	" " "	38.00	35.75	"	11	2	8.50	11	13	22.25	673.75	500	0.74211	1724.49	87066.5
36	" " "	41.00	35.75	"	11	23	38.25	11	34	26.00	647.75	450	0.69471	1816.71	85863.7
37	" " "	41.50	36.00	"	11	48	56.00	11	59	15.50	619.50	400	0.64568	1911.45	83965.5
38	February 20, P. M.			"	2	30	39.50	2	39	59.50	560.00	1000	1.78571	0.	0.
39	" " "	42.25	35.25	"	2	55	11.00	3	6	46.50	695.50	450	0.64702	1911.45	84139.1
40	" " "	41.75	35.75	"	3	21	56.50	3	30	16.50	500.00	300	0.60000	2011.52	82109.8
41	" " "	41.75	35.75	"	3	45	27.50	3	56	25.00	657.50	350	0.53232	2167.38	78492.2
42	" " "	41.50	36.00	"	4	9	29.00	4	18	39.50	550.50	250	0.45413	2367.88	73158.0
43	" " "			"	4	58	0.	4	59	30.00	90.00	0	0.	4213.38	0.
44	" " "			"	5	2	0.	5	4	30.00	150.00	0	0.	3946.38	0.
45	February 21, A. M.	36.25	35.50	"	9	22	57.50	9	32	18.25	560.75	1000	1.78333	0.	0.
46	" " "	36.25	35.75	"	9	38	3.00	9	49	4.00	661.00	550	0.83207	1565.21	88603.9
47	" " "	36.25	35.75	"	10	0	48.75	10	11	1.00	612.25	500	0.81666	1590.50	88367.8
48	" " "	35.75	36.25	"	10	25	38.50	10	37	3.25	684.75	550	0.80321	1614.79	88240.1
49	" " "	36.00	36.00	"	10	50	35.00	11	2	11.75	696.75	550	0.78938	1641.34	88146.2
50	" " "	35.75	36.25	"	11	14	54.00	11	25	44.25	650.25	500	0.76893	1679.62	878[illegible]5.8

11.

TREMONT MILLS, IN LOWELL, MASSACHUSETTS.

No. of the experiment.	11 Height of the water above the wheel, taken in the forebay, in feet.	12 Height of the water after passing the wheel, taken in the wheel-pit, in feet.	13 Total fall acting upon the wheel, in feet.	14 Depth of water on the weir, in feet.	15 Quantity of water which passed the weir, in cubic feet per second.	16 Total power of the water, in pounds avoirdupois raised one foot per second.	17 Ratio of the useful effect to the power expended.	18 Velocity due to the fall acting on the wheel, in feet per second.	19 Velocity of the interior circumference of the wheel, in feet per second.	20 Ratio of the velocity of the interior circumference of the wheel to the velocity due to the fall acting on the wheel.	21 Quantity of water which passed the wheel, reduced to a uniform fall of 13 feet, in cubic feet per second.	22 Ratio of the reduced quantity in column 21, to the reduced quantity in experiment 30.	23 Direction of the water leaving the wheel, as indicated by the vane.		24 Mean elevation of the pointer on the bell crank.
													deg.	m.	Feet.
1	15.082	2.218	12.864	1.8811	139.4206	111870.0	0.78449	28.7656	18.9525	0.65886	140.1557	1.01044	47	0	+0.002
2	15.079	2.219	12.860	1.8811	139.4206	111835.2	0.78300	28.7611	18.9107	0.65751	140.1775	1.01060	47	45	0.
3	15.087	2.218	12.869	1.8816	139.4676	111951.2	0.78378	28.7712	18.9491	0.65861	140.1756	1.01058	47	23	—0.001
4	15.036	2.482	12.554	2.0383	156.6470	122663.3	0.27187	28.4169	33.8553	1.19138	159.4053	1.14922			+0.013
5	15.042	2.431	12.611	2.0180	154.3891	121444.2	0.35318	28.4813	32.4909	1.14078	156.7522	1.13009			+0.008
6	15.051	2.398	12.653	1.9989	152.2682	120174.8	0.43005	28.5287	30.9921	1.08635	154.3421	1.11271			—0.003
7	15.061	2.365	12.696	1.9734	149.4653	118363.5	0.50726	28.5771	29.3167	1.02588	151.2442	1.09038	15	49	—0.001
8	15.069	2.349	12.720	1.9536	147.2942	116864.8	0.57199	28.6041	27.7653	0.97067	148.9066	1.07353	18	3	—0.010
9	15.089	2.312	12.777	1.9420	146.0204	116373.2	0.62298	28.6681	26.4041	0.92102	147.2891	1.06187	20	3	—0.018
10	15.102	2.302	12.800	1.9315	144.8734	115667.0	0.66704	28.6939	25.1203	0.87546	146.0009	1.05258	22	56	—0.013
11	15.100	2.281	12.819	1.9226	143.9088	115067.3	0.70350	28.7152	23.8602	0.83093	144.9212	1.04480	25	47	—0.001
12	15.028						0.		37.8336						
13	15.071	2.561	12.510	2.0989	163.4313	127527.3	0.	28.3670	37.8319	1.33366	166.6013	1.20110			
14	15.120	2.264	12.856	1.9098	142.5180	114284.2	0.73475	28.7566	22.6359	0.78716	143.3140	1.03321	29	49	+0.002
15	15.117	2.229	12.888	1.9054	142.0433	114187.1	0.74987	28.7924	21.7090	0.75398	142.6592	1.02849	33	30	+0.004
16	15.116	2.226	12.890	1.9048	141.9762	114150.8	0.75614	28.7946	21.1940	0.73604	142.5808	1.02792	35	37	—0.001
17	15.128	2.232	12.896	1.8983	141.2762	113640.9	0.76603	28.8013	20.5681	0.71414	141.8448	1.02262	38	20	—0.001
18	15.111	2.231	12.880	1.8908	140.4657	112848.8	0.77442	28.7835	20.0409	0.69626	141.1186	1.01738	41	26	—0.003
19	15.114	2.231	12.883	1.8873	140.0845	112568.7	0.77920	28.7868	19.5691	0.67979	140.7192	1.01450	44	26	+0.001
20	15.114	2.228	12.886	1.8866	140.0066	112532.3	0.78040	28.7902	19.3219	0.67113	140.6246	1.01382	46	18	—0.002
21	15.128	2.229	12.899	1.8856	139.9040	112563.3	0.78147	28.8047	19.1278	0.66405	140.4507	1.01257	47	26	—0.005
22	15.123	2.225	12.898	1.8834	139.6678	112364.5	0.78384	28.8036	19.0243	0.66048	140.2190	1.01090	48	26	0.
23	14.965	2.536	12.429	2.0834	161.6944	125355.0	0.	28.2750	37.7662	1.33567	165.3669	1.19220			
24	15.124	2.219	12.905	1.8773	139.0070	111893.5	0.78876	28.8114	18.9168	0.65657	139.5177	1.00584	49	28	—0.006
25	15.118	2.219	12.899	1.8775	139.0291	111859.4	0.78840	28.8047	18.7662	0.65150	139.5724	1.00623	50	37	+0.004
26	15.102	2.209	12.893	1.8750	138.7601	111591.0	0.78641	28.7980	18.5527	0.64424	139.3347	1.00452	51	40	—0.057
27	15.116	2.214	12.902	1.8758	138.8489	111740.3	0.78997	28.8080	18.6616	0.64779	139.3752	1.00481	51	18	—0.018
28	15.117	2.211	12.906	1.8760	138.8711	111792.9	0.78934	28.8125	18.5141	0.64257	139.3759	1.00482	52	25	—0.018
29	15.118	2.212	12.906	1.8727	138.5134	111504.9	0.79225	28.8125	18.3732	0.63768	139.0169	1.00223	53	52	—0.017
30	15.111	2.208	12.903	1.8697	138.1892	111218.1	0.79375	28.8092	18.0474	0.62645	138.7076	1.00000	58	10	—0.018
31	15.084	2.545	12.539	2.0891	162.3283	126960.2	0.	28.3999	37.9521	1.33635	165.2853	1.19161			
32	15.119	2.204	12.915	1.8704	138.2668	111384.0	0.79294	28.8225	17.7330	0.61525	138.7211	1.00010	61	54	—0.011
33	15.134	2.200	12.934	1.8701	138.2335	111521.1	0.79243	28.8437	17.2475	0.59796	138.5858	0.99912	66	5	—0.008
34	15.129	2.188	12.941	1.8687	138.0869	111463.0	0.78903	28.8515	16.6254	0.57624	138.4013	0.99779	86	12	—0.004
35	15.123	2.184	12.939	1.8652	137.7076	111139.7	0.78840	28.8493	15.7371	0.54549	138.0318	0.99513	99	25	—0.014
36	15.128	2.184	12.944	1.8539	136.4917	110200.9	0.77916	28.8549	14.7319	0.51055	136.7866	0.98615	115	48	—0.001
37	15.117	2.177	12.940	1.8412	135.1415	109077.1	0.76978	28.8504	13.6922	0.47459	135.4545	0.97655	131	18	—0.015
38	15.036	2.536	12.500	2.0834	161.6944	126071.1	0.	28.3557	37.8674	1.33544	164.8966	1.18881			
39	15.143	2.180	12.963	1.8431	135.3423	109433.4	0.76886	28.8761	13.7205	0.47515	135.5354	0.97718	130	51	—0.019
40	15.136	2.163	12.973	1.8380	134.7976	109077.0	0.75277	28.8872	12.7235	0.44045	134.9378	0.97282	139	45	—0.010
41	15.136	2.159	12.977	1.8282	133.7538	108265.8	0.72499	28.8916	11.2882	0.39071	133.8728	0.96514	147	25	—0.005
42	15.133	2.185	12.948	1.8252	133.4330	107764.7	0.67887	28.8593	9.6302	0.33370	133.7007	0.96390			—0.015
43	15.116	2.319	12.797	1.8460	135.6536	108280.4	0.	28.6906	0.	0.	136.7253	0.98571			
44	15.096	2.322	12.774	1.8457	135.6205	108059.4	0.	28.6648	0.	0.	136.8150	0.98635			
45	15.012	2.541	12.471	2.0864	162.0237	126034.8	0.	28.3228	37.8168	1.33521	165.4245	1.19261			
46	15.156	2.202	12.954	1.8737	138.6244	112008.3	0.79104	28.8660	17.6447	0.61126	138.8708	1.00117	60	52	—0.012
47	15.134	2.202	12.932	1.8726	138.5023	111720.6	0.79097	28.8415	17.3179	0.60045	138.8669	1.00114	63	16	—0.008
48	15.142	2.194	12.948	1.8723	138.4690	111832.0	0.78904	28.8593	17.0327	0.59020	138.7468	1.00028	66	27	—0.011
49	15.144	2.193	12.951	1.8714	138.3692	111777.2	0.78859	28.8627	16.7394	0.57997	138.6307	0.99945	81	48	—0.007
50	15.144	2.192	12.952	1.8694	138.1559	111613.5	0.78723	28.8638	16.3058	0.56492	138.4117	0.99787	89	44	—0.010

TABLE

EXPERIMENTS UPON THE TURBINE AT THE

1	2	3		4	5						6	7	8	9	10
No. of the experiment.	DATE, 1851.	Temperature of the atmosphere in degrees of Fahrenheit's thermometer.		Height of the regulating gate, in inches.	TIME.						Duration of the experiment, in seconds.	Total number of revolutions of the wheel during the experiment.	Number of revolutions of the wheel per second.	Weight in the scale, in pounds avoirdupois.	Useful effect, or the friction of the brake, in pounds avoirdupois, raised one foot per second.
		External air in the shade.	In the wheelpit.		Beginning of the experiment.			Ending of the experiment.							
					H.	min.	sec.	H.	min.	sec.					
51	February 21, P. M.	34.75	36.50	8.55	2	17	11.50	2	23	56.00	404.50	600	1.48331	390.95	39452.4
52	" " "	34.50	36.25	"	2	24	31.00	2	32	28.00	477.00	600	1.25786	775.61	66378.6
53	" " "	34.50	36.50	"	2	33	10.00	2	41	55.50	525.50	600	1.14177	968.30	74827.2
54	" " "	34.50	36.50	"	2	41	55.50	2	50	25.00	509.50	550	1.07949	1069.05	78512.0
55	" " "	34.50	37.00	"	2	50	25.00	2	58	33.00	488.00	500	1.02459	1150.77	80215.4
56	" " "	34.50	36.75	"	3	8	34.00	3	16	20.50	466.50	450	0.96463	1242.98	81572.6
57	" " "	34.50	37.00	"	3	17	13.50	3	25	17.00	483.50	450	0.93071	1293.63	81911.6
58	" " "	34.25	37.00	"	3	25	17.00	3	33	37.00	500.00	450	0.90000	1345.47	82382.7
59	" " "	34.25	37.25	"	3	33	37.00	3	42	17.00	520.00	450	0.86538	1396.11	82195.5
60	" " "	34.25	36.75	"	3	43	15.50	3	52	14.25	538.75	450	0.83527	1444.03	82057.9
61	" " "	34.25	36.50	"	4	10	26.00	4	19	45.50	559.50	450	0.80429	1494.68	81786.2
62	" " "	34.00	36.00	"	4	31	37.50	4	40	15.50	518.00	400	0.77220	1548.69	81360.6
63	" " "	34.00	36.00	"	4	51	21.50	4	59	36.00	494.50	350	0.70779	1656.98	79788.1
64	" " "			"	5	8	37.00	5	19	17.25	640.25	1100	1.71808	0.	0.
65	February 22, A. M.			5.65	8	57	1.00	9	7	47.00	646.00	1000	1.54799	0.	0.
66	" " "	35.50	35.75	"	9	15	39.00	9	21	12.00	333.00	450	1.35135	316.03	29054.7
67	" " "	35.50	36.50	"	9	21	51.25	9	28	0.50	369.25	450	1.21869	519.69	43087.9
68	" " "	35.50	36.75	"	9	28	45.00	9	36	26.00	461.00	500	1.08460	720.20	53142.4
69	" " "	35.50	37.25	"	9	36	26.00	9	43	4.50	398.50	400	1.00376	832.26	56834.2
70	" " "	35.75	37.25	"	9	43	57.00	9	50	16.00	379.00	350	0.92348	934.74	58727.1
71	" " "	35.75	37.25	"	9	51	14.00	9	58	9.00	415.00	350	0.84337	1033.30	59287.8
72	" " "	36.75	37.25	"	9	59	12.50	10	7	48.75	516.25	400	0.77482	1115.02	58776.2
73	" " "	39.00	35.75	"	10	21	10.00	10	30	50.25	580.25	450	0.77553	1115.02	58830.1
74	" " "	38.25	36.00	"	10	42	44.50	10	51	14.50	510.00	350	0.68627	1204.84	56253.1
75	" " "	38.25	36.25	"	10	58	28.00	11	6	35.00	487.00	300	0.61602	1277.98	53559.4
76	" " "	38.25	36.25	"	11	13	59.00	11	21	17.00	438.00	200	0.45662	1482.56	46056.1
77	February 22, A. M.	37.50	35.75	5.96	11	33	20.00	11	40	9.50	409.50	350	0.85470	1482.56	86207.6
78	" " "	38.00	36.00	"	11	46	21.00	11	53	27.00	426.00	350	0.82160	1544.87	86351.4
79	" " P. M.	40.75	35.75	"	0	0	42.50	0	8	6.00	443.50	350	0.78918	1604.85	86164.4
80	February 22, P. M.	42.75	36.00	2.875	2	33	11.00	2	38	31.25	320.25	400	1.24902	0.	0.
81	" " "	43.75	36.50	"	2	42	3.00	2	47	52.00	349.00	400	1.14613	118.59	9247.0
82	" " "	44.00	37.00	"	2	48	41.00	2	54	45.50	364.50	350	0.96022	325.39	21256.6
83	" " "	44.25	37.25	"	2	55	46.00	3	2	14.50	388.50	300	0.77220	519.86	27310.9
84	" " "	43.75	37.25	"	3	2	14.50	3	9	41.00	446.50	300	0.67189	612.22	27985.1
85	" " "	43.50	37.00	"	3	11	21.50	3	18	48.00	446.50	250	0.55991	704.44	26833.8
86	" " "	43.50	36.75	"	3	20	34.50	3	27	47.00	432.50	200	0.46243	777.58	24462.9
87	" " "	43.25	36.50	"	3	33	10.50	3	44	17.00	666.50	200	0.30007	882.02	18006.4
88	" " "	42.25	36.25	"	4	2	0.					0	0.	1195.06	0.
89	" " "			"	4	10	0.					0	0.	1054.25	0.
90	February 22, P. M.	42.00	37.75	1.00	4	27	3.50	4	33	47.00	403.50	250	0.61958	118.59	4998.8
91	" " "			"	4	35	1.00	4	42	16.50	435.50	300	0.68886	73.14	3427.7
92	" " "	41.00	37.25	"	4	45	42.00	5	2	54.00	1032.00	400	0.38760	296.39	7815.6

II. — Continued.

TREMONT MILLS, IN LOWELL, MASSACHUSETTS.

	11	12	13	14	15	16	17	18	19	20	21	22	23		24
No. of the experiment.	Height of the water above the wheel, taken in the forebay, in feet.	Height of the water after passing the wheel, taken in the wheel-pit, in feet.	Total fall acting upon the wheel, in feet.	Depth of water on the weir, in feet.	Quantity of water which passed the weir, in cubic feet per second.	Total power of the water, in pounds avoirdupois raised one foot per second.	Ratio of the useful effect to the power expended.	Velocity due to the fall acting on the wheel, in feet per second.	Velocity of the exterior circumference of the wheel, in feet per second.	Ratio of the velocity of the exterior circumference of the wheel to the velocity due to the fall acting on the wheel.	Quantity of water which passed the wheel, reduced to a uniform fall of 13 feet, in cubic feet per second.	Ratio of the reduced quantity in column 21, to the reduced quantity in experiment 30.	Direction of the water leaving the wheel, as indicated by the vane.		Mean elevation of the pointer on the bell crank.
													deg.	m.	Feet.
51	15.095	2.337	12.758	1.9173	143.8319	114060.7	0.34589	28.6468	31.4548	1.09802	144.6849	1.04309	12	0	+0.002
52	15.128	2.255	12.873	1.8792	139.2094	111778.7	0.59379	28.7756	26.6739	0.92696	139.8945	1.00856	17	32	+0.001
53	15.134	2.225	12.909	1.8656	137.7518	110917.6	0.67462	28.8159	24.2121	0.84023	138.2365	0.99660	22	19	+0.007
54	15.134	2.194	12.940	1.8660	137.7962	111219.8	0.70592	28.8504	22.8914	0.79345	138.1153	0.99573	25	0	+0.013
55	15.139	2.189	12.950	1.8586	137.0026	110664.7	0.72485	28.8616	21.7272	0.75281	137.2668	0.98961	28	56	−0.011
56	15.138	2.187	12.951	1.8450	135.5434	109494.5	0.74499	28.8627	20.4557	0.70873	135.7996	0.97903	33	58	−0.016
57	15.143	2.178	12.965	1.8408	135.0974	109252.2	0.74975	28.8783	19.7365	0.68344	135.2797	0.97529	37	47	−0.014
58	15.144	2.168	12.976	1.8336	134.3304	108724.1	0.75772	28.8905	19.0852	0.66060	134.4546	0.96934	42	43	−0.011
59	15.151	2.152	12.999	1.8240	133.3014	108082.5	0.76049	28.9161	18.3511	0.63463	133.3066	0.96106	47	30	−0.005
60	15.153	2.139	13.014	1.8186	132.7344	107746.9	0.76158	28.9328	17.7125	0.61219	132.6630	0.95642	54	37	−0.004
61	15.155	2.129	13.026	1.8117	131.9960	107246.3	0.76260	28.9461	17.0556	0.58922	131.8642	0.95066	59	39	−0.042
62	15.162	2.122	13.040	1.8022	130.9913	106544.4	0.76363	28.9617	16.3751	0.56541	130.7903	0.94292	76	36	−0.021
63	15.162	2.134	13.028	1.8013	130.8932	106366.6	0.75012	28.9484	15.0091	0.51848	130.7525	0.94265	94	4	−0.022
64	15.079	2.359	12.720	1.9742	149.5470	118652.1	0.	28.6041	36.4332	1.27370	151.1840	1.08995			
65	15.139	1.969	13.170	1.7160	121.9685	100194.6	0.	29.1057	32.8262	1.12783	121.1788	0.87363			
66	15.148	2.071	13.077	1.6829	118.5511	96699.5	0.30046	29.0028	28.6564	0.98806	118.2016	0.85216	6	30	+0.001
67	15.159	2.025	13.134	1.6590	116.0987	95112.0	0.45302	29.0659	25.8432	0.88912	115.5050	0.83272	8	30	+0.002
68	15.164	1.988	13.176	1.6409	114.2599	93904.9	0.56592	29.1128	22.9997	0.79003	113.4942	0.81823	12	32	−0.020
69	15.171	1.956	13.215	1.6309	113.2448	93346.0	0.60885	29.1554	21.2856	0.73007	112.3198	0.80976	16	5	−0.027
70	15.173	1.920	13.253	1.6139	111.5197	92188.4	0.63703	29.1973	19.5831	0.67072	110.4502	0.79628	21	17	−0.006
71	15.179	1.897	13.282	1.5959	109.7180	90893.3	0.65228	29.2292	17.8844	0.61187	108.5420	0.78252	29	56	−0.014
72	15.183	1.872	13.311	1.5793	108.0452	89707.1	0.65520	29.2611	16.4306	0.56152	106.7756	0.76979	39	2	+0.001
73	15.179	1.869	13.310	1.5783	107.9493	89620.7	0.65643	29.2600	16.4457	0.56205	106.6848	0.76913	40	13	−0.029
74	15.159	1.833	13.326	1.5541	105.5341	87720.9	0.64127	29.2776	14.5530	0.49707	104.2353	0.75147	69	27	−0.028
75	15.174	1.812	13.362	1.5371	103.8516	86555.6	0.61879	29.3171	13.0631	0.44558	102.4352	0.73850	95	0	−0.024
76	15.183	1.771	13.412	1.5034	100.5410	84110.0	0.54757	29.3719	9.6830	0.32966	98.9847	0.71362	144	56	−0.029
77	15.079	2.196	12.883	1.8620	137.3618	110380.8	0.78100	28.7868	18.1246	0.62961	137.9842	0.99478	55	52	−0.037
78	15.079	2.183	12.896	1.8583	136.9694	110176.6	0.78375	28.8013	17.4226	0.60492	137.5206	0.99144	64	0	−0.029
79	15.087	2.175	12.912	1.8544	136.5469	109973.0	0.78350	28.8192	16.7351	0.58069	137.0114	0.98777	74	28	−0.018
80	14.774	1.427	13.347	1.2914	80.4534	66979.0	0.	29.3006	26.4865	0.90396	79.4007	0.57243	0	30	
81	14.769	1.400	13.369	1.2737	78.8433	65746.8	0.14065	29.3248	24.3046	0.82881	77.7476	0.56051	1	30	+0.008
82	14.772	1.377	13.395	1.2492	76.6213	64018.1	0.33204	29.3533	20.3622	0.69369	75.4831	0.54419	4	32	+0.010
83	14.783	1.348	13.435	1.2206	74.0590	62062.9	0.44006	29.3971	16.3751	0.55703	72.8501	0.52521	11	30	+0.002
84	14.793	1.315	13.478	1.1960	71.8750	60424.6	0.46314	29.4441	14.2480	0.48390	70.5889	0.50890	20	9	−0.001
85	14.806	1.293	13.513	1.1748	70.0063	59006.4	0.45476	29.4823	11.8733	0.40273	68.6646	0.49503	41	34	−0.022
86	14.820	1.264	13.556	1.1497	67.8158	57342.0	0.42661	29.5292	9.8061	0.33208	66.4105	0.47878	81	40	−0.025
87	14.803	1.244	13.559	1.1113	64.5053	54554.9	0.33006	29.5324	6.3633	0.21547	63.1616	0.45536			−0.026
88	14.762	1.246	13.516	1.0623	60.3593	50886.6	0.	29.4856	0.	0.	59.1959	0.42677			
89	14.771	1.240	13.531	1.0630	60.4190	50993.4	0.	29.5020	0.	0.	59.2216	0.42695			
90	14.806	0.821	13.985	0.7798	38.2210	33340.8	0.14993	29.9928	13.1386	0.43806	36.8505	0.26567			+0.004
91	14.815	0.814	14.001	0.7846	38.5699	33683.5	0.10176	30.0099	14.6079	0.48677	37.1655	0.26794			+0.030
92	14.832	0.812	14.020	0.7653	37.1733	32508.0	0.24042	30.0303	8.2193	0.27370	35.7956	0.25806			−0.006

DESCRIPTION OF THE DIAGRAM REPRESENTING THE EXPERIMENTS.

75. For the purpose of presenting a general view of the experiments, the coefficients of effect, at different velocities, are plotted at figure 1, plate VI., on a system of coördinates. The ratios of the velocities of the interior circumference of the wheel, to the velocities due the fall acting upon the wheel, given in column 20, table II., are taken to represent the velocities; these ratios are here called the velocities, and are taken on the axis of abscissas AX; the corresponding coefficients of effect given in column 17, table II., are taken upon the axis of ordinates AY.

76. The line CD represents the experiments made with the regulating gate fully raised;—to avoid confusion a portion of the experiments are omitted;—the experiments represented are those numbered from 4 to 42, inclusive, which were made in regular sequence, with gradually increasing weights. It will be observed in the table of experiments, that several trials were made with the brake entirely removed; these were made, generally, after the wheel had been left for some time, for the purpose of seeing if it was in as good running order as usual; if any material change had taken place, it would have been indicated by a change in the velocity of the wheel.

The experiments thus made, omitting experiment 12, in which the height in the wheelpit was not observed, are collected together in the following table.

Number of the experiment.	Ratio of the velocity of the interior circumference of the wheel, to the velocity due the fall acting upon the wheel.
13	1.33366
23	1.33567
31	1.33635
38	1.33544
45	1.33521
Mean . . .	1.33527

The greatest variation in these velocities is in experiment 13, which is $\frac{1}{[illegible]}$ part below the mean; the running condition of the wheel must, consequently, have been nearly uniform.

In all the experiments with the brake removed, the coefficient of effect, of course, is nothing, and they would be represented on the diagram by points on the axis of abscissas; for the sake of distinctness, only one of those tried when the gate was at its full height, is represented on the diagram.

There is a small irregularity in the line *CD*, at numbers 26 and 27; both these experiments were made with the same weight in the scale, and under similar circumstances, except that in 26, water was used to lubricate the friction pulley, and in 27 oil was used.

It has been stated, that, with heavy loads, the brake operates much more steadily with oil as a lubricator, than with water, and the change in the lubricator at experiment 27, was made in consequence of the difficulty experienced by the operator, in regulating the tension of the brake screws. In experiment 26, nearly his whole strength, applied to the extremity of a wrench about three feet long, was required to move the nuts, whereas, in experiment 27, the same operation was performed with great ease. Experiment 26 was of much shorter duration than experiment 27, and a portion of the discrepancy may be due to a proportionally less perfect observation of the data in 26.

The line *CD* shows that, with a velocity of the interior circumference of the wheel not less than 44 or more than 75 per cent. of that due to the fall, the useful effect is 75 per cent. or more, of the total power expended. Beyond these points, the change in the coefficient of effect is nearly equal for equal and opposite variations of speed; thus, the diagram indicates that the coefficient of effect is 70 per cent. of the power expended, at the velocities 0.360 and 0.834.

$$0.436 - 0.360 = 0.076$$
$$0.834 - 0.750 = 0.084.$$

Taking the mean of the extreme velocities, that is, of 0, when the wheel was still, and 1.335, when the brake was removed, we have

$$\frac{1.335 + 0}{2} = 0.6675.$$

which is not far from the velocity giving the maximum coefficient of effect; that is to say, *when the gate is fully raised, the coefficient of effect is a maximum when the wheel is moving with about half its maximum velocity.*

77. Experiments 43 and 44 were both made with the gate fully raised, but the wheel at rest, the brake being screwed up sufficiently tight to prevent the wheel from revolving; — they were made for the purpose of ascertaining the total effort that could be exercised by the wheel.

By reference to column 9, of the table of experiments, it will be seen that, in experiment 43, the weight sustained was 4213.38 pounds, and in 44, the weight was 3946.38 pounds. These experiments were made under circumstances nearly identical, except that in 43, the weight preponderated, and in 44, the power of the wheel preponderated. In 43, the weight was the least that would cause the scale to lower when the bell crank was placed horizontally, and then left free; on the other hand, in experiment 44, the weight was the greatest that would allow the scale to be raised under the same circumstances; that is to say, in 43, the weight represents the force exercised by the water against the wheel, *plus* the friction of the entire apparatus, and in 44, the weight represents the same thing, *minus* the friction; the difference of the weights, or $4213.38 - 3946.38 = 267$ pounds, represents double the friction, and the true force exercised by the water against the wheel, is represented by the weight

$$\frac{4213.38 + 3946.38}{2} = 4079.88 \text{ pounds.}$$

This weight acted at a distance from the centre of the wheel, equal to the effective length of the brake, or 10.827778 feet (art. 50).

The radius of the turbine, at the outer extremities of the buckets, is 4.146 feet (art. 35), consequently, the equivalent force acting tangentially at the outer extremities of the buckets, was

$$\frac{4079.88 \times 10.827778}{4.146} = 10655.1 \text{ pounds.}$$

78. The line *EF* represents the experiments numbered 77, 78, and 79, made with the gate raised 9.96 inches, or about 87 per cent. of the full height. By a reference to the table of experiments, it will be seen that, although the regulating gate was lowered 13 per cent., the quantity of water discharged by the wheel was diminished less than one per cent.

79. The line *GH* represents the experiments numbered from 51 to 64, inclusive, made with the gate raised 8.55 inches, or about three fourths of the full height.

80. The line *IK* represents the experiments numbered from 65 to 76, inclusive, made with the gate raised 5.65 inches, or nearly a half of the full height.

81. The line *LM* represents the experiments numbered from 80 to 87, inclusive, made with the gate raised 2.875 inches, or one fourth of its full height. Experiments 88 and 89 were made with the same height of gate, but with the wheel held fast by the brake; the force exerted by the wheel at the distance 10.827778 feet, independent of friction, was

$$\frac{1195.06 + 1054.25}{2} = 1124.65 \text{ pounds.}$$

82. The line $N O$ represents the three experiments numbered 90, 91, and 92, made with the regulating gate raised one inch.

An examination of the diagram will show that the velocity corresponding to the maximum coefficient of effect, diminishes with the height of the gate. For heights not less than one fourth of the whole height, this diminution is sufficiently regular; for heights less than one fourth, the experiments are not sufficient to indicate the velocity giving the best effect, but the diminution is evidently more rapid than for greater heights of gate.

PATH DESCRIBED BY A PARTICLE OF WATER IN PASSING THROUGH THE WHEEL.

83. As in many other problems in hydraulics, resort is here had to a particular hypothesis, which, at best, is only an approximation to the truth, nevertheless, it may be the means of throwing some light upon the mode in which the water acts upon the wheel.

The particular hypothesis here assumed is this; *every particle of water contained in the wheel, situated at the same distance from the axis, moves in the same direction relative to the radius, and with the same velocity.* According to this hypothesis, the successive sections in which the same particles of water are found, are in cylindrical surfaces, concentric with the wheel.

Applying this hypothesis to experiment 30, on the Tremont Turbine, let us suppose

Q' = the mean quantity of water discharged through each aperture of the wheel, in cubic feet per second.

ω = the angular velocity of the wheel.

R = the radius of the circle inscribing the inner edges of the buckets, or $O A$, figure 3, plate VI.

R' = the radius $O B$.

t = the time occupied by a particle of water in passing from the section $A D$ to the section $B C$, or, which is the same thing, through the radial distance $R' - R$.

A = the area of $A B C D$, in square feet.

H = the mean height, in feet, between the crowns of the wheel, between the sections $A D$ and $B C$.

We have

$A H$ = the volume of water contained between the sections $A D$ and $B C$.

t is the time occupied by a particle of water in passing from the section AD to the section BC, and it will evidently be the time required for the discharge of the volume AH. We find t by the proportion

$$Q'' : 1 :: AH : t = \frac{AH}{Q''}.$$

If the wheel was at rest, a particle of water at A would arrive at B in the time t, but the wheel is moving with the angular velocity ω, therefore the point B, in the time t, will have advanced to E, and

$$BE = R' \omega t = \frac{R' \omega AH}{Q''},$$

consequently, a particle of water at A, instead of being at B, at the end of the time t, will have arrived, by some path, at the point E. In this manner, by taking successive values of R', sufficiently near to each other, the entire path of a particle of water, from its entrance into the wheel, up to the moment of its discharge, may be traced; and as, by the hypothesis, all the particles at the same distance from the axis move with the same velocity, and in the same relative direction, the path of the entire stream, from its entrance into the wheel to its discharge, will be determined.

In experiment 30, we have the total quantity discharged by the wheel equal to 138.1892 cubic feet per second; as the wheel has forty-four apertures,

$$Q'' = \frac{138.1892}{44} = 3.14066 \text{ cubic feet per second.}$$

The velocity of the interior circumference of the wheel was 18.0474 feet per second, and the interior radius of the wheel being 3.375 feet, we have

$$\omega = \frac{18.0474}{3.375} = 5.3474 \text{ feet per second,}$$

consequently,

$$BE = \frac{5.3474\, R' AH}{3.14066} = 1.7026\, R' AH.$$

84. The successive steps in the calculation for the entire path, are given in table III.

The arcs of circles FG, HI, etc. are drawn on a plan of the buckets, figure 2, plate VI., with the radii contained in the first column.

Column 2 contains the entire areas of these circles.

Column 3 contains the areas of the rings comprised between these circles, which are obtained by taking the differences of the successive areas in column 2.

Column 4 contains the areas reduced to square feet, of that part of each ring corresponding to a single aperture in the wheel, including also the area occupied by the thickness of the corresponding part of one bucket.

Column 5. Corrections for the thickness of the buckets;—these are deduced from measurements taken on a full sized plan of the buckets.

Column 6. True areas of the partial rings, being the differences of the corresponding areas in columns 4 and 5.

Column 7. Mean heights of the partial rings;—these are also taken from a full sized drawing of the wheel.

Column 8. Volumes of the partial rings, or the products of the corresponding numbers in columns 6 and 7.

Column 9. Volumes between the radius R, and the successive values of the radius R'. These are obtained by adding together the volumes of the partial rings, up to the corresponding radius;—they are the successive values of $A\,H$.

Column 10. The ordinates;—these are successive values of

$$1.7026\,R'A\,H,$$

the successive values of R' being taken in feet, instead of inches, as they are given in column 1.

TABLE III.

1	2	3	4	5	6	7	8	9	10
Value of R, and successive values of R'. Inches.	Areas in square inches, of circles of the radii in the last column.	Areas in square inches of the complete rings.	$\frac{1}{44}$ of the areas of the rings in the last column. Square feet.	Correction for the thickness of the bucket, in square feet.	True areas of the partial rings, in square feet.	Mean height of the partial rings, in feet.	Volumes of the partial rings, in cubic feet.	Volumes between R and the successive values of R'. Cubic feet.	Ordinates in feet, to be measured on arcs of the corresponding radii in column 1.
40.5	5152.997								
41.5	5410.608	257.611	0.04066	0.00091	0.03975	0.9264	0.03682	0.03682	0.2168
42.5	5674.502	263.894	0.04165	0.00099	0.04066	0.9080	0.03692	0.07374	0.4447
43.5	5944.679	270.177	0.04264	0.00106	0.04158	0.8940	0.03717	0.11091	0.6845
44.5	6221.139	276.460	0.04363	0.00115	0.04248	0.8840	0.03755	0.14846	0.9373
45.5	6503.882	282.743	0.04462	0.00128	0.04334	0.8775	0.03803	0.18649	1.2039
46.5	6792.909	289.027	0.04562	0.00146	0.04416	0.8755	0.03866	0.22515	1.4854
47.5	7088.218	295.309	0.04661	0.00174	0.04487	0.8800	0.03949	0.26464	1.7835
48.5	7389.811	301.593	0.04760	0.00212	0.04548	0.8920	0.04057	0.30521	2.1003
49.0	7542.964	153.153	0.02417	0.00138	0.02279	0.9055	0.02064	0.32585	2.2654
49.25	7620.129	77.165	0.01218	0.00078	0.01140	0.9145	0.01042	0.33627	2.3498
49.50	7697.687	77.558	0.01224	0.00087	0.01137	0.9210	0.01047	0.34674	2.4352
49.75	7775.638	77.951	0.01230	0.00081	0.01149	0.9277	0.01066	0.35740	2.5228

85. The arcs FG, HI, etc., figure 2, plate VI., are taken equal to the ordinates 0.2168, 0.4447 etc., in column 10 of the table; the points Q, G, I, etc. K, are joined by a line, which is the limit of the stream on one side. The limit on the other side is found by making the arcs $GL = FN$, $IM = HO$, etc.; the points R, L, M, etc. P, being joined by a line, give the limits of the stream on this side.

86. By an inspection of the figure, it is plain that, in experiment 30, the path of the water through the wheel must have been a continuation of the direction given to it by the fixed guides VW, and that there was no sudden change of direction or velocity, up to a point near where the water was discharged from the wheel. The abrupt change at this point, indicated by the figure, could not, in reality, have taken place, as we know by the direction assumed by the vane, which is represented at ST in its mean position during the experiment.

87. The foregoing hypothesis will evidently lead to results more nearly correct, the nearer the buckets are to each other, until, in the case in which the spaces between them are infinitely small, it will give the path accurately. In applications like the above, where the spaces are very considerable, it is assumed by the hypothesis that the water passes through in curved laminæ, superimposed on each other, the first of which, in contact with the concavity of the bucket, is constrained by it and the rotation of the wheel, to move in a particular path; this, in its turn, constrains the next lamina to move in a similar path; and so on.

By an inspection of figure 2, plate VI., it is reasonable to suppose, that a lamina, far removed from the concavity of the bucket, will take a path differing from that of a lamina near it; the abruptness in the curve near its extremity, will be diminished, somewhat in proportion to the distance of the lamina from the concavity of the bucket, the water passing out from the wheel more nearly in the direction in which it was moving, during its approach to the circumference of the wheel. These views go far to explain the discrepancy between the path determined by the hypothesis, and the direction assumed by the vane.

88. Whatever objection may be made to the method by which the path, given in figure 2, plate VI., is obtained, it cannot be denied that its general course must have been nearly as represented; this being admitted, it is difficult to see how centrifugal force can operate in the important manner that is commonly assigned to it. The path is concave to the axis only in a very slight degree, and through a part only of its course; nevertheless, it is only in con-

sequence of a concavity in the path, that centrifugal force can have any existence. With the gate only partially raised, this force may act powerfully in increasing the discharge, and a similar effect may be produced; at high velocities, with the gate fully raised; but in experiment 30, giving the maximum coefficient of effect, it can have had only a slight action.

RULES FOR PROPORTIONING TURBINES.

89. In making the designs for the Tremont, and other turbines, the author has been guided by the following rules, which he has been led to by a comparison of several turbines designed by Mr. Boyden, which have been carefully tested and found to operate well.

Rule 1st. The sum of the shortest distances between the buckets, should be equal to the diameter of the wheel.

Rule 2d. The height of the orifices at the circumference of the wheel, should be equal to one tenth of the diameter of the wheel.

Rule 3d. The width of the crowns should be four times the shortest distance between the buckets.

Rule 4th. The sum of the shortest distances between the curved guides, taken near the wheel, should be equal to the interior diameter of the wheel.

The turbines, from a comparison of which the above rules were derived, varied in diameter from twenty-eight inches to nearly one hundred inches, and operated on falls from thirty feet to thirteen feet. The author believes that they may be safely followed for all falls between five feet and forty feet, and for all diameters not less than two feet, and, with judicious arrangements in other respects, and careful workmanship, a useful effect of seventy-five per cent. of the power expended, may be relied upon. For falls greater than forty feet, the second rule should be modified, by making the height of the orifices smaller in proportion to the diameter of the wheel.

90. Taking the foregoing rules as a basis, we may, by aid of the experiments on the Tremont Turbine, establish the following formulas.

Let $D=$ the diameter of the wheel at the outer extremities of the buckets.

$d=$ the diameter of the wheel, at the interior extremities of the buckets.

$H=$ the height of the orifices of discharge, at the outer extremities of the buckets.

$W=$ the width of the crowns occupied by the buckets.

N = the number of buckets.

n = the number of guides.

P = the horse-power of the turbine; a horse-power being 550 pounds avoir. raised one foot per second.

h = the fall acting upon the wheel.

Q = the quantity of water expended by the turbine, in cubic feet per second.

V = the velocity due the fall acting upon the wheel.

V' = the velocity of the water passing the narrowest sections of the wheel.

v = the velocity of the interior circumference of the wheel: all the velocities being in feet per second.

C = the coefficient of V, or the ratio of the real velocity of the water passing the narrowest sections of the wheel, to the theoretical velocity due the fall acting upon the wheel.

The unit of length is the English foot.

It is assumed that the useful effect is seventy-five per cent. of the total power of the water expended.

According to rule 1, we have the sum of the widths of the orifices of discharge, equal to D. Then the sum of the areas of all the orifices of discharge, is equal to DH.

By the fundamental law of hydraulics we have

$$V = \sqrt{2gh}.$$

therefore

$$V' = C\sqrt{2gh}.$$

We can find the value of C in the last equation by experiment 30, on the Tremont Turbine. In that wheel we have for the sum of the widths of the orifices of discharge, $44 \times 0.18757 = 8.25308$ feet, and the height of the orifices of discharge $= 0.9314$ feet. Then we have, for the sum of the areas of all the orifices of discharge,

$$HD = 8.25308 \times 0.9314 = 7.68692 \text{ square feet.}$$

By experiment 30, we have

$$Q = 138.1892 \text{ cubic feet per second,}$$
$$h = 12.903 \text{ feet,}$$
$$\sqrt{2g} = 8.0202 \text{ feet,}$$

consequently,

$$138.1892 = 7.68692 \times 8.0202 \sqrt{12.903}\ C,$$
$$\text{or } C = 0.624.$$

By rule 2, we have $H = 0.10\,D$:

$$\text{then } HD = 0.10\,D^2,$$
$$\text{and } Q = HDV' = 0.10\,D^2\,C\sqrt{2gh},$$
$$\text{or } Q = 0.5\,D^2\sqrt{h}.$$

Calling the weight of a cubic foot of water 62.33 pounds avoir. we have

$$P = \frac{0.75 \times 62.33}{550}\,Qh,$$
$$\text{or } P = 0.085\,Qh;$$

or, substituting the value of Q just found,

$$P = 0.0425\,D^2 h\sqrt{h},$$

from which we may deduce

$$D = 4.85\sqrt{\frac{P}{h\sqrt{h}}}.$$

91. The number of buckets is, to a certain extent, arbitrary, and would usually be determined by practical considerations: some of the ideas to be kept in mind are the following.

The pressure on each bucket is less, as the number is greater; the greater number will therefore permit of the use of the thinner iron, which is important, in order to obtain the best results. The width of the crowns will be less for a greater number of buckets: a narrow crown appears to be favorable to the useful effect, when the gate is only partially raised. As the spaces between the buckets must be proportionally narrower for a larger number of buckets, the liability to become choked up, either with anchor ice, or other substances, is increased. The amount of power lost by the friction of the water against the surfaces of the buckets, will not be materially changed, as the total amount of rubbing surface on the buckets, will be nearly constant for the same diameter: there will be a little less on the crown, for the larger number. The cost of the wheel will probably increase with the number of buckets. The thickness and quality of the iron, or other metal intended to be used for the buckets, will sometimes be an element. In some waters, wrought iron is rapidly corroded.

The author is of opinion that a general rule cannot be given for the number of buckets; among the numerous turbines working satisfactorily in Lowell, there are examples in which the shortest distance between the buckets is as small as 0.75 inches, and in others as large as 2.75 inches.

As a guide in practice, to be controlled by particular circumstances, the following is proposed; to be limited to diameters of not less than two feet;

$$N = 3\,(D + 10).$$

Taking the nearest whole number for the value of N.

The Tremont Turbine is $8\frac{1}{3}$ feet in diameter, and, according to the proposed rule, should have fifty-five buckets, instead of forty-four. With fifty-five buckets, the crowns should have a width of 7.2 inches, instead of 9 inches; with the narrower width, it is probable that the useful effect, in proportion to the power expended, would have been a little greater when the gate was partially raised.

92. By the 3d rule, we have for the width of the crowns,

$$W = \frac{4\,D}{N};$$

and for the interior diameter of the wheel

$$d = D - \frac{8\,D}{N}.$$

By the 4th rule, d is also equal to the sum of the shortest distances between the guides, where the water leaves them.

93. The number n, of the guides, is, to a certain extent, arbitrary; the practice at Lowell has been, usually, to have from a half to three fourths of the number of the buckets; exactly half would probably be objectionable, as it would tend to produce pulsations, or vibrations.

94. The proper velocity to be given to the wheel, is an important consideration. Experiment 30, on the Tremont Turbine, gives the maximum coefficient of effect for that wheel; in that experiment the velocity of the interior circumference of the wheel, is 0.62645 of the velocity due to the fall acting upon the wheel. By reference to the other experiments with the gate fully raised, it will be seen, however, that the coefficient of effect varies only about two per cent. from the maximum, for any velocity of the interior circumference, between fifty per cent. and seventy per cent. of that due to the fall acting upon the wheel. By reference to the experiments in which the gate is only partially raised, it will be seen that the maximum corresponds to slower velocities; and as turbines,

to admit of being regulated in velocity for variable work, must, almost necessarily, be used with a gate not fully raised, it would appear proper to give them a velocity such, that they will give a good effect under these circumstances.

With this view, the following is extracted from the experiments in table II.

Number of the experiment.	Height of the regulating gate, in inches.	Ratio of the velocity of the interior circumference of the wheel, to the velocity due the fall acting upon the wheel, corresponding to the maximum coefficient of effect.
30	11.49	0.62645
62	8.55	0.56541
73	5.65	0.56205
84	2.875	0.48390

By this table it would appear, that, as turbines are generally used, a velocity of the interior circumference of the wheel, of about fifty-six per cent. of that due to the fall acting upon the wheel, would be most suitable. By reference to the diagram at plate VI., it will be seen that, at this velocity when the gate is fully raised, the coefficient of effect will be within less than one per cent. of the maximum.

Other considerations, however, must usually be taken into account, in determining the velocity; the most frequent is the variation of the fall under which the wheel is intended to operate. If, for instance, it was required to establish a turbine of a given power, on a fall liable to be diminished to one half, by backwater, and, that the turbine should be of a capacity to give the requisite power at all times; in this case, the dimensions of the turbine must be determined for the smallest fall; but if it has assigned to it a velocity, to give the maximum effect at the smallest fall, it will evidently move too slow for the greatest fall; and this is the more objectionable, as, usually, when the fall is greatest, the quantity of water is the least, and it is of the most importance to obtain a good effect. It would then be usually, the best arrangement, to give the wheel a velocity corresponding to the maximum coefficient of effect, when the fall is the greatest. To assign this velocity, we must first find the proportional height of gate, when the fall is greatest; this may be determined approximately by aid of the experiments on the Tremont Turbine.

We have seen that $P = 0.085\, Q h$.

Now, if h is increased to $2\,h$, the velocity, and, consequently, the quantity of water discharged, will be increased in the proportion of $\sqrt{h}$ to $\sqrt{2\,h}$; that is to say, the quantity for the fall $2\,h$, will be $\sqrt{2}\; Q$.

Calling P' the total power of the turbine on the double fall, we have

$$P' = 0.085 \sqrt{2}\, Q\, 2h,$$

or

$$P' = 0.085 \times 2.8284\, Qh.$$

Thus, the total power of the turbine is increased 2.8284 times, by doubling the fall; on the double fall, therefore, in order to preserve the effective power uniform, the regulating gate must be shut down to a point that will give only $\frac{1}{2.8284}$ part of the total power of the turbine.

In experiment 15, the fall acting upon the wheel was 12.888 feet, and the total useful effect of the turbine was 85625.3 pounds raised one foot per second; $\frac{1}{2.8284}$ part of this is 30273.4 lbs.; consequently, the same opening of gate that would give this last power, on a fall of 12.888 feet, would give a power of 85625.3 lbs. raised one foot per second, on a fall of 2×12.888 feet $= 25.776$ feet. To find this opening of gate, we must have recourse to some of the other experiments.

In experiment 73, the fall was 13.310 feet, the height of gate 5.65 inches, and the useful effect 58830.1 pounds. In experiment 83, the fall was 13.435 feet, the height of gate 2.875 inches, and the useful effect, 27310.9 pounds. Reducing both these useful effects to what they would have been, if the fall was 12.888 feet, —

the useful effect in experiment 73, $58830.1 \left(\frac{12.888}{13.310}\right)^{\frac{3}{2}} = 56054.5,$

" " " 83, $27310.9 \left(\frac{12.888}{13.435}\right)^{\frac{3}{2}} = 25660.1.$

By a comparison of these useful effects with the corresponding heights of gate, we find, by simple proportion of the differences, that a useful effect of 30273.4 pounds raised one foot high per second, would be given when the height of the regulating gate was 3.296 inches.

By another mode: —

$$\text{as } 25660.1 : 2.875 :: 30273.4 : 2.875 \times \frac{30273.4}{25660.1} = 3.392 \text{ inches},$$

a little consideration will show, that the first mode must give too little, and the second, too much; taking a mean of the two results, we have for the height of the gate, giving $\frac{1}{2.8284}$ of the total power of the turbine, 3.344 inches. Referring to table II., we see that, with this height of gate, in order to obtain the best coefficient of useful effect, the velocity of the interior circumference of the wheel, should be about one half of that due to the fall acting upon the wheel; and by comparison of experiments 74 and 84, it will be seen that, with this height of gate, and with this velocity, the coefficient of useful effect must be near 0.50.

This example shows, in a strong light, the well-known defect of the turbine, viz., giving a diminished coefficient of useful effect, at times when it is important to obtain the best results. One remedy for this defect would be, to have a spare turbine, to be used when the fall is greatly diminished; this arrangement would permit the principal turbine to be made nearly of the dimensions required for the greatest fall. As at other heights of the water, economy of water is usually of less importance, the spare turbine might generally be of a cheaper construction.

95. *To lay out the curve of the buckets*, the author makes use of the following method.

Referring to plate III., figure 1, the number of buckets, N, having been determined by the preceding rules, set off the arc $gi = \frac{\pi D}{N}$.

Let $\omega = gh$, the shortest distance between the buckets;
$t =$ the thickness of the metal forming the buckets.

Make the arc $gk = 5\omega$. Draw the radius Ok, intersecting the interior circumference of the wheel at l; the point l will be the inner extremity of the bucket. Draw the directrix lm tangent to the inner circumference of the wheel. Draw the arc on, with the radius $\omega + t$, from i, as a centre; the other directrix, gp, must be found by trial, the required conditions being, that, when the line ml is revolved round to the position gt, the point m being constantly on the directrix gp, and another point at the distance $mg = rs$, from the extremity of the line describing the bucket, being constantly on the directrix ml, the curve described shall just touch the arc no. A convenient line for a first approximation, may be drawn by making the angle $Ogp = 11°$. After determining the directrix according to the preceding method, if the angle Ogp should be greater than 12°, or less than 10°, the length of the arc gk should be changed, to bring the angle within these limits.

The curve $gss's''l$, described as above, is nearly the quarter of an ellipse, and would be precisely so, if the angle gml was a right angle; the curve may be readily described, mechanically, with an apparatus similar to the elliptic trammel; there is, however, no difficulty in drawing it by a series of points, as is sufficiently obvious.

96. The trace adopted by the author, for the corresponding guides, is as follows.

The number n having been determined, divide the circle, in which the extremities of the guides are found, into n equal parts, $v\,w$, $w\,x$, etc.

Put ω' for the width between two adjoining guides,

and t' for the thickness of the metal forming the guides.

We have by rule 4, $\omega' = \frac{d}{n}$.

With w as a centre, and the radius $\omega' + t'$, draw the arc $y\,z$; and with x as a centre, and the radius $2(\omega' + t')$, draw the arc $a'\,b'$. Through v draw the portion of a circle $v\,c'$, touching the arcs $y\,z$ and $a'\,b'$; this will be the curve for the essential part of the guide. The remainder of the guide, $c'd'$, should be drawn tangent to the curve $c'v$; a convenient radius is one that would cause the curve $c'd'$, if continued, to pass through the centre O. This part of the guide might be dispensed with, except that it affords great support to the part $c'v$, and thus permits the use of much thinner iron than would be necessary, if the guide terminated at c', or near it.

97. Collecting together the foregoing formulas for proportioning turbines, which, it is understood, are to be limited to falls not exceeding forty feet, and to diameters not less than two feet; we have

for the horse-power,

$$P = 0.0425\,D^2 h\sqrt{h};$$

for the diameter,

$$D = 4.85\sqrt{\frac{P}{h\sqrt{h}}};$$

for the quantity of water discharged per second,

$$Q = 0.5\,D^2\sqrt{h};$$

for the velocity of the interior circumference of the wheel, when the fall is not very variable,

$$v = 0.56\sqrt{2gh},$$

or,

$$v = 4.491\sqrt{h};$$

for the height of the orifices of discharge,

$$H = 0.10\,D;$$

for the number of buckets,

$$N = 3\,(D + 10)\,;$$

for the shortest distance between two adjacent buckets,

$$w = \frac{D}{N}\,;$$

for the width of the crown occupied by the buckets,

$$W = \frac{4\,D}{N}\,;$$

for the interior diameter of the wheel,

$$d = D - \frac{8\,D}{N}\,;$$

for the number of guides,

$$n = 0.50\,N \text{ to } 0.75\,N\,;$$

for the shortest distance between two adjacent guides,

$$w' = \frac{d}{n}.$$

Table IV. has been computed by these formulas.

For falls greater than forty feet, the height of the orifices in the circumference of the wheel, should be diminished; the foregoing formulas may, however, still be made use of; thus, supposing that for a high fall, it is determined to make the orifices three fourths of that given by the formula; divide the given power, or quantity of water to be used, by 0.75, and use the quotient in place of the true power, or quantity, in determining the dimensions of the turbine; no modification of the dimensions will be necessary, except that $\frac{1}{10}$ of the diameter of the turbine should be diminished to $\frac{3}{40}$ of the diameter, to give the height of the orifices in the circumference.

98. It is plain, from the method by which the preceding formulas have been obtained, that they cannot be considered as established, but should only be taken as guides in practical applications, until some more satisfactory are proposed, or the intricacies of the turbine have been more fully unravelled. The turbine has been an object of deep interest to many learned mathematicians, but, up to this time, the results of their investigations, so far as they have been published, have afforded but little aid to Hydraulic Engineers.

TABLE IV.

Table for Turbines of different diameters, operating on different falls; assuming that the useful effect is seventy-five per cent. of the power expended; also that the velocity of the interior circumference is fifty-six per cent. of the velocity due the fall; and also that the height between the crowns is $\frac{1}{10}$ of the outside diameter.

Fall in feet.	Outside diameter 2.000 feet. Inside " 1.556 " Number of buckets 26.			Outside diameter 3.000 feet. Inside " 2.386 " Number of buckets 33.			Outside diameter 4.000 feet. Inside " 3.238 " Number of buckets 42.			Outside diameter 5.000 feet. Inside " 4.111 " Number of buckets 45.			Outside diameter 6.000 feet. Inside " 5.000 " Number of buckets 48.		
	Quantity of water discharged in cubic feet per second.	Number of horse-power.	Number of revolutions per minute.	Quantity of water discharged in cubic feet per second.	Number of horse-power.	Number of revolutions per minute.	Quantity of water discharged in cubic feet per second.	Number of horse-power.	Number of revolutions per minute.	Quantity of water discharged in cubic feet per second.	Number of horse-power.	Number of revolutions per minute.	Quantity of water discharged in cubic feet per second.	Number of horse-power.	Number of revolutions per minute.
5	4.47	1.90	123.3	10.06	4.28	80.4	17.88	7.60	59.2	27.95	11.88	46.7	40.25	17.11	38.4
6	4.90	2.50	135.1	11.02	5.62	88.1	19.60	9.99	64.9	30.62	15.61	51.1	44.09	22.49	42.0
7	5.29	3.15	145.9	11.91	7.08	95.2	21.17	12.59	70.1	33.07	19.68	55.2	47.62	28.34	45.4
8	5.66	3.85	156.0	12.73	8.66	101.7	22.63	15.39	74.9	35.35	24.04	59.0	50.91	34.62	48.5
9	6.00	4.59	165.4	13.50	10.33	107.9	24.00	18.36	79.5	37.50	28.69	62.6	54.00	41.31	51.5
10	6.32	5.38	174.4	14.23	12.10	113.7	25.30	21.50	83.8	39.53	33.60	66.0	56.92	48.38	54.2
11	6.63	6.20	182.9	14.92	13.95	119.3	26.53	24.81	87.9	41.46	38.76	69.2	59.70	55.82	56.9
12	6.93	7.07	191.0	15.59	15.90	124.6	27.71	28.27	91.8	43.30	44.17	72.3	62.36	63.60	59.4
13	7.21	7.97	198.8	16.23	17.93	129.7	28.84	31.87	95.5	45.07	49.80	75.2	64.90	71.72	61.9
14	7.48	8.90	206.3	16.84	20.04	134.6	29.93	35.62	99.1	46.77	55.66	78.1	67.35	80.15	64.2
15	7.75	9.88	213.5	17.43	22.22	139.3	30.98	39.50	102.6	48.41	61.72	80.8	69.71	88.88	66.4
16	8.00	10.88	220.5	18.00	24.48	143.9	32.00	43.52	106.0	50.00	68.00	83.5	72.00	97.92	68.6
17	8.25	11.92	227.3	18.55	26.80	148.3	32.99	47.66	109.2	51.54	74.47	86.0	74.22	107.24	70.7
18	8.49	12.98	233.9	19.09	29.21	152.6	33.94	51.93	112.4	53.03	81.14	88.5	76.37	116.84	72.8
19	8.72	14.08	240.3	19.61	31.68	156.8	34.87	56.32	115.5	54.49	87.99	90.9	78.46	126.71	74.8
20	8.94	15.21	246.6	20.12	34.21	160.9	35.78	60.82	118.5	55.90	95.03	93.3	80.50	136.84	76.7
21	9.17	16.36	252.7	20.62	36.81	164.8	36.66	65.44	121.4	57.28	102.25	95.6	82.49	147.24	78.6
22	9.38	17.54	258.6	21.11	39.47	168.7	37.52	70.17	124.2	58.63	109.64	97.9	84.43	157.88	80.5
23	9.59	18.75	264.4	21.58	42.19	172.5	38.37	75.01	127.0	59.95	117.20	100.1	86.32	168.76	82.3
24	9.80	19.99	270.1	22.04	44.97	176.2	39.19	79.95	129.8	61.24	124.92	102.2	88.18	179.89	84.0
25	10.00	21.25	275.7	22.50	47.81	179.8	40.00	85.00	132.4	62.50	132.81	104.3	90.00	191.25	85.8
26	10.20	22.54	281.1	22.95	50.71	183.4	40.79	90.15	135.1	63.74	140.86	106.4	91.78	202.84	87.5
27	10.39	23.85	286.5	23.38	53.66	186.9	41.57	95.40	137.6	64.95	149.06	108.4	93.53	214.65	89.1
28	10.58	25.19	291.8	23.81	56.67	190.3	42.33	100.75	140.2	66.14	157.42	110.4	95.25	226.69	90.8
29	10.77	26.55	296.9	24.23	59.73	193.7	43.08	106.20	142.6	67.31	165.93	112.4	96.93	238.94	92.4
30	10.95	27.93	302.0	24.65	62.85	197.0	43.82	111.74	145.1	68.46	174.59	114.3	98.59	251.41	94.0
31	11.14	29.34	307.0	25.05	66.02	200.3	44.54	117.37	147.5	69.60	183.39	116.2	100.22	264.08	95.5
32	11.31	30.77	311.9	25.46	69.24	203.5	45.25	123.09	149.8	70.71	192.33	118.0	101.82	276.96	97.0
33	11.49	32.23	316.7	25.85	72.51	206.6	45.96	128.91	152.2	71.81	201.42	119.9	103.40	290.04	98.5
34	11.66	33.70	321.5	26.24	75.83	209.7	46.65	134.81	154.5	72.89	210.64	121.7	104.96	303.33	100.0
35	11.83	35.20	326.2	26.62	79.20	212.8	47.33	140.80	156.7	73.95	220.00	123.4	106.49	316.81	101.5
36	12.00	36.72	330.8	27.00	82.62	215.8	48.00	146.88	158.9	75.00	229.50	125.2	108.00	330.48	102.9
37	12.17	38.26	335.4	27.37	86.09	218.8	48.66	153.04	161.1	76.03	239.13	126.9	109.49	344.34	104.3
38	12.33	39.82	339.9	27.74	89.60	221.7	49.32	159.29	163.3	77.05	248.89	128.6	110.96	358.40	105.7
39	12.49	41.40	344.3	28.10	93.16	224.6	49.96	165.62	165.4	78.06	258.78	130.3	112.41	372.64	107.1
40	12.65	43.01	348.7	28.46	96.77	227.5	50.60	172.03	167.5	79.06	268.79	132.0	113.84	387.06	108.5

TABLE IV. — Continued.

Fall in feet.	Outside diameter 7.000 feet. Inside " 5.902 " Number of buckets 52.			Outside diameter 8.000 feet. Inside " 6.815 " Number of buckets 54.			Outside diameter 9.000 feet. Inside " 7.737 " Number of buckets 57.			Outside diameter 10.000 feet. Inside " 8.667 " Number of buckets 60.		
	Quantity of water discharged, in cubic feet per second.	Number of horse-power.	Number of revolutions per minute.	Quantity of water discharged, in cubic feet per second.	Number of horse-power.	Number of revolutions per minute.	Quantity of water discharged, in cubic feet per second.	Number of horse-power.	Number of revolutions per minute.	Quantity of water discharged, in cubic feet per second.	Number of horse-power.	Number of revolutions per minute.
5	54.78	23.28	32.5	71.55	30.41	28.1	90.56	38.49	24.8	111.80	47.52	22.1
6	60.01	30.61	35.6	78.38	39.97	30.8	99.20	50.59	27.2	122.47	62.46	24.2
7	64.82	38.57	38.4	84.67	50.37	33.3	107.15	63.76	29.3	132.29	78.71	26.2
8	69.30	47.12	41.1	90.51	61.55	35.6	114.55	77.90	31.4	141.42	96.17	28.0
9	73.50	56.23	43.6	96.00	73.44	37.8	121.50	92.95	33.3	150.00	114.75	29.7
10	77.47	65.86	46.0	101.19	86.02	39.8	128.07	108.86	35.1	158.11	134.40	31.3
11	81.26	75.97	48.2	106.13	99.23	41.7	134.32	125.59	36.8	165.83	155.05	32.8
12	84.87	86.57	50.3	110.85	113.07	43.6	140.30	143.10	38.4	173.21	176.67	34.3
13	88.34	97.61	52.4	115.38	127.49	45.4	146.03	161.36	40.0	180.28	199.21	35.7
14	91.67	109.09	54.4	119.73	142.48	47.1	151.53	180.33	41.5	187.08	222.63	37.0
15	94.89	120.98	56.3	123.94	158.02	48.7	156.86	199.99	42.9	193.65	246.90	38.3
16	98.00	133.28	58.1	128.00	174.08	50.3	162.00	220.32	44.3	200.00	272.00	39.6
17	101.02	145.97	59.9	131.94	190.65	51.9	166.99	241.29	45.7	206.16	297.89	40.8
18	103.94	159.03	61.7	135.76	207.72	53.4	171.83	262.89	47.0	212.13	324.56	42.0
19	106.79	172.47	63.3	139.48	225.27	54.9	176.53	285.10	48.3	217.94	351.98	43.1
20	109.57	186.26	65.0	143.11	243.28	56.3	181.12	307.91	49.6	223.61	380.13	44.3
21	112.27	200.41	66.6	146.64	261.75	57.7	185.60	331.28	50.8	229.13	408.99	45.4
22	114.91	214.89	68.2	150.09	280.67	59.0	189.96	355.23	52.0	234.52	438.55	46.4
23	117.50	229.71	69.7	153.47	300.03	60.4	194.23	379.72	53.2	239.79	468.79	47.5
24	120.02	244.85	71.2	156.77	319.81	61.7	198.41	404.76	54.3	244.95	499.70	48.5
25	122.50	260.31	72.7	160.00	340.00	62.9	202.50	430.31	55.4	250.00	531.25	49.5
26	124.93	276.09	74.1	163.17	360.60	64.2	206.51	456.39	56.5	254.95	563.44	50.5
27	127.30	292.17	75.5	166.28	381.61	65.4	210.45	482.97	57.6	259.81	596.26	51.4
28	129.64	308.55	76.9	169.33	403.00	66.6	214.31	510.05	58.7	264.58	629.69	52.4
29	131.93	325.22	78.3	172.32	424.78	67.8	218.09	537.61	59.7	269.26	663.72	53.3
30	134.19	342.19	79.6	175.27	446.94	68.9	221.83	565.66	60.7	273.86	698.35	54.2
31	136.41	359.44	80.9	178.17	469.47	70.1	225.50	594.18	61.7	278.39	733.55	55.1
32	138.59	376.97	82.2	181.02	492.37	71.2	229.10	623.16	62.7	282.84	769.33	56.0
33	140.74	394.78	83.5	183.82	515.63	72.3	232.66	652.59	63.7	287.23	805.67	56.9
34	142.86	412.86	84.7	186.59	539.24	73.4	236.16	682.48	64.6	291.55	842.57	57.7
35	144.94	431.21	86.0	189.31	563.21	74.5	239.60	712.82	65.6	295.80	880.02	58.5
36	147.00	449.82	87.2	192.00	587.52	75.5	243.00	743.58	66.5	300.00	918.00	59.4
37	149.03	468.69	88.4	194.65	612.17	76.6	246.35	774.77	67.4	304.14	956.51	60.2
38	151.03	487.82	89.6	197.26	637.15	77.6	249.66	806.40	68.3	308.22	995.55	61.0
39	153.00	507.20	90.8	199.84	662.47	78.6	252.92	838.44	69.2	312.25	1035.11	61.8
40	154.95	526.83	91.9	202.39	688.12	79.6	256.15	870.89	70.1	316.23	1075.17	62.6

EXPERIMENTS ON A MODEL OF A CENTRE-VENT WATER-WHEEL, WITH STRAIGHT BUCKETS.

99. The author was led to this design by the consideration of the path of the water in passing through the wheel, according to the hypothesis in art. 83. It is a wheel well suited for low falls, in which the water, over the wheel, may stand at its natural height, without requiring a vertical shaft of great length. Its simplicity and cheapness, combined with its other good qualities as a hydraulic motor, must recommend it for many such situations.

100. Plate VII., figure 1, is a general plan, and figure 2, a vertical section of the apparatus.

Figure 3 is a vertical section through the apertures in the guides and wheel; the guides and buckets are omitted to avoid confusion in the figure.

Figure 4 is a horizontal section of part of the guides and buckets, showing, also, the path of the water in experiment 3, according to the hypothesis in art. 83.

A is the wheel; the exterior diameter is $22\frac{1}{4}$ inches; the interior diameter is $19\frac{3}{4}$ inches; the height between the crowns, or *B C*, figure 3, is $2\frac{13}{16}$ inches; it carries thirty-six buckets, *E E*, figure 4, of steel, about $\frac{1}{32}$ of an inch in thickness, fastened to the wheel by means of the wooden cushions *F F*, figure 3; the upper cushions are screwed to the disc *D*, and the lower ones to the crown *G*. The disc *D* is of cast-iron, $\frac{3}{8}$ inch thick, with a suitable hub by which it is connected with the vertical shaft.

H H are guides of cast-iron, which direct the water into the wheel, and also support the plate *I*, which protects the wheel from pressure on its upper surface; the contraction of the streams entering the apertures between the guides, is diminished by the curved wooden garniture *K*; there are twenty-four guides. The mean shortest distance between the buckets at *a b*, figure 4, is 0.0339 feet; the mean shortest distance between the guides *c d*, figure 4, is 0.0437 feet; and the height of both is $2\frac{13}{16}$ inches $= 0.2344$ feet; we have, therefore, for the sum of the areas of the smallest sections between the guides,

$$0.0437 \times 0.2344 \times 24 = 0.24584 \text{ square feet.}$$

Similarly, the sum of the areas of the smallest sections between the buckets is

$$0.0339 \times 0.2344 \times 36 = 0.28606 \text{ square feet.}$$

The water is admitted into the forebay *L*, by the pipes *MM*; the diaphragm *N* is to diminish the agitation of the water.

101. The apparatus for gauging the water discharged by the wheel, consisted of the weir *O*, which had sharp edges; the depth on the weir was measured by a hook gauge, in the box *P*, which communicated, by a small aperture, with the surrounding water; the height of the water above the wheel was taken at a gauge in the box *Q*; this box was made sloping on one side, in order to permit a better view of the gauge. The zeros of both gauges were at the level of the top of the weir; consequently, the difference in the readings of the gauges gave at once the fall acting upon the wheel.

102. The apparatus for measuring the power, consisted of the Prony dynamometer *R*, attached to the upper part of the vertical shaft; the weights were applied by means of the bell crank *S*, figures 1, 2, and 5; the oscillations of the brake were diminished by the hydraulic regulator *T*, and the extent of the oscillations was limited by the stops *UU*. The speed of the wheel was obtained by means of a counter, driven by the worm *V*, attached to the top of the upright shaft; this was so arranged as to strike a bell once in fifty revolutions of the wheel.

In order to diminish the passive resistances, the weight, bearing upon the step *W*, was counterbalanced, in part, by other weights, one of which is represented at *y*, figure 2; these were attached to the brakes at the points *XX*, by vertical cords passing over pulleys; the weight, resting on the step when the wheel was immersed, and the dynamometer attached, was found to be 170 pounds; the counterbalance was 160 pounds, leaving 10 pounds bearing upon the step. The entire apparatus for measuring the power, was in equilibrium when there were no weights in the scale.

103. In all the experiments, except experiment 10, the brake was lubricated with oil; in experiment 10 water was used for this purpose; experiments 9 and 10 were identical in all other respects. It was noticed in experiment 10 that the whole apparatus trembled very much; this must have consumed some power, which is perceptible in the coefficients of effect. Experiment 9, in which oil was used, and in which the trembling of the apparatus was very slight, gives a coefficient of effect of 0.6922; while experiment 10, in which water was used to lubricate the brake, and in which the trembling of the apparatus was very distinct, gave 0.6886 as the coefficient of effect.

104. All the apparatus was constructed with great care and precision; the surfaces of the cast-iron guides were ground smooth; and the cast-iron disc and lower crown of the wheel were turned true, and polished, in order to diminish, as much as possible, the resistance of the water to the motion of the wheel.

105. In table V., the quantity of water discharged has been calculated by the formula

$$Q = 3.33\,(l - 0.1\,n h)\,h^{\frac{3}{2}},$$

in which Q = the quantity in cubic feet per second; l = the length of the weir = 3.003 feet; n = the number of end contractions = 2; h = the depth upon the weir. The weights were obtained for the purpose from Mr. O. A. Richardson, the official sealer of weights and measures for the City of Lowell. The effective length of the lever of the dynamometer, was two feet. The temperature of the water was 63½° Fahrenheit. Temperature of the air at 8^h, 35′ A. M., 63° Fahrenheit. The weight of a cubic foot of water is taken at 62.3128 pounds, which is deduced from table I.

If, in any experiment, the brake touched, even momentarily, either of the stops *U U*, it was rejected; with the use, however, of a regular and sufficient quantity of oil to lubricate the brake, and a properly constructed hydraulic regulator, there is seldom any difficulty from this cause, except at very low velocities.

TABLE V.

EXPERIMENTS ON A MODEL OF A CENTRE-VENT WATER-WHEEL.

1	2						3	4	5	6	7	8	9	10	11	12	13	14	15	16
No. of the experiment.	Time. June 19, 1847. Beginning of the experiment.			Ending of the experiment.			Duration of the experiment, in seconds.	Total number of revolutions of the wheel, during the experiment.	Weight in the scale, in pounds avoirdupois.	Useful effect, or the friction of the brake in pounds avoirdupois, raised one foot per second.	Total fall acting upon the wheel, in feet.	Depth of water on the weir, in feet.	Quantity of water passing the weir, in cubic feet per second.	Total power of the water, in pounds raised one foot per second.	Ratio of the useful effect to the power expended.	Velocity due the fall acting on the wheel, in feet per second.	Velocity of the exterior circumference of the wheel, in feet per second.	Ratio of the velocity of the exterior circumference of the wheel, to the velocity due the fall acting on the wheel.	Quantity of water discharged by the wheel, reduced to a uniform fall of 2½ feet, in cubic feet per second.	Ratio of the reduced quantity in the preceding column to the reduced quantity in experiment 3.
	H.	min.	sec.	H.	min.	sec.														
1	9	41	9	9	51	23	614	700	16	229.22	2.5198	0.3649	2.1507	337.69	0.6788	12.731	6.827	0.5363	2.1422	0.9884
2	9	59	57	10	9	16½	559½	750	14	235.83	2.4609	0.3662	2.1620	331.54	0.7113	12.581	8.028	0.6381	2.1791	1.0054
3	10	21	5	10	30	16½	551½	850	12½	242.10	2.5017	0.3662	2.1681	337.98	0.7163	12.685	9.230	0.7276	2.1674	1.0000
4	10	52	10	10	52	26½	16½	still	25½	0.	2.5160	0.3440	1.9714	309.07	0.	12.722	0.	0.	1.9651	0.9067
5	10	56	45	10	56	58	13	"	19½	0.	2.5074	0.3426	1.9596	306.17	0.	12.700	0.	0.	1.9567	0.9028
6	11	4	0	11	4	26	26	"	19¼	0.	2.4405	0.3395	1.9334	294.02	0.	12.529	0.	0.	1.9568	0.9029
7	11	9	40	11	9	58½	18½	"	24¼	0.	2.3735	0.3365	1.9082	282.23	0.	12.356	0.	0.	1.9584	0.9036
8	11	21	44	11	28	42	418	700	12	252.53	2.5977	0.3720	2.2127	358.17	0.7051	12.926	10.029	0.7759	2.1707	1.0015
9	11	45	32	11	52	16½	404½	700	11¼	250.08	2.6059	0.3734	2.2250	361.29	0.6922	12.947	10.364	0.8005	2.1795	1.0055
10	3	3	57½	3	9	18½	321	550	11½	247.61	2.5986	0.3729	2.2206	359.57	0.6886	12.929	10.261	0.7936	2.1781	1.0049
11	3	23	53	3	31	56½	483½	900	10	233.91	2.6001	0.3737	2.2276	360.84	0.6482	12.932	11.147	0.8620	2.1843	1.0078
12	3	45	1	3	50	11	310	700	6	170.25	2.6331	0.3750	2.2390	367.37	0.4634	13.014	13.523	1.0391	2.1817	1.0066
13	4	45	10½	4	49	50	279	750	0	0.	2.6287	0.3641	2.1437	351.15	0.	13.003	16.070	1.2359	2.0906	0.9646

106. In the foregoing table, experiments 4, 5, 6, and 7, were made with the wheel still; the brake was screwed up tight, and the pressure of the water upon the buckets, was measured by weights in the scale. In experiments 4 and 7, the weights were sufficient to balance the effect of the pressure of the water on the buckets, and also to overcome the friction of the apparatus; in other words, the weights were the least that would cause the scale to preponderate over the active and passive forces. In experiments 5 and 6, the weights in the scale were the greatest that the pressure upon the buckets would raise, and overcome the friction of the apparatus; consequently, the force of the water acting upon the buckets, may be considered as balanced by the average of the weights in the fourth and fifth experiments, and, also, by the average in the sixth and seventh experiments.

To obtain the true weight that would balance the pressure, we must reduce the weights in the different experiments to what they would have been, if the fall acting upon the wheel had been constant.

The following table shows the weights reduced to a uniform fall of 2.5 feet, obtained by simple proportion; thus, in the fourth experiment,

2.5160 : 25.75 :: 2.500 : 25.586.

The quantities discharged are also given for a uniform fall of 2.5 feet.

Number of experiment.	Actual fall acting upon the wheel, in feet.	Weight in scale by experiment, in pounds.	Weight reduced to a uniform fall of 2.5 feet.	Quantity of water discharged, reduced to a uniform fall of 2.5 feet, in cubic feet per second.
4	2.5160	25.750	25.586	1.9651
5	2.5074	19.375	19.318	1.9567
6	2.4405	19.250	19.719	1.9568
7	2.3735	24.125	25.411	1.9584
Means			22.5085	1.9592

The mean reduced weight, when the weights preponderated, is 25.4985 pounds.
and when the pressure on the buckets preponderated, . 19.5185 "
Difference, ... 5.9800 pounds.

Half of this difference, or 2.99 pounds, may be considered as the measure of the passive resistances, or, rather, of the friction of the apparatus.

107. In experiment 13, the brake was entirely removed, and the wheel allowed to run without load; with the brake, the counterbalance was necessarily

removed, consequently the passive resistance arising from the friction of the step, was much greater than in the other experiments.

108. Fig. 6, plate VII., is a diagram representing the experiments; the abscissas represent the ratios of the velocities of the exterior circumference of the wheel, to the velocities due to the falls acting upon the wheel, as given in column 14, of table V.; the ordinates represent the ratios of the useful effects to the powers expended, as given in column 11; the points, representing experiments 12 and 13, are connected by a broken line, because the latter experiment is not strictly comparable with the others, in consequence of the removal of the counterbalance.

109. The following table contains the successive steps of the calculation for the ordinates of the path of the water in experiment 3, represented at figure 4, plate VII.; the operations are all similar to those explained in articles 83 and 119. The ordinates in column 10 are obtained by the formula

$$O = \frac{R' \omega A H}{Q''},$$

in which

O is the ordinate,

R' the corresponding value of the radius in column 1,

ω, the angular velocity $= \frac{850 \times 2\pi}{551.5} = 9.684$,

$A H$, the corresponding volume in column 9,

Q'', the mean quantity discharged by each aperture in the wheel $= \frac{2.1681}{36} = 0.06022$

1	2	3	4	5	6	7	8	9	10
Values of R and successive values of R', in inches.	Areas in square inches, of circles of the radii in column 1.	Areas in square inches, of the complete rings.	$\frac{1}{36}$ of the areas of the rings in column 3, in square feet.	Correction for the thickness of the bucket, in square feet.	True areas of the partial rings, in square feet.	Height of the partial rings, in feet.	Volumes of the partial rings, in cubic feet.	Volumes between R and the successive values of R', in cubic feet.	Ordinates in feet, measured on arc of the radii in column 1.
11.437	410,936								
11.000	380,133	30,803	0,005942	0,000262	0,005680	0.2344	0,001331	0,001331	0.1962
10,500	346,361	33,772	0,006515	0,000350	0,006165	"	0,001445	0,002776	0,3906
10,250	330,064	16,297	0,003144	0,000198	0,002946	"	0,000691	0,003467	0,4762
10,000	314,159	15,905	0,003068	0,000228	0,002840	"	0,000666	0,004133	0,5538
9,875	306,354	7,805	0,001506	0,000156	0,001350	"	0,000316	0,004449	0,5887
9.750	298,648	7.706	0.001486	0,000175	0,001311	"	0,000307	0,004756	0,6214

EXPERIMENTS ON THE POWER OF A CENTRE-VENT WATER-WHEEL, AT THE BOOTT COTTON-MILLS IN LOWELL, MASSACHUSETTS.

110. This wheel is one of a pair constructed from the designs of the author by the Lowell Machine Shop, for the Boott Cotton-Mills, in 1849. During a considerable portion of the year, the fall, on which these wheels operate, is about nineteen feet; with this fall, and with the regulating gates raised to the full height, they each furnish an effective power of about 230 horse-power.

A patent for the term of fourteen years was issued, July 26, 1838, by the Government of the United States of America, to Samuel B. Howd, of Geneva, in the State of New York, for a water-wheel resembling, in some respects, the wheels at the Boott Cotton-Mills.* Under this patent, a large number of wheels have been constructed, and a great many of them are now running in different parts of the country; they are known in some places as the *Howd wheel*, in others as the *United States wheel;* they have uniformly been constructed in a very simple and cheap manner, in order to meet the demands of a numerous class of millers and manufacturers, who must have cheap wheels if they have any.

111. Figures 3 and 4, plate IX., are a plan and vertical section of one of the Howd wheels, constructed by the owners of the patent right for a portion of New England. *A*, the wooden guides by which the water is directed on to the buckets; *B*, buckets of cast-iron, fastened to the upper and lower crowns of the wheel, by bolts; the upper crown is connected with the vertical shaft *E*, by the arms *C*. *D*, the regulating gate, placed *outside* of the guides; this is made of wood; the apparatus by which it is moved is not represented; it is a simple arrangement of levers. The upright shaft *E* runs on a step at the bottom. This wheel is usually placed in the bottom of a rectangular forebay, which, in high falls, may be closed at the top, so as to avoid the necessity of using a vertical shaft of great length. The peculiarly shaped projections on one side of the buckets, it is said, increase the efficiency of the wheel, by diminishing the

* A wheel similar, in its essential features, was proposed in France, in 1826, by *Poncelet*.

waste of water; it is possible that some such effect may be produced by them. The author is not aware that any exact experiments have been made on the power of these wheels; from their form and construction, however, it is plain that they cannot be classed among those using water with very great economy. In the design for the Boott wheel, the author has so modified the form and arrangement of the whole, as to produce a wheel essentially different from the Howd wheel, as above described, although it may, possibly, be technically covered by the patent for that wheel.

112. Figures 1 and 2, plate VIII., are a vertical section, and a plan of the Boott centre-vent wheel, showing, also, the apparatus used in the experiments. *A*, the lower end of a pipe, about one hundred and thirty feet long, and eight feet in diameter, by which the water is conducted into the forebay *B*; this pipe is constructed of plate iron, three eighths of an inch in thickness, riveted together in the usual manner of making steam-boilers. For local reasons, the top of the forebay *B* is closed, so as to prevent the water from rising to its natural level, by about six or seven feet. *C*, the surface of the water in the Merrimack River, represented at about its medium height during the experiments. *D*, the wheel; *E*, the guides; *F*, the regulating gate, the apparatus for moving which, is not represented; *G*, the disc, which relieves the wheel from the vertical pressure of the water, and which also supports the lower bearing of the vertical shaft. The leather packing of the regulating gate *F*, slides against the circumference of the disc, which is turned smooth and cylindrical for that purpose, and the disc itself is supported by means of four brackets, two of which are represented at *H H*, by the columns *I I*. The vertical shaft *K* is of wrought iron, and it passes through the stuffing box *L*, and is supported by the box *M*, which has a series of recesses lined with babbit metal, fitted to receive a corresponding series of projections in the vertical shaft. The wheel, the vertical shaft, and the bevel gear usually on the latter, have a total weight of about 15,200 pounds; the bearing surface in the box *M* is about 331 square inches, consequently, the weight, per square inch, of bearing surface, is about 46 pounds.

Figures 3 and 4, plate VIII., represent the wheel and guides on a larger scale. The buckets and guides are equal in number, there being forty of each; the buckets are of plate iron, $\frac{1}{4}$ of an inch in thickness; the guides are of the same material, $\frac{5}{16}$ of an inch in thickness. The following dimensions were taken after the parts were finished:—

Mean shortest distance between adjacent buckets, or *a b* figure 4, ... 0.1384 feet.

Mean height between the crowns, at the inner extremities of the buckets, or *c d*, figure 3, ...	1.2300	feet.
Mean height between the crowns, at the outer extremities of the buckets, or *e f*, figure 3, ...	0.9990	"
Mean shortest distance between the adjacent guides, or *g h*, figure 4, ...	0.1467	"
Mean height of the orifices between the guides, or *i k*, figure 3,	1.0066	"
Diameter of the wheel at the outside of the buckets, ...	9.338	"
Diameter of the wheel at the inside of the buckets, ...	7.987	"

113. Several of the peculiar features of this design are covered by patents issued by the Government of the United States to U. A. Boyden. His patents cover the arrangement of the regulating gate, by placing it between the guides and the wheel, and having it detached from the garniture; making the height between the crowns of the wheel greater where the water is discharged, than where it enters; they also cover the self-adjusting apparatus on which the box *M* is supported.

114. Returning to figures 1 and 2, plate VIII., *N* is the friction pulley of the dynamometer, which is attached to the part of the shaft intended to receive the hub of the bevel gear, for the transmission of the power; *O*, the brake of maple wood; *P*, the bell crank, and *Q*, the hydraulic regulator; the friction pulley and the brake were subsequently used in the experiments on the Tremont Turbine, in the account of which they are more particularly described, (see arts. 37 and 38). *R*, the weir at which the water discharged by the wheel was gauged; *S*, a grating for the purpose of equalizing the flow of the water towards the weir; *T*, the gauge box in which the depths on the weir were observed. The communication between the water inside the box, and that surrounding it, was maintained by means of an aperture in the bottom of the box, (which extended 1.06 feet below the top of the weir,) and which was 4.12 feet from the weir. It may be thought, at first sight, that the depths on the weir were taken so near it, as to be affected by the curvature in the surface, caused by the discharge over the weir, but the experiments at the Lower Locks, (art. 173,) prove, conclusively, that when the communication between the water inside the box, and that outside of it, is maintained, by means of a pipe opening near the bottom of the canal, the depths are not affected in any appreciable degree, by the curvature in the surface. If any such effect was produced in this case, it must have been very slight. *U* and *V* are the gauge boxes at which the heights of the water, below and above the wheel, were observed, in order to

obtain the fall acting upon the wheel. The velocity of the wheel was obtained by means of the counter *W*. The apparatus for lubricating the brake is not represented on the plate; in some of the experiments, water was used, and in others, linseed oil.

The experiments were made according to the method of continuous observations, which has been sufficiently described in the account of the experiments on the Tremont Turbine.

115. The experiments on the Boott centre-vent water-wheel, are given in detail in table VI., which will be intelligible, without much further explanation than is contained in the respective headings of the several columns.

116. Column 10. *Useful effect, or the friction of the brake, in pounds avoirdupois raised one foot per second.* The brake was connected with the vertical arm of the bell crank, by a link, which was horizontal when the brake was in its normal position. When in this position, the length of a perpendicular, from the centre of the vertical shaft, to the line joining the points of the brake and bell crank to which the link was attached, was 9.743 feet; the effective length of the vertical arm of the bell crank, was 4.5 feet, and of the horizontal arm to which the scale was attached, 5 feet; consequently, the effective length of the brake was

$$\frac{9.743 \times 5}{4.5} = 10.826 \text{ feet.}$$

117. Column 15. *Quantity of water passing the wheel, in cubic feet per second.* This quantity was gauged at the weir. The length of the weir was 13.998 feet; the width of the raceway on the upstream side of the weir, was 17 feet; the crest of the weir was 11.14 feet above the bottom of the raceway. The quantity has been computed by the formula

$$Q = 3.33\,(l - 0.1\,n\,h)\,h^{\frac{3}{2}},$$

determined from the experiments made, in 1852, at the Lower Locks. (See art. 258.) In this formula

Q = the quantity in cubic feet per second.
l = the length of the weir = 13.998 feet.
n = the number of end contractions = 2.
h = the depth on the weir, given in column 14.

TABLE

EXPERIMENTS ON THE BOOTT

1	2	3	4	5						6	7	8	9	10
				TIME.										
No. of the experiment.	DATE, 1849.	Temperature of the water in degrees of Fahrenheit's thermometer.	Height of the regulating gate, in inches.	Beginning of the experiment.			Ending of the experiment.			Duration of the experiment, in seconds.	Total number of revolutions of the wheel during the experiment.	Number of revolutions of the wheel per second.	Weight in the scale, in pounds avoirdupois.	Useful effect, or the friction of the brake, in pounds avoirdupois, raised one foot per second.
				H.	min.	sec.	H.	min.	sec.					
1	October 17, A.M.	54	3	10	13	19	10	17	32	253	150	0.59289	575.56	23211.8
2	" " "	"	"	11	30	3.5	11	36	42	398.5	350	0.87829	202.09	12073.5
3	" " "	"	"	11	46	11	11	54	15	484	350	0.72314	407.25	20032.3
4	" 29, "	49.5	"	11	2	17	11	19	7	1010	550	0.54455	606.00	22447.2
5	" " "	"	"	11	33	41	11	45	22	701	350	0.49929	666.34	22630.5
6	" " "	"	"	11	59	24	0	6	49	445	200	0.44944	720.50	22096.8
7	November 5, P.M.	44	"	4	14	47	4	21	20	393	100	0.25445	931.87	16129.1
8	" 7, A.M.	45	"	9	40	19	9	48	32	493	500	1.01420	0.	0.
9	October 17, P.M.	53	6	2	34	15.5	2	44	59	643.5	650	1.01010	334.06	22952.9
10	" " "	"	"	2	56	6	3	5	39	573	550	0.95986	441.22	28807.9
11	" " "	"	"	3	17	6	3	26	56	590	550	0.93220	501.72	31814.1
12	" " "	"	"	4	7	3	4	15	22.5	499.5	450	0.90090	562.59	34476.0
13	" " "	"	"	4	50	36	5	4	13.5	817.5	700	0.85627	656.59	38243.0
14	" 29, "	50	"	2	4	10	2	14	58	648	450	0.69444	955.50	45135.3
15	" " "	"	"	2	41	41	2	52	50	669	400	0.59791	1140.94	46402.8
16	November 7, A.M.	45	"	9	29	27	9	37	57	510	600	1.17647	0.	0.
17	October 29, P.M.	50	9	3	20	41	3	27	28	407	450	1.10565	263.00	19779.8
18	" " "	"	"	3	33	18	3	35	49	151	150	0.99338	531.75	35931.0
19	" " "	"	"	3	36	44	3	44	8	444	400	0.90090	786.75	48212.7
20	" " "	"	"	3	45	5	3	54	10.5	545.5	450	0.82493	1001.47	56195.7
21	" " "	"	"	3	55	14	4	6	53.5	699.5	550	0.78628	1107.37	59226.4
22	" " "	"	"	4	21	14	4	30	10	536	400	0.74627	1205.00	61168.8
23	" " "	"	"	4	31	19	4	40	31	552	400	0.72464	1259.16	62065.4
24	" " "	"	"	4	41	42	4	51	4.5	562.5	400	0.71111	1297.31	62752.2
25	" " "	"	"	4	54	35	5	4	7.5	572.5	400	0.69869	1329.78	63199.3
26	November 7, A.M.	45	"	9	19	24	9	27	22.5	478.5	600	1.25392	0.	0.
27	November 5, A.M.	44	12	9	4	34.5	9	13	58.5	564	400	0.70922	1554.22	74979.3
28	" " "	"	"	9	15	10	9	21	7.5	357.5	250	0.69930	1584.00	75347.2
29	" " "	"	"	9	33	15	9	39	23	368	250	0.67935	1613.94	74580.9
30	" " "	"	"	9	40	37	9	48	3.5	446.5	300	0.67189	1644.37	75153.2
31	" " "	"	"	10	0	3	10	7	37.5	454.5	300	0.66007	1675.06	75208.3
32	" " "	"	"	10	8	54.5	10	16	37	462.5	300	0.64865	1705.47	75249.2
33	" " "	"	"	10	32	31	10	41	41	550	350	0.63636	1735.94	75142.9
34	" " "	"	"	10	43	0	10	51	0.5	480.5	300	0.62435	1768.41	75103.3
35	" " "	"	"	11	1	53	11	10	2	489	300	0.61350	1802.06	75202.0
36	" " "	"	"	11	11	24	11	18	26	422	250	0.59242	1836.19	73993.4
37	" 6, P.M.	"	"	3	31	12	3	35	20	248	0	0.	3155.34	0.
38	" " "	"	"	3	40	16	3	42	22	126	0	0.	2797.27	0.
39	" 7, A.M.	45	"	9	10	57	9	18	5	428	550	1.28505	0.	0.

In experiments Nos. 8, 16, 26, and 39, the brake was removed.

In experiment No. 37, the weight preponderated. In No. 38, the wheel preponderated (art. 77.)

VI.

CENTRE-VENT WATER-WHEEL

	11	12	13	14	15	16	17	18	19	20
No. of the experiment.	Height of the water above the wheel.	Height of the water below the wheel, taken in the wheelpit.	Total fall acting upon the wheel.	Depth of water on the weir.	Quantity of water passing the wheel, in cubic feet per second.	Total power of the water, in pounds avoirdupois, raised one foot per second.	Ratio of the useful effect to the power expended.	Velocity due to the fall acting on the wheel, in feet per second.	Velocity of the exterior circumference of the wheel, in feet per second.	Ratio of the velocity of the exterior circumference of the wheel, to the velocity due to the fall acting on the wheel.
	Feet.	Feet.	Feet.	Feet.						
1	16.013	1.410	14.603	1.2964	67.532	61493.4	0.37747	30.648	17.393	0.56750
2	16.036	1.364	14.672	1.2619	64.887	59364.4	0.20338	30.721	25.766	0.83871
3	15.955	1.387	14.568	1.2821	66.432	60347.0	0.33195	30.612	21.214	0.69301
4	15.558	1.400	14.158	1.2845	66.614	58821.6	0.38161	30.178	15.975	0.52937
5	15.607	1.410	14.197	1.2899	67.029	59351.7	0.38129	30.219	14.647	0.48470
6	15.563	1.420	14.143	1.2881	66.889	59002.2	0.37332	30.162	13.185	0.43714
7	15.604	1.360	14.244	1.2943	67.368	59858.1	0.26946	30.269	7.465	0.24661
8	15.573	1.273	14.300	1.2115	61.083	54486.4	0.	30.329	29.753	0.98101
9	15.956	1.668	14.288	1.5145	84.998	75732.8	0.30308	30.316	29.633	0.97746
10	15.930	1.704	14.226	1.5308	86.355	76608.0	0.37604	30.250	28.159	0.93086
11	15.914	1.717	14.197	1.5395	87.080	77093.2	0.41267	30.219	27.347	0.90496
12	15.923	1.730	14.193	1.5467	87.685	77607.2	0.44424	30.215	26.429	0.87470
13	15.944	1.750	14.194	1.5539	88.285	78143.8	0.48939	30.216	25.120	0.83134
14	15.581	1.803	13.778	1.5762	90.166	77480.4	0.58254	29.770	20.372	0.68433
15	15.481	1.875	13.606	1.5943	91.697	77812.1	0.59634	29.584	17.540	0.59291
16	15.451	1.506	13.945	1.4180	77.112	67076.7	0.	29.950	34.513	1.15237
17	15.408	1.890	13.518	1.6418	95.762	80736.3	0.24499	29.488	32.436	1.09997
18	15.323	1.950	13.373	1.6734	98.490	82145.2	0.43741	29.329	29.142	0.99362
19	15.352	1.983	13.369	1.6955	100.418	83728.2	0.57582	29.325	26.429	0.90125
20	15.413	2.017	13.396	1.7184	102.421	85571.4	0.65671	29.354	24.200	0.82442
21	15.426	2.047	13.379	1.7230	102.825	85800.0	0.69029	29.336	23.066	0.78629
22	15.418	2.076	13.342	1.7308	103.517	86138.0	0.71013	29.295	21.893	0.74731
23	15.424	2.102	13.322	1.7337	103.769	86218.8	0.71986	29.273	21.258	0.72620
24	15.465	2.131	13.334	1.7328	103.689	86229.3	0.72774	29.286	20.861	0.71232
25	15.464	2.160	13.304	1.7380	104.229	86483.7	0.73077	29.253	20.497	0.70067
26	15.417	1.715	13.702	1.5981	92.018	78648.0	0.	29.688	36.785	1.23907
27	15.398	1.998	13.400	1.8316	112.525	94057.5	0.79716	29.359	20.806	0.70868
28	15.434	2.003	13.431	1.8367	112.987	94662.2	0.79596	29.393	20.515	0.69796
29	15.321	1.990	13.331	1.8320	112.562	93603.9	0.79677	29.283	19.929	0.68058
30	15.369	1.991	13.378	1.8368	112.996	94296.4	0.79699	29.335	19.711	0.67193
31	15.367	1.981	13.386	1.8377	113.071	94415.2	0.79657	29.343	19.364	0.65990
32	15.369	1.986	13.383	1.8387	113.164	94471.1	0.79653	29.340	19.029	0.64856
33	15.336	1.980	13.356	1.8379	113.090	94219.0	0.79753	29.311	18.668	0.63692
34	15.362	1.981	13.381	1.8443	113.673	94881.9	0.79154	29.338	18.316	0.62431
35	15.385	1.980	13.405	1.8511	114.293	95571.2	0.78687	29.364	17.998	0.61291
36	15.292	1.971	13.321	1.8476	113.969	94703.1	0.78132	29.272	17.379	0.59371
37	15.442	1.905	13.537	1.8087	110.454	93270.1	0.	29.508	0.	0.
38	15.477	1.902	13.575	1.8072	110.325	93422.4	0.	29.550	0.	0.
39	15.415	1.819	13.596	1.6884	99.795	84635.0	0.	29.573	37.698	1.27477

118. The results of the experiments in table VI., are represented by a system of coördinates at figure 1, plate IX.; — the relative velocities, given in column 20, are taken for the abscissas, and the corresponding ratios of the useful effects to the powers expended, given in column 17, are taken for the ordinates. The numbers on the figure refer to the experiments in table VI., which the several points represent; — the points not numbered represent some experiments not reported, in consequence of an imperfection in the gauge of the quantity of water discharged, owing to a defective arrangement of the grating. These experiments have been corrected by a comparison with those that are reported; notwithstanding this correction, however, they ought not to be considered as of equal value with those reported in table VI. In the figure, the points representing the latter experiments, are connected by full lines; the points representing the experiments considered imperfect, are connected by broken lines. The line *AB* represents the experiments reported, that were made with the regulating gate fully raised; the line *CD*, the experiments with the gate raised three quarters of its full height; *EF*, the experiments with the gate raised a half, and *GH*, the experiments with the gate raised one quarter of its full height. It will be seen that the maximum coefficient of effect, with the gate fully raised, is given, when the outside of the wheel is moving with a velocity equal to about sixty-seven per cent. of that due to the fall acting upon the wheel, at which velocity, the useful effect is very nearly eighty per cent. of the total power of the water. The coefficient of effect diminishes rapidly as the regulating gate is lowered, and the maximum is also found at a slower speed; thus, when the gate is raised three inches, or one quarter of its full height, the maximum coefficient of effect is thirty-eight per cent. of the power expended; which is given when the outside of the wheel is moving with a velocity about one half of that due to the fall acting upon the wheel.

119. *ABCD*, figure 2, plate IX., represents the path of the water as it passed through one of the apertures of the wheel, in experiment 30, according to the hypothesis in art. 83; the steps in the calculation for which, are given in table VII. In the formula

$$O=\frac{R'\omega A H}{Q'},$$

we have for this case,

$O=$ the ordinate measured on the arc of a circle the radius of which is R'; its several values are given in column 10.

$R'=$ the distance from the centre of the wheel for which the ordinate is computed; — its several values are given in inches, in column 1; — to compute the value of O in feet, R' must be taken in feet.

$\omega =$ the angular velocity. In experiment 30, the velocity of the outside of the wheel was 19.711 feet per second, and the radius of the outside of the wheel is 4.669 feet, consequently,

$$\omega = \frac{19.711}{4.669} = 4.2217.$$

$A\,H =$ the volume of that part of the space between two adjacent buckets, included between the outside of the wheel and the radius R'; — its several values are given in column 9.

$Q'' =$ the quantity of water discharged, per second, by each orifice in the wheel. In experiment 30, we have, by table VI., the total quantity discharged $= 112.996$ cubic feet per second, and as there are forty orifices, we have

$$Q'' = \frac{112.996}{40} = 2.8249.$$

In figure 2, plate IX., the buckets and guides are drawn to a scale one fourth the full size; — the radius of the arc $AB = R = 56.028$ inches. To find the limit of the stream on the side $B\,C$, the arcs $I\,F$, $K\,H$, etc., $N\,C$, are drawn with the radii 55 inches, 54 inches, etc., 47.922 inches; — the arcs $E\,F$, $G\,H$, etc., $O\,C$, being taken from column 10, equal to 0.415 feet, 0.796 feet, etc., 2.748 feet; the points B, F, H, etc., C, being connected by suitable lines, determine the limit of the stream on that side. The limit of the stream on the other side is found by making the arcs $F\,L = E\,I$, $H\,M = G\,K$, etc., $C\,D = O\,N$; — the points A, L, M, etc., D, being connected by suitable lines, determine the limit of the stream on that side.

By an examination of figure 2, it will be seen, that the section of the stream just after it has entered the wheel, is sensibly greater than the section of the stream as it leaves the guides, and that, consequently, if the stream flowed according to the hypothesis, there must have been a sudden change in the velocity of the water, causing a shock, which, according to the common theory, implies a loss of power. This indicates a defect in the design; nevertheless, the success attending this first essay, on a large scale, of a centre-vent water-wheel, in which due regard has been paid to accuracy of construction and perfection of workmanship, guided by such light as the present imperfect theories can afford, ought to encourage us to hope, that, when it has received the same degree of attention as the turbine, it will not be much behind that celebrated motor, in its economical use of water.

TABLE VII.

1	2	3	4	5	6	7	8	9	10
Value of R and successive values of R', in inches.	Areas in square inches, of circles of the radii in the preceding column.	Areas in square inches, of the complete rings.	$\frac{1}{4}a$ of the areas of the complete rings in the preceding column, in square feet.	Correction for the thickness of the bucket, in square feet.	Corrected areas of the partial rings, in square feet.	Mean height of the partial rings, in feet.	Volumes of the partial rings, in cubic feet.	Volumes between R and the successive values of R', in cubic feet.	Ordinates in feet, to be measured on arcs of the corresponding radii in column 1.
56.028	9861.890								
55.000	9503.318	358.572	0.06225	0.00168	0.06057	1.001	0.06063	0.06063	0.415
54.000	9160.884	342.434	0.05945	0.00210	0.05735	1.008	0.05781	0.11844	0.796
53.000	8824.734	336.150	0.05836	0.00227	0.05609	1.021	0.05727	0.17571	1.160
52.000	8494.866	329.868	0.05727	0.00262	0.05465	1.042	0.05695	0.23266	1.507
51.000	8171.282	323.584	0.05618	0.00304	0.05314	1.070	0.05686	0.28952	1.839
50.000	7853.982	317.300	0.05509	0.00386	0.05123	1.105	0.05661	0.34613	2.155
49.000	7542.964	311.018	0.05400	0.00561	0.04839	1.147	0.05550	0.40163	2.451
48.750	7466.191	76.773	0.01333	0.00168	0.01165	1.177	0.01371	0.41534	2.522
48.500	7389.811	76.380	0.01326	0.00181	0.01145	1.190	0.01363	0.42897	2.591
48.250	7313.824	75.987	0.01319	0.00202	0.01117	1.204	0.01345	0.44242	2.659
47.922	7214.723	99.101	0.01721	0.00252	0.01469	1.221	0.01794	0.46036	2.748

PART II.

EXPERIMENTS ON THE FLOW OF WATER OVER WEIRS, AND IN SHORT RECTANGULAR CANALS.

EXPERIMENTS ON THE FLOW OF WATER OVER WEIRS.

120. The laws governing the flow of water over weirs, have received the attention of several distinguished engineers and men of science, among whom may be named Smeaton and Brindley in England; Du Buat, Navier, D'Aubuisson, Castel, Poncelet, Lesbros, and Boileau, in France; and Eytelwein and Weisbach in Germany. A great number of experiments have been made and recorded; the earlier ones rude and imperfect; the later ones, particularly those by Poncelet, Lesbros, and Boileau, with a perfection of apparatus previously unknown.

There has been in this branch of hydraulics, as well as in others, a steady advance with the accumulation of experiments and the improvement of the means of observation; the result, however, of these numerous labors, is far from satisfactory to the practical engineer. On a careful review of all that has been done, he finds that the rules given for his use, are founded on the single natural law governing the velocity of fluids, known as the theorem of Torricelli; omitting, in consequence of the extreme complexity of the subject, all consideration of many other circumstances, which, it is well known, materially affect the flow of water through orifices. He finds also that it has been attempted to correct the theoretical expression thus found, by coefficients obtained by comparing the results derived from it, with those furnished by experiment; but when he comes to investigate these experiments, even after rejecting all excepting those made with the greatest care, and with apparatus capable of insuring the greatest precision, he finds such discordances in the resulting coefficients, that he loses all hope of arriving at correct results when he applies them on the great scale. They will undoubtedly furnish sufficiently accurate results, if the apparatus used is a reproduction, both in form and dimension, of that used in the experiments; but this is seldom attainable, the experiments having been made on such a minute scale. Boileau,* in discussing the various formulas that have been proposed, points out many of their defects, and has himself proposed a new one, coupled, however, with some special conditions in the form of the weir, and the mode of taking the depth upon the sill.

No correct formula for the discharge of water over weirs, founded upon natural laws, and including the secondary effects of these laws, being known, we must rely entirely upon experiments, taking due care in the application of any formula deduced from them, not to depart too far from the limits of the experiments on which it is founded.

Engineers have generally agreed that the most convenient form of weir for gauging streams of water, is one which is cut in a vertical plane side of a reservoir, the sill being horizontal, the sides vertical, and the contraction complete. In order that the contraction may be complete, the sill and sides of the weir must be so far removed from the bottom and lateral sides of the reservoir, that they may produce no more effect upon the discharge, than if they were removed a distance indefinitely great; also, the aperture must be effectively the same, as if cut in a plate having no sensible thickness. The condition relating to the distance of the bottom and sides of the reservoir, can seldom be strictly complied with, when gauging large streams of water; it is found, however, that, when the sill is at a height above the bottom of the reservoir not less than twice the height of the water above the sill, and the sides are removed a distance at least equal to the height above the sill, a correction free from serious error can usually be made for the effect of the velocity of the water approaching the weir. The condition that the aperture shall be effectively the same as if cut in a plate having no sensible thickness, is usually more easily complied with. The effect of the contraction is such, that the water has a strong tendency to leave the bottom and sides of the aperture for a certain distance, and to touch the aperture only at the upstream edge; if, however, the thickness of the plank or other material, exceeds a certain amount, (depending upon the depth flowing over,) the water will follow the top of the plank; in this case, all that is requisite is, to cut away the downstream side of the weir at an angle of, say, forty-five or sixty degrees with the horizontal; leaving horizontal, only a small part of the thick-

* *Jaugeage des cours d'eau a faible ou a moyenne section* by M. P. Boileau (Paris: 1850); or *Journal de l'Ecole Polytechnique*, No. xxxiii.

ness of the sill. It is essential, however, that the corners of the sill and sides of the weir presented to the stream, should be full and sharp, and not rounded or bevelled in any degree.

121. Two modes present themselves for studying, experimentally, the laws governing the discharge of water over weirs. *First*, that which has been uniformly adopted heretofore, namely, to obtain by direct measurement the quantity of water discharged in a given time, through an aperture of known dimensions; this is evidently the only mode of resolving the question completely. To perform the experiments, however, upon a scale of magnitude corresponding to the ordinary practical applications, usually requires an apparatus of great cost, and such as is beyond the reach of most experimenters. The great difficulty is, to obtain a suitable basin, in which to make the direct measurement of the quantity discharged by the weir.

The *second* mode dispenses with a direct measurement of the quantity. If we have two weirs of the same form, but of different lengths, and we know that the quantities of water discharged by them, in certain circumstances, are equal; knowing also the depth upon the sill of each weir, we have the data for an equation by which one unknown quantity may be determined. Neither the coefficient of contraction, nor the absolute discharge can, however, be obtained by such an equation.

122. The discharge over weirs is commonly assumed to vary as the square root of the third power of the depth; let us suppose it to be unknown, and equal to a.

Suppose also l the length, and h the depth, on one of the weirs; and l' and h' the corresponding dimensions for the other weir; C, a constant coefficient; Q, the quantity which, by hypothesis, is the same for both weirs. Assuming, according to the common formula, that the quantity is proportional to the length of the weir, we have

$$Q = C l h^a;$$
$$Q = C l' h'^a;$$

consequently,

$$C l h^a = C l' h'^a;$$
$$\left(\frac{h}{h'}\right)^a = \frac{l'}{l};$$

taking the logarithms, we have

$$a\,(\text{Log.}\,h - \text{Log.}\,h') = \text{Log.}\,l' - \text{Log.}\,l;$$

therefore,

$$a = \frac{\text{Log.}\, l' - \text{Log.}\, l}{\text{Log.}\, h - \text{Log.}\, h'}.$$

We can thus, by means of two experiments, determine the power of the depth which will lead to identical quantities in the computed discharge of the two weirs.

123. It is assumed in the above equations, that the quantity discharged by a weir is directly proportioned to its length; this, in weirs having complete contraction, is, however, known not to be true, in consequence of the contraction which takes place at the ends of the weir. This contraction diminishes the discharge. When the weir is of considerable length in proportion to the depth of the water flowing over, this diminution is evidently a constant quantity, whatever may be the length, provided the depth is the same; we may, therefore, assume that the end contraction effectively diminishes the length of such weirs, by a quantity depending only upon the depth upon the weir. It is evident that the amount of this diminution must increase with the depth; we are unable, however, in the present state of the science, to discover the law of its variation; but experiment has proved that it is very nearly in direct proportion to the depth. As it is of great importance, in practical applications, to have the formula as simple as possible, it is assumed in this work that the quantity to be subtracted from the absolute length of a weir having complete contraction, to give its effective length, is directly proportional to the depth. It is also assumed that the quantity discharged by weirs of equal effective lengths, varies according to a constant power of the depth. There is no reason to think that either of these assumptions is perfectly correct; it will be seen, however, that they lead to results agreeing very closely with experiment.

124. The formula proposed for weirs of considerable length in proportion to the depth upon them, and having complete contraction, is

$$Q = C(l - bnh)h^{n};*$$

in which

Q = the quantity discharged in cubic feet per second.

C = a constant coefficient.

l = the total length of the weir in feet.

b = a constant coefficient.

---

* This formula was first suggested to the author by Mr. Boyden, in 1846.

n = the number of end contractions. In a single weir having complete contraction, n always equals 2, and when the length of the weir is equal to the width of the canal leading to it, $n = 0$.

h = the depth of water flowing over the weir, taken far enough upstream from the weir, to be unaffected by the curvature in the surface caused by the discharge.

a = a constant power.

The coefficient C can be determined only from experiments in which the actual discharge is known; the constants, a and b can, however, be determined without knowing the actual discharge in any particular case.

It has been stated that the proposed formula is applicable only to weirs having a considerable length in proportion to the depth of water running over them. It is found by experiment that, when the length equals or exceeds three times the depth, the formula applies; but in lengths less than this in proportion to the depth, the formula cannot be used with safety; the error increasing as the relative length of the weir diminishes.

It is evident, from the construction of the formula, that it cannot be of general application. The factor $l - bnh$ represents the *effective* length of the weir; if $l = bnh$ this effective length becomes 0, and the formula would give 0 for the discharge, which is absurd; similarly, if $bnh > l$, the discharge given by the formula would be negative. In weirs of very short length in proportion to the depth, the effect of the end contraction cannot be considered as independent of the length. The end contraction influences the discharge to a certain distance, A, from the end of a weir; if the whole length of the weir is greater than $2A$, the effect of the end contraction is independent of the length; but if the length is less than $2A$, the whole breadth of the stream is affected in its flow by the end contractions, and, consequently, the proposed formula would not apply.

In practical applications, this will seldom be an inconvenience, as it is nearly always practicable so to proportion the weir, that the length may not be less than three times the depth upon it; if, however, there is no end contraction, the proportion of the length to the depth is not material.

125. The author has made numerous experiments on the discharge of water over weirs, according to each of the methods described above.

First, those at the Tremont Turbine, and at the centre-vent water-wheel for moving the guard gates of the Northern Canal. In none of these experiments has any attempt been made to measure the absolute quantities flowing over the weirs; but simply to cause quantities of water known to be equal, to pass over weirs of different dimensions, noting the depth of water and length of weir in each case. From these data, as is explained above, certain factors in the formula can be determined.

Second, those at the Lower Locks, in which the absolute quantities passing over weirs of known dimensions, were measured directly.

As each of these three sets of experiments were made with different apparatus, they will be described separately.

EXPERIMENTS MADE AT THE TREMONT TURBINE, ON THE FLOW OF WATER OVER WEIRS.

126. The apparatus constructed to gauge the water discharged by the Tremont Turbine, with some modifications, was used for the experiments on the discharge over the weir; for a general description of this apparatus, see arts. 44, 45, and 46.

The experiments consisted in allowing a quantity of water, of unknown volume, to enter the wheelpit, through the turbine, the regulating gate of which was sufficiently opened for the purpose. This volume of water was then caused to flow over weirs of different dimensions, and the corresponding depth on the weir, assumed by the water in each experiment, was noted after the water had arrived at a uniform state.

The experiments are divided into series, in each of which the regulating gate was unchanged throughout, so that the apertures through which the water entered the wheelpit remained constant during each series.

Some variations necessarily occurred in the head acting upon these orifices; they were small, however, when compared to the whole head. The depths on the weir have been reduced, according to well-known principles, to what they would have been if the head had been constant. The leakage of the wheelpit also rendered another small correction necessary. After the corrections are made, we have in each series a collection of experiments in which the quantity discharged is the same, and we have also the requisite dimensions of the different weirs. These data, if perfectly accurate, are sufficient to enable us to determine, in the proposed formula for the discharge, the values of the constants a and b. It is not to be presumed, however, that the data are perfectly correct, but we can, at any rate, find the values of a and b that will give the most uniform results to the computed discharges in all the experiments in a series; the actual discharge being, by hypothesis, a constant quantity.

127. Some additions to the apparatus used in the experiments on the turbine were made for the weir experiments. The partitions, represented by figures

5, 6, and 7, plate V., were provided for the purpose of shortening or subdividing the weir. They were made of wood, faced on part of one side with plates of sheet-iron a, $\frac{3}{16}$ of an inch in thickness; the width bc was about 1.5 feet; the iron plate was two inches less. One side of the timber P, figure 2, was in the same vertical plane as the upstream edge of the weir H. When the partitions were placed upon the weir, the top of them was supported by the timber P, and the bottom by the plate of iron a, which rested against the weir. Flashboards, represented by figures 8, 9, and 10, plate V., were also provided to close up portions of the weir; these, together with the partitions, were maintained in their respective positions, simply by the pressure of the water against them. Wherever leaks appeared at the joints of the partitions or flashboards, they were stopped with great ease and effect, by a little dough made of unbolted Indian meal, a handful of which was drawn over the upstream side of the joints; of course the orifices closed in this manner were very minute. In plate X., all the modifications of the weir produced by changing the partitions and flashboards, are represented; the several figures are referred to in column 8, table X. In the greater number of the experiments, two or more spaces were used at the same time; they were always of very nearly equal length, so that the length of each may be obtained by dividing the whole length of the weir given in column 6 by half the number of end contractions given in column 7.

The brackets N, figures 1 and 2, plate V., were placed on the downstream side of the weir, to support a board on which to stand for the purpose of adjusting the partitions and flashboards. The top of the board was about 9.5 inches below the top of the weir. In some of the experiments, a part of the sheet of water fell upon this board; in experiment 50 it was moved nearer to the weir, so that the entire sheet of water fell upon it, but without producing any sensible effect upon the discharge. In experiment 51, a three inch plank was placed on the top of the board, as is represented by the dotted lines at O, figure 2, plate V.; the effect of this obstruction, as indicated by the increased depth on the weir as measured by the hook gauge, was, to diminish the discharge, with the same depth on the weir, about $\frac{1}{1000}$.

It is to be regretted that the casting forming the sill of the weir, was not planed on its whole height on the side $H\,Q$, figure 4, plate V. When the weir was erected no thought was entertained of using it for these experiments, requiring, as they do, to be of value, to be free from all disturbing causes. The disturbance caused by the projection at I, can, however, have been scarcely sensible.

128. The data furnished by observation, together with the necessary reductions, and the results deduced from them, are contained in table X. Most of

the columns are sufficiently explained by the respective headings; several of them, however, require further explanation.

129. COLUMN 11. *Fall affecting the leakage of the wheelpit.* This is obtained by adding together the corresponding numbers in columns 9 and 10.

130. COLUMN 12. *Depth of water on the weir corrected for the leakage of the wheelpit.* This is obtained in the following manner.

It was clear, from the construction of the wheelpit, (art. 23,) that nearly the whole of the leakage passed through the wooden flooring, and that all the orifices through which it passed were constantly below the surface of the lower canal. In the construction of the wheelpit, no particular precautions were taken to prevent a free communication from the bottom of the wooden flooring to the lower canal; and as the amount of the leakage was very small, and the material, fine sand free from large springs, it is clear that the water could have had no appreciable obstruction after passing through the flooring, except from the pressure of the water in the lower canal. This being the case, the amount of the leakage would depend upon the head; or, in other words, upon the height from the surface of the water in the wheelpit, to the surface of the water in the lower canal. Let

$L=$ the quantity of water leaking out of the wheelpit, in cubic feet per second.

A, A', A'', etc. $=$ the areas of the several orifices through which the water passed.

C, C', C'', etc. $=$ the corresponding coefficients of contraction.

$h=$ the head, or the height from the surface of the water in the wheelpit, to the surface of the water in the lower canal. This head applies to all the orifices, as they are all below the surface of the water in the lower canal.

$$L = CA\sqrt{2gh} + C'A'\sqrt{2gh} + C''A''\sqrt{2gh} + \text{etc.};$$

$$L = (CA + C'A' + C''A'' + \text{etc.},)\sqrt{2gh}.$$

The areas A, A', A'', etc., are constant, as are also the coefficients C, C', C'', etc., the variations in the head not being very great. Let

$$c = CA + C'A' + C''A'' + \text{etc.}:$$

then

$$L = c\sqrt{2gh} = c\sqrt{2g}\sqrt{h}.$$

The factor $c\sqrt{2g}$, being constant, can be determined by an experiment in which L and h are known. To determine this constant, the following experiment was made.

The weir was closed up by the flashboards, and made tight in the usual manner, so that no appreciable quantity passed over the weir; the head gate was closed, and the small quantity leaking through it was caught in the leak box and carried over the weir in the leak pipe (art. 24). The water in the wheelpit having then no supply, its surface began to lower, in consequence of the leakage through the floor; while thus falling, the following observations were made.

February 5, 1851, at 10^h, $20'$, $30''$, A. M., the height of the water in the wheelpit above the top of the weir, was . . . . 0.596 feet.
And at 11^h, $1'$, $46''$, A. M., the height was . . . . 0.396 "
Consequently the surface of the water in the wheelpit lowered in $2476''$ . . . . 0.200 feet.

The area of the surface of the water in the wheelpit, after making the proper deductions, was about 506 square feet; consequently,

$$L = \frac{506 \times 0.2}{2476} = 0.0409 \text{ cubic feet per second.}$$

During the interval of 2476 seconds, the mean height of the water in the lower canal was 1.2316 feet below the top of the weir, and the mean height in the wheelpit, during the same period, was 0.496 feet above the top of the weir, then

$$h = 1.2316 + 0.4960 = 1.7276 \text{ feet.}$$

Substituting these values of L and h in the equation

$$L = c\sqrt{2g}\sqrt{h},$$

we have

$$c\sqrt{2g} = 0.03112:$$

consequently,

$$L = 0.03112\sqrt{h}.$$

To find the depth on the weir, corrected for the leakage of the wheelpit, let

$h' =$ the depth on the weir by observation,

h'' = the depth on the weir corrected for the leakage,

l = length of the weir,

Q = the quantity passing over the weir, the dimensions being all in feet.

We have $Q+L$ = the total quantity entering the wheelpit, and which would have passed over the weir, if there had been no leakage out of the wheelpit.

To determine the corrected depth, it is necessary to assume some formula giving nearly the relations between the quantities h', l, and Q. Let us use that given by Lesbros* for a depth of 0.20 metres and complete contraction, which, when reduced to the English foot as the unit, and adopting our own notation, is

$$Q = 3.12\, l h'^{\frac{3}{2}};$$

we shall have also

$$Q + L = 3.12\, l h''^{\frac{3}{2}};$$

by subtraction

$$L = 3.12\, l h''^{\frac{3}{2}} - 3.12\, l h'^{\frac{3}{2}};$$

from which we derive

$$h'' = \left(h'^{\frac{3}{2}} + \frac{L}{3.12\, l}\right)^{\frac{2}{3}};$$

or substituting for L its value $0.03112\,\sqrt{h}$, we have

$$h'' = \left(h'^{\frac{3}{2}} + \frac{0.03112\,\sqrt{h}}{3.12\, l}\right)^{\frac{2}{3}}.$$

By this formula, the reduced heights given in column 12 have been obtained.

131. COLUMN 15. *Fall from the surface of water in the forebay, to the surface of the water in the wheelpit.* This is obtained by taking the difference of the corresponding numbers in columns 13 and 14.

132. COLUMN 16. *Uniform fall from the forebay to the wheelpit, to which the depths on the weir in each series are reduced.* The fall in the same series given in column 15, which is the nearest to the mean fall in all the experiments in the series, is assumed for this purpose; it is unimportant what fall is taken, provided it is near the mean.

133. COLUMN 17. *Depth on the weir corrected for the leakage of the wheelpit, and the variation in the fall.* It must be recollected that all the experiments of each

* *Experiences Hydrauliques sur les lois de l'ecoulement de l'eau*, by M. Lesbros, Paris: 1851. Table XXXIX.

series, were made with the same opening of the regulating gate of the turbine; that is, the areas of the orifices through which the water entered the wheelpit, were the same in each. In all the experiments, a small quantity of the water entering the wheelpit, passed between the gate and the lower curb, in consequence of the leather packing not being perfectly adjusted; this did not affect the results, however, as these orifices were also submerged in the wheelpit. Under these circumstances, if the head had been constant, the quantity of water entering the wheelpit, would also have been constant; but the head was subject to a variation, comparatively small certainly, but sufficient to produce a material change in the quantity of water entering the wheelpit, and, consequently, in the depth on the weir.

To clear the results from this source of irregularity, it will be necessary to ascertain what the depths on the weir would have been, if the head had been constant. For this purpose, let

$H =$ the constant head to which the depths on the weir, in any particular series, are to be reduced, and which varies but little from the actual heads in the same series;

$H' =$ the actual head in the particular experiment to be reduced;

$h''' =$ the depth on the weir, corrected for the variation of the head, or corresponding to the constant head H;

$h'' =$ the depth on the weir corresponding to the head H', and which is the depth given by observation, corrected for the leakage of the wheelpit;

$q =$ the quantity of water, in cubic feet per second, that would have entered the wheelpit, if the head had been H;

$q' =$ the quantity of water corresponding to the head H', and which is the same as $Q + L$ (art. 130);

$l =$ the length of the weir;

$C =$ the coefficient of the formula for the discharge over weirs;

a, a', a'', etc. $=$ the areas of the several orifices through which the water entered the wheelpit, all of them being submerged in the wheelpit;

c, c', c'', etc. $=$ the corresponding coefficients of contraction;

$$q' = ca\sqrt{2gH'} + c'a'\sqrt{2gH'} + c''a''\sqrt{2gH'} + \text{etc.};$$

$$q' = (ca + c'a' + c''a'' + \text{etc.})\sqrt{2gH'};$$

similarly

$$q=(ca+c'a'+c''a''+\text{etc.})\sqrt{2gH};$$

by division,

$$\frac{q'}{q}=\sqrt{\frac{H'}{H}};$$

also

$$q'=Clh''^{\frac{3}{2}} \text{ and } q=Clh'''^{\frac{3}{2}};$$

whence,

$$\frac{q'}{q}=\left(\frac{h''}{h'''}\right)^{\frac{3}{2}};$$

therefore,

$$\left(\frac{H'}{H}\right)^{\frac{1}{2}}=\left(\frac{h''}{h'''}\right)^{\frac{3}{2}} \text{ or } \left(\frac{H'}{H}\right)^{\frac{1}{3}}=\frac{h''}{h'''};$$

whence, we derive

$$h'''=h''\left(\frac{H}{H'}\right)^{\frac{1}{3}}.$$

By this last formula, the corrected depths given in column 17 have been computed.

By an inspection of column 13, it will be seen that the level of the water above the wheel was maintained throughout each series with great uniformity, excepting in a few experiments in which it was intentionally altered, as will be seen presently. The height of the water in the wheelpit necessarily varied with the depth upon the weir, and this is the principal cause of the variations in the fall.

Several of the experiments given in table X., were made for the express purpose of testing the accuracy of the method of reduction just described. Thus, in experiments 41 and 42, the weir was in the same state as in experiment 40, but the height of the water above the wheel was lowered, and the differences in the observed depths upon the weir, given in column 9, are to be attributed entirely to the diminution in the quantity of water entering the wheelpit, in consequence of the diminished head. If the method of reduction is accurate, however, the corrected depths in these three experiments, given in column 17, should be the same.

In table VIII., are collected all the experiments made for this object, together with the other experiments forming part of the corresponding series, with which they may be compared, the weir having been in the same state.

TABLE VIII.

Number of the experiment.	Fall from the forebay to the wheelpit. Feet.	Corrected depth upon the weir, in Feet.	Variation in the fall from the initial experiment. Feet.	Variation in the corrected depth, from the initial experiment. Feet.
40	14.088	0,79096		
41	13.554	0,79049	—0,534	—0,00047
42	13.149	0.78976	—0,939	—0,00120
49	13,904	0,95477		
52	13,436	0,95380	—0,468	—0,00097
53	12.962	0,95097	—0,942	—0,00380
63	13.719	1.13177		
64	12.806	1.12508	—0.913	—0,00669
72	13.816	0,92170		
73	13,315	0,92145	—0,501	—0,00025
74	12.665	0,92153	—1.151	—0,00017

It will be perceived that the variations in the fall, to which the method of reduction is applied in these experiments, are, nearly all of them, much greater than any that occur in the regular experiments. This was arranged for the purpose of applying an extreme test to the method. Several of the variations in the corrected depths, are not within the limits of ordinary observation; several of them, however, are sensible, and being all in the same direction, they cannot be attributed entirely to errors of observation, but, in part at least, to either a slight defect in the method of reduction, or to the instability of the apparatus.

It was observed during the course of the experiments, that the quantity of water entering the wheelpit, sometimes diminished sensibly, although no change had been made in the height of the regulating gate; the precaution having been taken to fix, in a secure manner, the apparatus by which the gate was moved. At the time the experiments were made, this change was attributed to a minute lowering of the gate, taking place very slowly, and arising from a defect in the stiffness of the apparatus, aided by a slight, but not totally insensible vibration of the whole apparatus, caused by the passage of the water through the apertures. To show how minute a change in the height of the regulating gate, would produce the observed changes in the quantity, let us take the two first experiments given in table IX. The regulating gate was raised to a height not exceeding 0.01 feet; supposing it to have been at just that height, and that any change in its height would have produced an equal proportional change in the discharge, the observed proportional change in the quantity was 0.00046; consequently, the absolute change in the height of the gate must have been 0.0000046 feet.

In order to prevent this source of irregularity from affecting the experiments, the regulating gate was usually set some hours before the experiments were made. This probably obviated the difficulty in part, but not entirely, as will be seen by table IX., in which are collected all the experiments that were repeated under identical circumstances.

TABLE IX.

Number of the experiment.	Number of the series.	Corrected depth upon the weir, in feet.	Variation in the depth from the initial experiment, in feet.	Proportional change in the quantities that entered the wheelpit.	Time that had elapsed when the experiment was made, since the gate was set.	
					Hours.	Minutes.
3	I.	0.19583				
7	"	0.19577	—0.00006	—0.00046		
8	II.	0.23386	. . . .	. . . .	0	22
12	"	0.23505	+0.00119	+0.00764	1	31
16	III.	0.29223	. . . .	. . . .	4	33
20	"	0.29166	—0.00057	—0.00292	5	39
24	"	0.29210	—0.00013	—0.00067	6	39
26	IV.	1.06532	. . . .	. . . .	16	58
30	"	1.06548	+0.00016	+0.00023	17	51
35	V.	0.79190	. . . .	. . . .	2	25
40	"	0.79096	—0.00094	—0.00178	3	34
44	VI.	0.95656	. . . .	. . . .	5	20
49	"	0.95477	—0.00179	—0.00281	6	40
55	VII.	1.13356	. . . .	. . . .	2	26
58	"	1.13306	—0.00050	—0.00066	3	31
63	"	1.13177	—0.00179	—0.00237	4	39
66	VIII.	1.06358	. . . .	. . . .	3	08
69	"	1.06272	—0.00086	—0.00121	3	46
Mean proportional change in the quantity, neglecting the signs, ...				0.00208		

Although the variations in the depths given in the preceding table are very small, the fact that they are nearly all negative precludes the idea that they are entirely due to errors of observation; we must, therefore, attribute to some other cause a portion of the irregularity.

134. COLUMN 19. *Combination of experiments used to determine the value of a.* It has been shown (art. 122) how, by means of two experiments in which the quantities passing over different weirs are equal, we may determine a in the formula

$$Q = Clh^a.$$

We now propose to show how, by means of two such experiments, the value of a may be found in the proposed formula

$$Q = C(l - bnh)h^a.$$

In this equation, we have b and a constant quantities to be determined; we have also C a constant, which we may here consider as indeterminate; the same may be said of Q, as limited to the experiments in the same series.

Let l, n, and h, represent the length of the weir, the number of end contractions, and the depth upon the weir in one experiment; and l_1, n_1, and h_1, the corresponding quantities in another experiment of the same series; we have

$$Q = C(l - bnh)h^a;$$

and

$$Q = C(l_1 - bn_1h_1)h_1^a:$$

since for the same series Q is constant, we have

$$(l - bnh)h^a = (l_1 - bn_1h_1)h_1^a:$$

taking the logarithms,

$$a \operatorname{Log.} h + \operatorname{Log.}(l - bnh) = a \operatorname{Log.} h_1 + \operatorname{Log.}(l_1 - bn_1h_1):$$

whence we derive

$$a = \frac{\operatorname{Log.}(l_1 - bn_1h_1) - \operatorname{Log.}(l - bnh)}{\operatorname{Log.} h - \operatorname{Log.} h_1}.$$

This equation is still indeterminate, but can be rendered determinate, by assuming a value for b.

If the formula represents the true law, and the experiments from which the values of the constants are to be derived are perfectly accurate, the particular combination of experiments to be used is evidently unimportant. As such an assumption would be very unreasonable, we have combined the experiments, with a view of obtaining the best approximation from imperfect data; and this we have accomplished by selecting experiments the most remote from each other in the values of the respective data they furnish; thus, in series I., the combinations are made by combining experiment 6, in which l has the least, and, consequently, h the greatest value, with each of the others, omitting entirely all the experiments which, for any reason, appear to be unsuitable.

Generally, in each series, one experiment has been repeated as a test, in order to show if any change had taken place in the apparatus; thus, in series III., experiments 16, 20, and 24, were made, so far as is known, under identical circumstances; in such cases, means deduced from the repeated experiments have been used instead of making a separate combination with each.

135. COLUMNS numbered 20 to 25. Values of a when $b = 0.07$, $b = 0.065$, etc. The object is, to find the values of a and b, in the formula

$$Q = C(l - bnh)h^a,$$

that will give to the computed discharges in each series the most uniform results. For this purpose, successive values of b are assumed, and the corresponding values of a, determined. The value of b leading to values of a, having the least variation among themselves, will evidently be that most nearly fulfilling this condition. To aid in the selection of the proper value of b, the table gives the differences between the values of a deduced from each combination, and the mean value of a deduced from all the combinations, with the same value of b, and the sums of these differences (having no regard to the sign) are also given. It will be seen that the sum of the differences is least when the value of $b = 0.05$, the corresponding mean value of a being 1.46994, or 1.47 very nearly; consequently, to represent the whole of the experiments with the most uniformity, the formula becomes

$$Q = C(l - 0.05nh)h^{1.47}.$$

TABLE X.

EXPERIMENTS ON THE FLOW OF WATER OVER WEIRS, MADE AT THE TREMONT TURBINE.

1	2	3		4				5	6	7	8	9	10	11
Number of the series and of the experiment.	Date of the experiment 1851.	Temperature of the atmosphere in degrees of Fahrenheit's thermometer.		TIME.				Duration of the experiment, in minutes.	Total length of the weir, in feet. l.	No of the end contractions. n.	Reference to the figures on plate X.	Depth of water on the weir by observation; in feet. h'	Height of the water in the lower canal, below the top of the weir, in feet.	Fall affecting the leakage of the wheel-pit; in feet. h.
		External air in the shade.	Near the weir.	Beginning of the experiment.		Ending of the experiment.								
				H.	min.	H.	min.							
Series I.														
Exp. 1	January 30, A. M.			10	0	10	12	12	6.987	2	Fig. 1	0.3125	1.16	1.47
" 2	" " "			10	18	10	25	7	13.978	4	" 2	0.1948	1.16	1.35
" 3	" " "	6.50		10	39	10	46.5	7.5	13.978	8	" 3	0.1952	1.16	1.36
" 4	" " "	6.25	31.50	11	2	11	6	4	10.482	6	" 4	0.2389	1.16	1.40
" 5	" " "	5.75	31.00	11	20	11	26	6	7.000	4	" 5	0.3149	1.17	1.48
" 6	" " "	6.25		11	40	11	45	5	3.500	2	" 6	0.5028	1.25	1.75
" 7	" " "	5.75		11	52	11	55	3	13.978	8	" 3	0.1951	1.22	1.42
Series II.														
Exp. 8	January 30, P. M.		30.75	2	22	2	26	4	13.978	8	Fig. 3	0.2330	1.10	1.33
" 9	" " "	4.50	30.50	2	35.5	2	41	5.5	10.482	6	" 4	0.2842	1.10	1.38
" 10	" " "	4.50	30.50	2	54	3	0	6	7.000	4	" 5	0.3738	1.10	1.47
" 11	" " "	4.25		3	15	3	21	6	3.500	2	" 6	0.5973	1.20	1.80
" 12	" " "	4.00	30.75	3	31	3	38	7	13.978	8	" 3	0.2341	1.27	1.50
" 13	" " "	3.50	30.50	3	53	3	59	6	13.978	4	" 2	0.2330	1.22	1.45
" 14	" " "			4	11	4	16.5	5.5	6.987	2	" 1	0.3719	1.10	1.47
" 15	" " "	2.75	30.75	4	24	4	32	8	16.980	4	" 7	0.2046	1.09	1.29
Series III.														
Exp. 16	January 31, P. M.	5.00	31.00	2	23.5	2	32	8.5	13.978	4	Fig. 2	0.2916	1.17	1.46
" 17	" " "	5.00	31.25	2	41	2	49.5	8.5	6.987	2	" 1	0.4652	1.17	1.64
" 18	" " "	5.00	31.25	2	56.5	3	5	8.5	13.978	8	" 3	0.2932	1.16	1.45
" 19	" " "	5.00	31.00	3	12	3	18	6	10.484	6	" 4	0.3564	1.13	1.49
" 20	" " "	5.00	30.75	3	29.5	3	35.5	6	13.978	4	" 2	0.2910	1.12	1.41
" 21	" " "	4.50	30.50	3	46	3	53	7	6.989	4	" 5	0.4684	1.15	1.62
" 22	" " "	4.50		4	2.5	4	8.5	6	3.500	2	" 8	0.7478	1.12	1.87
" 23	" " "	4.25	31.00	4	14	4	20	6	16.980	4	" 7	0.2548	1.12	1.37
" 24	" " "	4.00		4	29.5	4	35.5	6	13.978	4	" 2	0.2914	1.16	1.45
Series IV.														
Exp. 25	February 1, A. M.	5.00	31.00	9	15	9	21	6	13.978	4	Fig. 2	0.4071	1.14	1.55
" 26	" " "	7.00		9	38.5	9	46	7.5	3.496	2	" 8	1.0447	1.12	2.16
" 27	" " "	8.50	30.00	9	52	9	57	5	6.989	4	" 5	0.6577	1.15	1.81
" 28	" " "	10.00		10	4	10	11	7	10.484	6	" 4	0.4977	1.10	1.60
" 29	" " "	10.00	30.50	10	16	10	21	5	13.978	8	" 3	0.4096	1.15	1.56
" 30	" " "	10.50		10	31.5	10	39.5	8	3.496	2	" 8	1.0456	1.17	2.22
" 31	" " "	14.25	31.50	10	57	11	1	4	3.496	2	" 8	1.0452	1.12	2.17
" 32	" " "	15.00		11	19	11	24.5	5.5	6.987	2	" 1	0.6494	1.17	1.82
" 33	" " "	15.50	30.75	11	29.5	11	36	6.5	16.980	4	" 7	0.3576	1.22	1.58

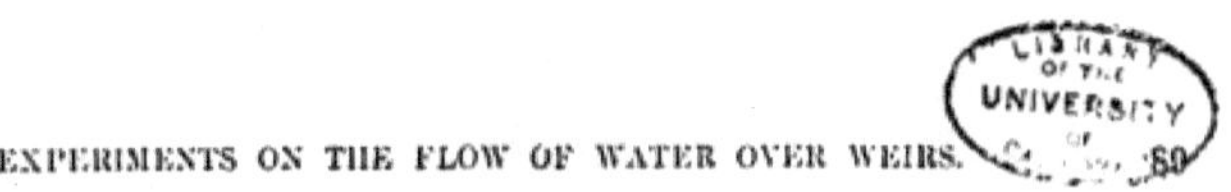

TABLE X—Continued.

EXPERIMENTS ON THE FLOW OF WATER OVER WEIRS, MADE AT THE TREMONT TURBINE.

Number of the series and of the experiment.	12 Depth of water on the weir, corrected for the leakage of the wheelpit, in feet. h''.	13 Height of water above the wheel, taken in the forebay; in feet.	14 Height of the water in the wheelpit; in feet.	15 Fall from the surface of the water in the forebay, to the surface of the water in the wheelpit; in feet. H'.	16 Uniform fall from the forebay to the wheelpit, to which the depths on the weir in each series are reduced; in feet. H.	17 Depth on the weir, corrected for the leakage of the wheelpit, and the variation in the fall; in feet. h'''.	18 REMARKS.
Series I.							
Exp. 1	0.31456	14.869	0.320	14.549	14.549	0.31456	In experiments 2, 3, and 7, the contraction was incomplete, as the water followed the top of the weir.
" 2	0.19605	14.896	0.201	14.695	"	0.19540	
" 3	0.19645	14.894	0.205	14.689	"	0.19583	
" 4	0.24043	14.881	0.247	14.634	"	0.23997	
" 5	0.31696	14.892	0.320	14.572	"	0.31679	
" 6	0.50634	14.876	0.510	14.366	"	0.50848	
" 7	0.19638	14.886	0.200	14.686	"	0.19577	
Series II.							
Exp. 8	0.23414	14.910	0.240	14.670	14.619	0.23386	In experiment 15 the contraction was incomplete, as the water followed the top of the weir.
" 9	0.28560	14.909	0.290	14.619	"	0.28560	
" 10	0.37568	14.915	0.380	14.535	"	0.37640	
" 11	0.60059	14.916	0.610	14.306	"	0.60494	
" 12	0.23530	14.906	0.240	14.666	"	0.23505	
" 13	0.23419	14.912	0.240	14.672	"	0.23390	
" 14	0.37379	14.910	0.380	14.530	"	0.37455	
" 15	0.20558	14.918	0.210	14.708	"	0.20517	
Series III.							
Exp. 16	0.29266	14.897	0.300	14.597	14.532	0.29223	
" 17	0.46698	14.900	0.470	14.430	"	0.46808	
" 18	0.29426	14.897	0.300	14.597	"	0.29382	
" 19	0.35770	14.897	0.365	14.532	"	0.35770	
" 20	0.29205	14.890	0.300	14.590	"	0.29166	
" 21	0.47017	14.887	0.477	14.410	"	0.47149	
" 22	0.75080	14.883	0.760	14.123	"	0.75798	
" 23	0.25571	14.878	0.260	14.618	"	0.25520	
" 24	0.29246	14.886	0.300	14.586	"	0.29210	
Series IV.							
Exp. 25	0.40803	14.904	0.420	14.484	14.537	0.40852	In experiments 26 and 30, the water flowing over the weir fell upon a board placed upon the brackets N, figures 1 and 2, plate V.; in experiment 31 the board was removed. So far as is known the three experiments were identical in all other respects. By comparing the corrected depths upon the weir given in column 17, it appears that the board offered no appreciable obstruction to the discharge.
" 26	1.04743	14.877	1.060	13.817	"	1.06532	
" 27	0.65928	14.871	0.670	14.201	"	0.66444	
" 28	0.49884	14.877	0.510	14.367	"	0.50080	
" 29	0.41053	14.893	0.420	14.473	"	0.41113	
" 30	1.04887	14.908	1.060	13.848	"	1.06548	
" 31	1.04794	14.886	1.060	13.826	"	1.06560	
" 32	0.65099	14.904	0.660	14.244	"	0.65543	
" 33	0.35842	14.307	0.370	14.537	"	0.35842	

TABLE X—Continued.

EXPERIMENTS ON THE FLOW OF WATER OVER WEIRS, MADE AT THE TREMONT TURBINE.

Number of the series and of the experiment.	Combination of experiments used to determine the value of a.	20 b = 0.07			21 b = 0.065.			22 b = 0.06.		
19		Values of a.	Differences from the mean value of a, or from 1.47797.		Values of a.	Differences from the mean value of a, or from 1.47595.		Values of a.	Differences from the mean value of a, or from 1.47394.	
			+	−		+	−		+	−
Series I.										
Exp. 1	1 and 6	1.4691		0.00887	1.4669		0.00905	1.4648		0.00914
" 2										
" 3										
" 4	4 " 6	1.4753		0.00267	1.4742		0.00175	1.4731		0.00084
" 5	5 " 6	1.4814	0.00343		1.4801	0.00415		1.4789	0.00496	
" 6										
" 7										
Series II.										
Exp. 8	8, 12, and 11	1.4768		0.00117	1.4756		0.00035	1.4744	0.00046	
" 9	9 " 11	1.4787	0.00073		1.4775	0.00155		1.4763	0.00236	
" 10	10 " 11	1.4805	0.00253		1.4791	0.00315		1.4777	0.00376	
" 11										
" 12	8, 12, " 11	1.4768		0.00117	1.4756		0.00035	1.4744	0.00046	
" 13	13 " 11	1.4781	0.00013		1.4766	0.00065		1.4751	0.00116	
" 14	14 " 11	1.4774		0.00057	1.4748		0.00115	1.4723		0.00164
" 15	15 " 11	1.4800	0.00203		1.4786	0.00265		1.4772	0.00326	
Series III.										
Exp. 16	16, 20, 24, and 22	1.4778		0.00017	1.4759		0.00005	1.4740	0.00006	
" 17	17 " 22	1.4784	0.00043		1.4752		0.00075	1.4721		0.00184
" 18	18 " 22	1.4811	0.00313		1.4797	0.00375		1.4782	0.00426	
" 19	19 " 22	1.4827	0.00473		1.4811	0.00515		1.4795	0.00556	
" 20	16, 20, 24, " 22	1.4778		0.00017	1.4759		0.00005	1.4740	0.00006	
" 21	21 " 22	1.4814	0.00343		1.4796	0.00365		1.4778	0.00386	
" 22										
" 23	23 " 22	1.4752		0.00277	1.4734		0.00255	1.4716		0.00234
" 24	16, 20, 24, " 22	1.4778		0.00017	1.4759		0.00005	1.4740	0.00006	
Series IV.										
Exp. 25	25 and 26, 30	1.4827	0.00473		1.4800	0.00405		1.4773	0.00336	
" 26										
" 27	27 " 26, 30	1.5023	0.02433		1.4997	0.02375		1.4971	0.02316	
" 28	28 " 26, 30	1.4857	0.00773		1.4834	0.00745		1.4812	0.00726	
" 29	29 " 26, 30	1.4838	0.00583		1.4817	0.00575		1.4796	0.00566	
" 30										
" 31										
" 32	32 " 26, 30	1.4878	0.00983		1.4832	0.00725		1.4786	0.00466	
" 33	33 " 26, 30	1.4853	0.00733		1.4828	0.00685		1.4802	0.00626	

TABLE X—Continued.

EXPERIMENTS ON THE FLOW OF WATER OVER WEIRS, MADE AT THE TREMONT TURBINE.

Number of the series and of the experiment.	23 $b = 0.055$			24 $b = 0.05$.			25 $b = 0.045$.		
	Values of a.	Differences from the mean value of a, or from 1.47194.		Values of a.	Differences from the mean value of a, or from 1.46994.		Values of a.	Differences from the mean value of a, or from 1.46795.	
		+	−		+	−		+	−
Series I.									
Exp. 1	1.4626		0.00934	1.4605		0.00944	1.4584		0.00955
" 2									
" 3									
" 4	1.4721	0.00016		1.4710	0.00106		1.4700	0.00205	
" 5	1.4778	0.00586		1.4765	0.00656		1.4754	0.00745	
" 6									
" 7									
Series II.									
Exp. 8	1.4733	0.00136		1.4722	0.00226		1.4710	0.00305	
" 9	1.4750	0.00306		1.4737	0.00376		1.4725	0.00455	
" 10	1.4762	0.00426		1.4749	0.00496		1.4734	0.00545	
" 11									
" 12	1.4733	0.00136		1.4722	0.00226		1.4710	0.00305	
" 13	1.4735	0.00156		1.4721	0.00216		1.4706	0.00265	
" 14	1.4697		0.00224	1.4671		0.00284	1.4646		0.00335
" 15	1.4758	0.00386		1.4744	0.00446		1.4730	0.00505	
Series III.									
Exp. 16	1.4721	0.00016		1.4702	0.00026		1.4683	0.00035	
" 17	1.4688		0.00314	1.4656		0.00434	1.4624		0.00555
" 18	1.4768	0.00486		1.4753	0.00536		1.4739	0.00595	
" 19	1.4780	0.00606		1.4764	0.00646		1.4748	0.00685	
" 20	1.4721	0.00016		1.4702	0.00026		1.4683	0.00035	
" 21	1.4760	0.00406		1.4742	0.00426		1.4725	0.00455	
" 22									
" 23	1.4699		0.00204	1.4681		0.00184	1.4663		0.00165
" 24	1.4721	0.00016		1.4702	0.00026		1.4683	0.00035	
Series IV.									
Exp. 25	1.4746	0.00266		1.4720	0.00206		1.4693	0.00135	
" 26									
" 27	1.4945	0.02256		1.4920	0.02206		1.4895	0.02155	
" 28	1.4789	0.00696		1.4767	0.00676		1.4745	0.00655	
" 29	1.4776	0.00566		1.4755	0.00556		1.4735	0.00555	
" 30									
" 31									
" 32	1.4741	0.00216		1.4695		0.00044	1.4651		0.00285
" 33	1.4777	0.00576		1.4752	0.00526		1.4728	0.00485	

TABLE X—Continued.

EXPERIMENTS ON THE FLOW OF WATER OVER WEIRS, MADE AT THE TREMONT TURBINE.

1	2	3		4				5	6	7	8	9	10	11
Number of the series and of the experiment.	Date of the experiment, 1851.	Temperature of the atmosphere in degrees of Fahrenheit's thermometer.		TIME.				Duration of the experiment, in minutes.	Total length of the weir, in feet. l.	No. of the end contractions. n.	Reference to the figures on plate X.	Depth of water on the weir by observation; in feet. h'.	Height of the water in the lower canal, below the top of the weir, in feet.	Fall affecting the leakage of the wheel-pit; in feet. λ.
		External air in the shade.	Near the weir.	Beginning of the experiment. H.	min.	Ending of the experiment. H.	min.							
Series V.														
Exp. 34	February 1, P. M.	20.75	31.50	2	11	2	15.5	4.5	13.978	4	Fig. 2	0.4937	1.12	1.61
" 35	" " "	20.00	31.50	2	25	2	33	8	6.987	2	" 1	0.7908	1.16	1.95
" 36	" " "	21.50	31.50	2	39	2	43	4	13.978	8	" 3	0.4981	1.17	1.67
" 37	" " "	22.25	31.50	2	49.5	2	54	4.5	10.484	6	" 4	0.6060	1.23	1.84
" 38	" " "	20.50		3	3.5	3	14	10.5	6.989	4	" 5	0.8000	1.21	2.01
" 39	" " "	21.50	30.00	3	19	3	23.5	4.5	16.980	4	" 7	0.4337	1.22	1.65
" 40	" " "	21.00	31.50	3	34	3	48	14	6.987	2	" 1	0.7896	1.24	2.03
" 41	" " "	18.50	31.25	4	16	4	26	10	6.987	2	" 1	0.7790	1.25	2.03
" 42	" " "	18.00		4	52	5	0	8	6.987	2	" 1	0.7704	1.35	2.12
Series VI.														
Exp. 43	February 3, P. M.	38.25		2	6	2	11	5	13.978	4	Fig. 2	0.5977	1.10	1.70
" 44	" " "	38.25	32.25	2	20.5	2	30	9.5	6.987	2	" 1	0.9561	1.17	2.13
" 45	" " "	38.25		2	37	2	43	6	6.987	4	" 9	0.9636	1.17	2.13
" 46	" " "	37.50	32.00	2	48.5	2	56	7.5	13.978	8	" 3	0.6023	1.16	1.76
" 47	" " "	37.50		3	6.5	3	13	6.5	10.488	6	" 4	0.7308	1.11	1.84
" 48	" " "	37.25	32.00	3	16.5	3	22.5	6	16.980	4	" 7	0.5238	1.13	1.65
" 49	" " "	37.00		3	40	3	45	5	6.987	2	" 1	0.9535	1.13	2.08
" 50	" " "			3	47	3	59	12	6.987	2	" 1	0.9531	1.13	2.08
" 51	" " "			4	2	4	4	2	6.987	2	" 1	0.9539	1.13	2.08
" 52	" " "			4	23	4	31	8	6.987	2	" 1	0.9415	1.13	2.07
" 53	" " "	33.75		4	46.5	5	0	13.5	6.987	2	" 1	0.9275	1.14	2.07
Series VII.														
Exp. 54	February 4, A. M.	26.00	31.75	9	6	9	12.5	6.5	16.980	4	Fig. 7	0.5233	1.12	1.64
" 55	" " "	26.50		9	26.5	9	37	10.5	5.487	2	" 10	1.1278	1.14	2.27
" 56	" " "	31.00	31.75	9	44	9	57	13	6.987	2	" 11	0.9544	1.15	2.10
" 57	" " "	30.00	31.75	10	17	10	22	5	8.489	2	" 12	0.8375	1.14	1.98
" 58	" " "	31.25	31.75	10	31	10	36	5	5.487	2	" 10	1.1269	1.17	2.30
" 59	" " "	33.50	31.75	10	42	10	46.5	4.5	6.987	4	" 13	0.9609	1.13	2.09
" 60	" " "	34.00	31.75	10	56.5	11	1	4.5	13.978	8	" 3	0.6017	1.13	1.73
" 61	" " "	35.00		11	10	11	15	5	10.489	6	" 4	0.7303	1.12	1.85
" 62	" " "	37.50	31.75	11	20	11	26	6	13.978	4	" 2	0.5971	1.15	1.75
" 63	" " "	38.00	31.75	11	39.5	11	55	15.5	5.487	2	" 10	1.1256	1.14	2.27
" 64	" " P. M.	40.75		0	16	0	25	9	5.487	2	" 10	1.0935	1.13	2.22
Series VIII.														
Exp. 65	February 4, P. M.	38.25	32.00	3	10.5	3	14	3.5	16.980	4	Fig. 7	0.2316	1.19	1.42
" 66	" " "	37.50		3	33.5	3	40	6.5	1.829	2	" 14	1.0581	1.17	2.23
" 67	" " "	37.00		3	45	3	52	7	3.658	4	" 15	0.6650	1.18	1.84
" 68	" " "	36.75		3	58	4	2	4	5.487	6	" 16	0.5066	1.24	1.75
" 69	" " "			4	11.5	4	16	4.5	1.829	2	" 14	1.0574	1.21	2.27
" 70	" " "	36.25	32.00	4	21	4	25	4	8.489	2	" 12	0.3706	1.19	1.56
" 71	" " "			4	32	4	37	5	5.487	2	" 10	0.4980	1.19	1.69
Series IX.														
Exp. 72	February 4, P. M.	34.25		4	53	5	2	9	16.980	4	Fig. 7	0.9206	1.18	2.10
" 73	" " "	34.25		5	17.5	5	24	6.5	16.980	4	" 7	0.9091	1.02	1.93
" 74	" " "	33.75		5	31	5	43	12	16.980	4	" 7	0.8941	1.17	2.06

TABLE X—Continued.

EXPERIMENTS ON THE FLOW OF WATER OVER WEIRS, MADE AT THE TREMONT TURBINE.

Number of the series and of the experiment.	12 Depth of water on the weir, corrected for the leakage of the wheelpit, in feet. h''.	13 Height of water above the wheel, taken in the forebay; in feet.	14 Height of water in the wheelpit; in feet.	15 Fall from the surface of the water in the forebay, to the surface of the water in the wheelpit; in feet. H'.	16 Uniform fall from the forebay to the wheelpit, to which the depths on the weir in each series are reduced; in feet. H.	17 Depth on the weir, corrected for the leakage of the wheelpit, and the variation in the fall; in feet. h'''.	18 REMARKS.
Series V.							
Exp. 34	0.49456	14.845	0.510	14.335	14.079	0.49160	In experiments 41 and 42 the weir was in the same state as in experiment 40; the height of the water above the wheel was reduced for the purpose of testing the method of reduction.
" 35	0.79229	14.910	0.810	14.100	"	0.79190	
" 36	0.49897	14.891	0.520	14.371	"	0.49557	
" 37	0.60710	14.891	0.620	14.271	"	0.60437	
" 38	0.80151	14.899	0.820	14.079	"	0.80151	
" 39	0.43446	14.908	0.450	14.458	"	0.43063	
" 40	0.79112	14.897	0.809	14.088	"	0.79096	
" 41	0.78054	14.352	0.798	13.554	"	0.79049	
" 42	0.77198	13.939	0.790	13.149	"	0.78976	
Series VI.							
Exp. 43	0.59850	14.903	0.620	14.283	13.907	0.59320	In experiments 50 and 51 the weir was in the same state as in experiment 49, excepting that in 50 a board was placed on the brackets *N*, figs. 1 and 2, plate V., on which the water fell; and in exp. 51, the plank *O*, fig. 2, plate V., was placed in the position represented; the top of the plank was 6.5 inches below the top of the weir. In exps. 52 and 53, the weir was in the same state as in exp. 49; the height of the water above the wheel was lowered for the purpose of testing the method of reduction.
" 44	0.95752	14.929	0.980	13.949	"	0.95656	
" 45	0.96501	14.915	0.992	13.923	"	0.96464	
" 46	0.60311	14.894	0.630	14.264	"	0.59804	
" 47	0.73181	14.887	0.755	14.132	"	0.72790	
" 48	0.52449	14.889	0.550	14.339	"	0.51917	
" 49	0.95471	14.884	0.980	13.904	"	0.95477	
" 50	0.95451	14.886	0.980	13.906	"	0.95453	
" 51	0.95530	14.887	0.980	13.907	"	0.95530	
" 52	0.94291	14.406	0.970	13.436	"	0.95380	
" 53	0.92892	13.914	0.952	12.962	"	0.95097	
Series VII.							
Exp. 54	0.52399	14.889	0.550	14.339	13.882	0.51837	In experiments 63 and 64 the weir was in the same state; in experiment 64 the height of the water above the wheel was lowered for the purpose of testing the method of reduction.
" 55	1.12952	14.884	1.150	13.734	"	1.13356	
" 56	0.95581	14.882	0.980	13.902	"	0.95535	
" 57	0.83870	14.875	0.865	14.010	"	0.83614	
" 58	1.12863	14.872	1.152	13.720	"	1.13306	
" 59	0.96230	14.872	0.990	13.882	"	0.96230	
" 60	0.60251	14.868	0.629	14.239	"	0.59743	
" 61	0.73131	14.865	0.754	14.111	"	0.72733	
" 62	0.59791	14.860	0.620	14.240	"	0.59286	
" 63	1.12732	14.869	1.150	13.719	"	1.13177	
" 64	1.09523	13.926	1.120	12.806	"	1.12508	
Series VIII.							
Exp. 65	0.23257	14.894	0.240	14.654	13.839	0.22817	In experiments 66, 67, 68, and 69, the lengths of the several bays of the weir were deemed to be too short relative to the depth flowing over, for the proposed formula to apply.
" 66	1.06337	14.899	1.068	13.831	"	1.06358	
" 67	0.66802	14.895	0.676	14.219	"	0.66202	
" 68	0.50885	14.902	0.515	14.387	"	0.50231	
" 69	1.06272	14.905	1.066	13.839	"	1.06272	
" 70	0.37221	14.905	0.380	14.525	"	0.36625	
" 71	0.50023	14.909	0.505	14.404	"	0.49360	
Series IX.							
Exp. 72	0.92119	14.864	1.048	13.816	13.839	0.92170	Experiments 72, 73, and 74 were made for the express purpose of testing the mode of reduction.
" 73	0.90967	14.350	1.035	13.315	"	0.92145	
" 74	0.89469	13.678	1.013	12.665	"	0.92153	

TABLE X—Continued.

EXPERIMENTS ON THE FLOW OF WATER OVER WEIRS, MADE AT THE TREMONT TURBINE.

Number of the series and of the experiment.	19	20			21			22		
	Combination of experiments used to determine the value of a.	b = 0.07			b = 0.065.			b = 0.06.		
		Values of a.	Differences from the mean value of a, or from 1.47797.		Values of a.	Differences from the mean value of a, or from 1.47595.		Values of a.	Differences from the mean value of a, or from 1.47394.	
			+	−		+	−		+	−
Series V.										
Exp. 34	34 and 38	1.4645		0.01347	1.4611		0.01485	1.4577		0.01624
" 35	35, 40, " 39	1.4737		0.00427	1.4726		0.00335	1.4716		0.00234
" 36	36 " 38	1.4679		0.01007	1.4660		0.00995	1.4641		0.00984
" 37	37 " 38	1.4651		0.01287	1.4630		0.01295	1.4610		0.01294
" 38										
" 39	39 " 38	1.4700		0.00797	1.4670		0.00895	1.4640		0.00994
" 40	35, 40, " 39	1.4737		0.00427	1.4726		0.00335	1.4716		0.00234
" 41										
" 42										
Series VI.										
Exp. 43	43 and 45	1.4827	0.00473		1.4785	0.00255		1.4744	0.00046	
" 44	44, 49, " 48	1.4706		0.00737	1.4694		0.00655	1.4681		0.00584
" 45										
" 46	46 " 45	1.4821	0.00413		1.4798	0.00385		1.4774	0.00346	
" 47	47 " 46	1.4890	0.01103		1.4870	0.01105		1.4850	0.01106	
" 48	48 " 47	1.4879	0.00993		1.4834	0.00745		1.4788	0.00486	
" 49	44, 49, " 48	1.4706		0.00737	1.4694		0.00655	1.4681		0.00584
" 50										
" 51										
" 52										
" 53										
Series VII.										
Exp. 54	54 and 55, 58, 63	1.4715		0.00647	1.4696		0.00635	1.4677		0.00624
" 55										
" 56	56 and 54	1.4699		0.00807	1.4686		0.00735	1.4674		0.00654
" 57	57 and 55, 58, 63	1.4881	0.01013		1.4843	0.00835		1.4806	0.00666	
" 58										
" 59	59 and 54	1.4850	0.00703		1.4814	0.00545		1.4778	0.00386	
" 60	60 and 55, 58, 63	1.4695		0.00847	1.4689		0.00705	1.4683		0.00564
" 61	61 " 55, 58, 63	1.4619		0.01607	1.4620		0.01395	1.4620		0.01194
" 62	62 " 55, 58, 63	1.4711		0.00687	1.4691		0.00685	1.4671		0.00684
" 63										
" 64										
Series VIII.										
Exp. 65	65 and 71	1.4755		0.00247	1.4747		0.00125	1.4739		0.00004
" 66										
" 67										
" 68										
" 69										
" 70	70 " 71	1.4845	0.00653		1.4829	0.00695		1.4813	0.00736	
" 71										
Series IX.										
Exp. 72										
" 73										
" 74										
Sums of the differences and mean values of a.		1.47797	0.26767		1.47595	0.25085		1.47394	0.23672	

TABLE X—Continued.

EXPERIMENTS ON THE FLOW OF WATER OVER WEIRS, MADE AT THE TREMONT TURBINE

Number of the series and of the experiment.	23 $b = 0.055$			24 $b = 0.05$.			25 $b = 0.045$.		
	Values of a.	Differences from the mean value of a, or from 1.47194.		Values of a.	Differences from the mean value of a, or from 1.46994.		Values of a.	Differences from the mean value of a, or from 1.46795.	
		+	−		+	−		+	−
Series V.									
Exp. 34	1.4543		0.01764	1.4510		0.01894	1.4476		0.02035
" 35	1.4706		0.00134	1.4695		0.00044	1.4684	0.00045	
" 36	1.4622		0.00974	1.4602		0.00974	1.4584		0.00955
" 37	1.4589		0.01304	1.4567		0.01324	1.4547		0.01325
" 38									
" 39	1.4610		0.01094	1.4581		0.01184	1.4551		0.01285
" 40	1.4706		0.00134	1.4695		0.00044	1.4684	0.00045	
" 41									
" 42									
Series VI.									
Exp. 43	1.4703		0.00164	1.4662		0.00374	1.4622		0.00575
" 44	1.4668		0.00514	1.4656		0.00434	1.4643		0.00365
" 45									
" 46	1.4751	0.00316		1.4728	0.00286		1.4705	0.00255	
" 47	1.4830	0.01106		1.4811	0.01116		1.4790	0.01105	
" 48	1.4744	0.00246		1.4699		0.00004	1.4654		0.00255
" 49	1.4668		0.00514	1.4656		0.00434	1.4643		0.00365
" 50									
" 51									
" 52									
" 53									
Series VII.									
Exp. 54	1.4657		0.00624	1.4638		0.00614	1.4619		0.00605
" 55									
" 56	1.4661		0.00584	1.4649		0.00504	1.4636		0.00435
" 57	1.4769	0.00496		1.4733	0.00336		1.4696	0.00165	
" 58									
" 59	1.4742	0.00226		1.4706	0.00066		1.4670		0.00095
" 60	1.4677		0.00424	1.4672		0.00274	1.4666		0.00135
" 61	1.4620		0.00994	1.4621		0.00784	1.4621		0.00585
" 62	1.4652		0.00674	1.4633		0.00664	1.4613		0.00665
" 63									
" 64									
Series VIII.									
Exp. 65	1.4731	0.00116		1.4722	0.00226		1.4714	0.00345	
" 66									
" 67									
" 68									
" 69									
" 70	1.4797	0.00776		1.4782	0.00826		1.4765	0.00855	
" 71									
Series IX.									
Exp. 72									
" 73									
" 74									
	1.47194	0.23124		1.46994	0.22900		1.46795	0.23945	

EXPERIMENTS ON THE FLOW OF WATER OVER WEIRS, MADE AT THE CENTRE-VENT WHEEL FOR MOVING THE GUARD GATES OF THE NORTHERN CANAL.

136. This centre-vent wheel usually operates under about ten feet fall, and is of about sixty horse-power under this fall. It was constructed from nearly the same designs as the model centre-vent wheel, described in art. 100, and represented on plate VII. For a general description of the Guard Gates, see vol. I., page 775, *Appleton's Dictionary of Machines, Mechanics*, etc., New York: D. Appleton & Co., 1852.

A set of experiments upon the power of this wheel was made in 1848, in which the water discharged by the wheel was gauged at a weir constructed for the purpose, below the wheel. The following experiments were made with the same apparatus.

The total length of the weir was 18.02 feet, which, for the purposes of these experiments, was diminished to 16.02 feet by two movable planks or partitions, one foot wide each, the upstream faces of which, when placed upon the weir, were in the same plane as the upstream face of the weir. The form of the weir was such as to give complete contraction; it was constructed of wood, with the upstream face vertical. The crest of the weir was formed of southern hard pine plank, four inches in thickness; the top was 0.53 inches wide, and bevelled off on the downstream side, at an angle of 40° with the vertical; the ends of the weir and the sides of the partitions were of the same form.

The bottom of the canal or basin, measured near the weir, was about 6.72 feet below the top of the weir. The water discharged by the wheel passed to the basin through an irregular and contracted channel, cut in rock, and confined by cement masonry. This basin was specially excavated in the rock, of large dimensions, in order that the water might reach the weir in a sufficiently quiet state to permit a satisfactory measurement to be made; and also, for the same object, two gratings were placed across the basin, parallel to each other, and about six feet apart, the downstream grating being about seventeen feet from the weir. The effect of these several precautions was such that, although the water escaped from the wheelpit in a rapid and turbulent current, in the basin between the downstream grating and the weir, the water was tranquil and free from perceptible irregularities in its motion towards the weir.

The depths upon the weir were measured by the hook gauge, described at art. 45 and represented by figures 9, 10, and 11, plate IV.; this was placed in the basin about eight feet from the weir, in a box, in which the communication with the surrounding water was maintained by a small aperture in the bottom; the box and hook gauge were firmly attached to a timber strongly bolted to the masonry forming one side of the basin.

The quantity of water discharged by the wheel is usually regulated by the head gate, admitting the water from the river into the forebay above the wheel. When it is desired to diminish the quantity discharged by the wheel, this gate is partially closed, the effect of which is to diminish the fall acting upon the wheel; but this method was unsuitable for these experiments, on account of the great agitation in the forebay, produced by the fall at the head gate. During these experiments, the head gate was fully opened, and the quantity of water discharged by the wheel was diminished by closing up a portion of the spaces between the guides, with pieces of wood.

The wheel was prevented from revolving by the brake of the Prony dynamometer. The entire apparatus about the wheel remained unchanged throughout the four experiments, except that the head gate was closed on several occasions, to enable the partitions on the weir to be moved. This gate was large (five feet square,) and care was taken to keep it open to its full extent, in all these experiments.

The apertures through which the water entered the wheelpit being the same, the quantity of water discharged must have been uniform, if the head acting upon the orifices had been constant; small variations, however, unavoidably occurred in the head, for which it was necessary to correct the depths upon the weir. This has been done in a manner precisely similar to that adopted in the experiments upon the weir at the Tremont Turbine, described at art. 133.

The apertures in the wheel and between the guides, were entirely submerged. The effective height of the water in the wheelpit was measured in a chamber constructed for the purpose, in the masonry. A free communication was maintained between the water in the wheelpit and in the chamber by an iron pipe about 3.5 inches diameter. The surface of the water in the chamber was, in all the experiments, above the level of the top of the apertures between the guides. The height above the wheel was taken in the forebay nearly over the wheel, the gauge being placed in a box in the usual manner; the zeros of the gauges, at which both these heights were taken, were at the same level, consequently, the difference in the readings gave the fall acting upon the apertures.

TABLE XI.

EXPERIMENTS ON THE FLOW OF WATER OVER WEIRS, MADE AT THE CENTRE-VENT WHEEL FOR MOVING THE GUARD GATES OF THE NORTHERN CANAL AT LOWELL, MASSACHUSETTS.

No. of the experiment.	TIME. November 14, 1848. Beginning. H.	min.	sec.	Ending. H.	min.	sec.	Length of the experiment, in seconds.	Mean height of the water above the wheel taken in the forebay, in feet.	Mean height of the water below the wheel, taken in the chamber communicating with the wheel-pit, in feet.	Fall acting upon the apertures in the wheel and guides, in feet.	Total length of the weirs, in feet.	No. of end contractions in the weir.	Depth on the weir as measured by the hook gauge, in feet.	Constant fall acting upon the apertures in the wheel and guides, to which the depths on the weir are reduced, in feet.	Depth on the weir reduced to the uniform fall in the preceding column, in feet.	REMARKS.
1	11	18	2	11	27	42	580	10.373	1.755	8.618	16.02	6	1.0160	8.612	1.0157	Three weirs, 5.34 feet long each, separated by partitions one foot wide each.
2	11	52	0	11	58	54	414	10.357	1.745	8.612	16.02	2	1.0072	8.612	1.0072	One weir, both of the partitions placed at one end.
3	2	1	49	2	16	27	878	10.368	1.746	8.622	16.02	4	1.0137	8.612	1.0133	Two weirs, 8.01 feet long each, separated by the two partitions placed close together, making a single partition two feet wide.
4	2	42	18	2	50	27	489	10.365	1.744	8.621	16.02	2	1.0071	8.612	1.0067	One weir, same as in experiment 2.

EXPERIMENTS ON THE EFFECT PRODUCED ON THE FLOW OF WATER OVER WEIRS, BY THE HEIGHT OF THE WATER ON THE DOWNSTREAM SIDE.

137. These were made at the weir at the centre-vent wheel for moving the guard gates, with the apparatus used in the preceding experiments.

A singular phenomenon was here produced, namely: *under particular circumstances, the flow of water over a weir may be increased by raising the height of the water on the downstream side of the weir.* Ordinarily, when water flows over a weir having contraction on the bottom, the under side of the sheet near the weir, is elevated above the level of the top of the weir, taking a curved form; representations of this curve are given in several works, the most perfect of which are by M. M. Poncelet and Lesbros,* who ascertained with great care the forms for several depths upon the weir. In such cases, the space between the sheet of water and the plank or other material of which the weir is composed, is filled with air which communicates more or less freely with the external atmosphere.

Suppose the sheet, after passing the weir, to fall into a body of water of considerable depth, in which the natural level of the surface is not very much below the top of the weir, but sufficiently so, as not sensibly to affect the discharge. The weir having complete contraction, the air will remain under the sheet, even if the weir is of very considerable length in proportion to the depth flowing over. Suppose now, that the communication of the air under the sheet, with the external atmosphere, is entirely cut off by placing boards on the downstream side of the weir in contact with each side of the sheet, or by other means, the effect will ordinarily be, that the air under the sheet will be wholly or partially driven out by the lateral communication of motion in fluids, and a partial vacuum will be produced, unless water takes the place of the air that is driven out. In either case, the equilibrium of the atmospheric pressure on the upper and lower sides of the sheet, will be destroyed, the pressure on the upper side preponderating, the effect will be to alter the form of the sheet, and to increase the discharge, by the operation of forces bearing some resemblance to the action in the well-known experiment with Venturi's tube.

In the following experiments, this effect was produced by raising the level

---

* *Expériences Hydrauliques sur les lois de l'écoulement*, etc. Paris: 1832. Plate 6.

of the water on the downstream side of the weir, to a height a little above the top of the weir, in consequence of which, by the lateral communication of motion, the air was driven out, and the flow over the weir facilitated.

During the following experiments, the apparatus was arranged in the same manner as in the preceding experiments, with these exceptions, namely: the partitions were not used; the quantity of water entering the wheelpit was diminished by closing up more of the spaces between the guides; the wheel was entirely removed; and means were provided for varying the height of the water on the downstream side of the weir.

The depths on the weir are reduced in the same manner to what they would have been, if the quantity of water entering the wheelpit, and flowing over the weir, had been uniform. The details of the experiments are given in table XII.

That the quantity of water entering the wheelpit changed only in a very small degree from any change in the apparatus, is proved by the depths upon the weir in experiments 1 and 9. The circumstances in both being the same, the corrected depth on the weir in experiment 9, is 0.0006 feet less than in experiment 1, corresponding to a change in quantity of about $\frac{1}{545}$ part. A mean of the depths on the weir in these two experiments has been taken, with which to compare the other experiments.

Measurements were also taken of the thickness of the sheet, in the plane of the upstream face of the weir. This was done by means of a graduated rod terminating in a fine point, and so arranged as to slide in a vertical groove, supported from one end of the weir. These measurements were not taken with the same precision as were the depths on the weir with the hook gauge, principally in consequence of the oscillations of the surface.

In consequence of the want of symmetry in the channel carrying off the water from the weir, the water on the downstream side did not assume the same height at both ends of the weir. Gauges were placed at both ends, protected in a considerable degree from the agitation of the water immediately below the weir, and placed so as to indicate the height of the water a short distance downstream from the sheet, but the heights taken at these gauges have not the exactness of those taken with the hook gauge. There were also much greater variations in the height of the water during the course of an experiment, than occurred on the upstream side of the weir; some of the heights given in column 9 may consequently be erroneous to the extent of 0.02 feet.

The differences given in column 10, indicate the effect produced on the discharge by the height of the water on the downstream side. When this height was about 3 inches below the top of the weir, the effect was insensible. When about level with the top of the weir, the obstruction was very minute and barely sensible. When the height on the downstream side was about $\frac{3}{4}$ of an inch above the top of the weir, (at which height the air did not remain under the sheet,) the increase in the discharge is quite sensible, the discharge with the same depth being increased about $\frac{1}{150}$. When the height on the downstream side is 1.25 inches above the top of the weir, the obstruction is quite distinct, and it increases rapidly with the increase of height.

TABLE XII.

EXPERIMENTS ON THE EFFECT PRODUCED ON THE FLOW OF WATER OVER WEIRS, BY THE HEIGHT OF THE WATER ON THE DOWNSTREAM SIDE.

1	2						3	4	5	6	7	8	9			10	11	12
No. of the experiment.	TIME November 17, 1848. Beginning.			Ending.			Length of the experiment. Seconds.	Mean height of the water above the wheel, taken in the forebay	Mean height of the water below the wheel, taken in the chamber communicating with the wheelpit.	Fall acting upon the apertures between the guides.	Depth on the weir as measured by the hook gauge.	Depth on the weir reduced to the uniform fall of 8.242 feet.	Height of the water on the downstream side of the weir, above or below the top of the weir, in feet.			Difference of the depth on the weir from the mean in experiments 1 and 9.	Actual depth on the weir at the upstream edge of the weir, 3.2 feet from the west end.	REMARKS.
	H.	min.	sec.	H.	min.	sec.		Feet.	Feet.	Feet.	Feet.	Feet.	East end.	West end.	Mean.	Feet.	Feet.	
1	2	49	0	2	51	30	150	9.762	1.520	8.242	0.8528	0.8528	—1.00	—1.10	—1.050		0.7280	Air under the sheet; form of the sheet not perceptibly changed by raising the water below.
2	2	59	0	3	4	30	330	9.753	1.523	8.230	0.8523	0.8527	—0.22	—0.25	—0.235	+0.0002	0.7277	
3	3	6	30	3	11	30	300	9.753	1.522	8.231	0.8528	0.8532	+0.03	+0.01	+0.020	+0.0007	0.7280	
4	3	18	30	3	24	0	330	9.745	1.519	8.226	0.8480	0.8485	+0.08	+0.05	+0.065	—0.0040	0.7215	No air under the sheet; a very slight lowering of the surface of the water on the downstream side, caused the air to pass under the sheet, and allow it to take the usual form.
5	3	29	0	3	34	30	330	9.746	1.525	8.221	0.8515	0.8522	+0.10	+0.07	+0.085	—0.0003	0.7247	No air under the sheet.
6	3	37	30	3	40	0	150	9.738	1.521	8.217	0.8562	0.8571	+0.12	+0.09	+0.105	+0.0046	0.7290	
7	3	45	0	3	49	30	270	9.730	1.528	8.202	0.8806	0.8820	+0.24	+0.20	+0.220	+0.0295	0.7565	
8	3	57	0	4	2	0	300	9.731	1.584	8.147	0.9667	0.9704	+0.50	+0.48	+0.490	+0.1179	0.8607	
9	4	4	0	4	10	0	360	9.725	1.518	8.207	0.8510	0.8522	—1.06	—1.23	—1.145		0.7260	Air under the sheet, same as in experiment 1.

EXPERIMENTS ON THE FLOW OF WATER OVER WEIRS, MADE AT THE LOWER LOCKS, IN LOWELL.

138. In the year 1852, the author, in connection with James F. Baldwin, Esq., the eminent engineer of Boston, Massachusetts, was employed to ascertain the amount of water-power used by the several manufacturing companies at Lowell. In order to be able to do this in a satisfactory manner, it was found necessary to determine anew the rules for computing the discharge of water over weirs of certain forms; and for this purpose an extensive series of experiments was made, with a very complete apparatus, and on a scale of unusual magnitude. The execution of these experiments was intrusted to the author; and *The Proprietors of the Locks and Canals on Merrimack River*, at whose expense they were made, have, with great liberality, given the author permission to publish an account of them.

139. The great difficulty in this kind of experiment, usually, is to obtain a suitable basin in which the water flowing over the weir for a certain period of time may be actually measured. Fortunately for our purpose, the Lower Locks at Lowell are seldom used, except during the high water in the spring, when rafts can pass over the rapids in the river below. These locks were rebuilt principally of wood, in 1841, and at the time when the experiments were made, they were still in good condition; they however required some alterations to adapt them to the requirements of the experiments; which alterations, together with the entire apparatus employed, and the mode of conducting the experiments, will now be described.

140. Plate XI., figure 1, is a general plan of the Lower Locks and the vicinity, on a scale of eighty feet to an inch. *A* is the lower level of Pawtucket Canal; *B*, the Eastern Canal; *C*, the Concord River, which enters the Merrimack River at about 1200 feet below the foot of the lock; *D* is the dam for discharging the surplus water from the Pawtucket Canal into Concord River, passing through the wasteway *E*; *F*, the Middlesex Mills, which are carried by water-power from the Pawtucket Canal, through the covered penstock *H*; *I*, an apparatus erected for the purpose of gauging the water drawn by the Middlesex Mills, which was removed before these experiments were made; *K*, the upper chamber of the lock, which was converted into the gauging basin for these experiments, and which is represented as it was before the alterations were made.

141. Plate XII. represents the gauging chamber subsequent to the alterations, on a scale of 10 feet to an inch. Figure 1 is a plan; figure 2, a longitudinal section; and figures 3 and 4 transverse sections. The side wall *A* was built in 1822, of large and small stones laid without mortar; in order to render the lock capable of holding water, it is lined with planks about three inches thick, secured by tree-nails and spikes to wooden frames, which are supported on the bottom by the earth and some rough walls, and on the sides by the side walls of the lock. As originally constructed, the planking was fastened to posts resting immediately against the side walls; but when reconstructed in 1841, the chambers, together with the gates, were narrowed to the width represented, which is about one half the former width; at that time, also, the parts *B B*, about the hollow quoins, were built anew in cut granite, laid in hydraulic cement.

To prepare the chamber for these experiments, the upper set of lock gates and the corresponding mitre sills were removed, and the weir *C*, plate XII., figures 1 and 2, constructed in place of them; the middle gates were also removed, and the lower end of the chamber closed with timbers and plank, as represented at *D*; in the lower part of this timber work the waste gate *K* was constructed, for the purpose of drawing off the water from the chamber, after each experiment.

The construction of the wooden sides of the chamber was such, that when the chamber was partially or wholly filled with water, they would yield a little to the pressure, and the capacity would, consequently, be increased beyond what it was when empty, which was necessarily the case when the dimensions were taken. To diminish, as much as practicable, this source of error, the braces *E* were placed across the chamber, just above the water-line *F F*, nearly up to which the chamber was filled in the experiments. These braces were placed opposite each side timber in the frame of the chamber, excepting at *G G*, where a flooring of thick plank, put in for another object, answered the same purpose; afterwards, every accessible timber in the sides was strongly braced and keyed up from the side walls, which was done with such force, that the ends of the braces *E* were indented into the planks forming the sides of the chamber. At *H H*, where the space between the walls and the planking was too small to admit of the bracing, the spaces between the timbers were filled with small stones, dropped and rammed in from the top. These operations stiffened the sides of the chamber so much, that the correction required for the enlargement of the capacity of the chamber, in consequence of the yielding of the sides, was very minute. All the leakages that could be detected were stopped by various contrivances; the depressions in the planks, about the heads of the spikes, were filled up with cement; the sides of the planking towards the chamber were scraped, and rendered as smooth and uniform as practicable.

A part of the wall *A* was removed at *I*, for the purpose of discharging the water, flowing over the weir, directly into the wasteway, whenever it was necessary to divert its flow from the chamber; the floor *GG* was continued through the wall, as represented in figures 1 and 4, plate XII.

142. Plate XIII., figure 1, is a longitudinal sectional elevation through the middle of the weir, showing most of the apparatus immediately connected with it. *A* is a plate of cast-iron forming the crest of the weir; it is ten feet long, thirteen inches wide, and an inch thick, accurately and smoothly planed in every part; the upper corner presented to the current is square and sharp, or as nearly so as cast-iron can be conveniently maintained; the horizontal part of the top is 0.25 inches wide; the remainder of the top is bevelled off at an angle of 45°; this plate is secured to the timber work by numerous screws with countersunk heads; the timber work is strongly bolted to the granite hollow quoins of the lock. The ends of the weir *B* are formed of plates of cast-iron, of similar section to the plate *A*. The whole upstream side of the weir forms a vertical plane 13.96 feet in length, and 4.60 feet in depth, from the top of the plate *A* to the top of the masonry *C*; the upstream side of the plates *B* are also in the same vertical plane.

D is the swing gate for admitting and diverting, at will, the stream of water flowing over the weir, into or from the measuring chamber *E*. *FFF* are leak boxes or troughs, to catch the leakage by the edges of the swing gate, when shut. The water thus caught is conveyed to openings *G*, cut through the planking on each side of the chamber, through which it is discharged, thus preventing any embarrassment from the leakage of the swing gate when shut, as it does not enter the chamber *E*. The swing gate is suspended from the pivots *H*; all its parts are made as light as practicable, consistent with the required stiffness, in order that the time occupied in opening or shutting it may be as short as possible. A very important part of the experiments consisted in determining the length of time during which the water flowed into the measuring chamber *E*; this was obtained by observing the time when the swing gate was opened and shut, which was done by an observer in the building *I*, by means of an electric telegraph and a marine chronometer, in the following manner. The break circuit apparatus *K* is fixed in such a position that, when one half only of the stream flowing over the weir passes into the chamber, the cam *L*, attached to the frame of the swing gate, depresses the knob as represented in the plate, and breaks the circuit of the electric current in the wire *M*; this causes a sound to be made by the call *N*, in the small building *I*, where sits the observer with his eye on the chronometer, who notes the time when the sound is made; the chronometer used beats half seconds, but, by employing a practised observer, the time was noted to tenths of a second, the error probably rarely exceeding two tenths of a second. The gate, with its accompanying apparatus, was balanced, so that it could be opened or shut with sensibly the same amount of force; this balancing was done with the water flowing over the weir, and was done anew for each material variation in the quantity. To each of the timbers *O* and *P*, plate XIII., figure 1, was attached, by a joint, a prop *L*, shown at figure 4, plate XII.; the prop at the timber *O*, for the purpose of retaining the swing gate in its position when open, and the other at the timber *L*, to retain it in position when shut. The movement of the gate was produced by placing weights upon the frame at *Q* and *R*, plate XIII., where the gate is represented as at the middle point of its motion while shutting; the motion being produced by the gravitation of the weights at *Q*. As soon as the gate is shut, the prop *L*, plate XII., figure 4, is placed under the frame at *R*, plate XIII., and keyed up tight; the weights are then taken off at *Q*, and about the same amount of weight is placed at *R*; then, when it is desired to open the gate, an assistant strikes the prop from under *R* with a sledge-hammer, when the weight at *R* causes the gate to open; the prop is then immediately placed under the frame at *Q*. To prevent injurious concussions from the action of the weights, thick pieces of India-rubber, operating as springs, were fastened on the under-side of the frame at *Q* and *R*, which, when the gate attained either of its extreme positions, struck upon the corresponding stops *S* and *T*.

From the foregoing description of the apparatus, the manner of operating the swing gate will be readily understood. Four assistants were employed for the purpose. Suppose that the chamber is nearly filled, and that it is required that, when the water reaches a certain height, the flow of the water shall be diverted from the chamber: one assistant, who has been watching the rise of the water, gives a signal when the water has reached the desired height, at which the prop under the frame at *Q* is immediately knocked away, the weights at *Q* cause the gate to move until it strikes the weir, or the India-rubber springs strike the stops *T*; at that moment another assistant places the prop under *R*, and the flow of the water is diverted from the chamber; another assistant then changes the weights, and the apparatus is ready for the reverse operation by which the gate is opened. Much time was occupied in adjusting this apparatus so that the cam *L*, plate XIII., would strike the break circuit when the gate was in such a position that one half of the water flowing over the weir passed into the chamber; and also, that the time in which the gate moved through each half of the thickness of the sheet, would be the same. It required a new adjustment for each depth upon the weir. Precise accuracy was not attained or attempted, in any of these adjustments, but such an approximation was made, that it is believed that the errors arising from want of complete exactness, are entirely insensible in the results.

143. The depths upon the weir were observed by means of the hook gauges *U* and *V*, plate XII., figures 1 and 2, and plate XIII., figure 1. One of these gauges is represented in detail by figures 2, 3, and 4, plate XIII., ¼ the full size. They were made by the Lowell Machine Shop. This valuable instrument has been sufficiently described in the account of the experiments on the Tremont Turbine (art 45). These gauges were placed in wooden boxes closed on all sides, excepting at the top; in the bottom of each of which was a hole about an inch in diameter, and in that part of the bottom projecting beyond the lines of the canal walls, due care being taken that the plugs, by which the holes were partially closed, did not project through the bottom. In the experiments on the weir in which the end contraction was suppressed, a communication was established between the gauge boxes and the canal leading to the weir, by pipes opening at *B*, figures 8, 9, and 10, plate XIV. The pipes opening near the bottom of the canal, six feet from the weir, forming part of the system for taking the heights at different distances from the weir, were also used in some of the experiments. The boxes were securely fastened to wooden posts in the angles of the gate recesses; and the posts were strongly fastened to the walls, by several iron bolts driven into holes drilled in the granite stones for the purpose. It was very important that these gauges should be immovably fixed, relatively to the weir. It is probable, however, that they were not perfectly firm. During the course of the experiments, two comparisons were made of the relative heights of the gauges and the top of the weir; one on October 26th; the other, November 8th, when there was found to be a sensible difference in them, the most probable cause of which was, that changes took place in the absolute height of the gauges, that did not affect the weir in the same degree. It is difficult to perceive how the weir could change from a settlement of the masonry, founded, as it is, on rock; the walls to which the gauges were attached, were much less substantially built and not founded on rock; it is not impossible that changes took place in the timber-work of the weir, by the absorption of water, notwithstanding it was

fixed in place several weeks before the experiments were made, with a view to its complete saturation.

If the apparatus is sufficiently stable, the comparison of the heights of the hook gauges with the top of the weir, can be made with any desired degree of precision. For making the comparison in these experiments, the following apparatus was devised. The water being drawn out of the canal, the top of the weir was inclosed in a water-tight trough, containing only a small quantity of water, but sufficient to cover the crest of the weir to a small depth; this trough was connected with the hook gauge boxes, by leaden pipes; the boxes were rendered water-tight by coating the joints with pitch, and plugging up the holes in the bottom; they were also carefully propped up. The communication being free, and the leakages very small, the water on the crest of the weir and in the boxes, would stand at the same level; consequently, all that remained to be done, was to measure the height of the water with the hook gauges, and, at the same time, the depths upon the crest of the weir. The measurement by the hook gauges presented no difficulty, as it required nothing more than the ordinary use of the instruments. To measure the depths upon the crest of the weir, had always been a difficulty in making similar comparisons; to meet it in this case, the instrument, represented at plate XIII., figure 5, was devised. The points were numbered from 1 to 10, and the exact height of each of them above a horizontal plane, on which the instrument stood, was ascertained. In using this instrument, the water in the trough was adjusted to a convenient level; the top of the weir was divided into ten equal spaces; the instrument was placed upon one of them, and when the water became quite tranquil, the number of the point that coincided with the surface was noted, and, at the same moment, the heights of the water in the boxes were observed with the hook gauges. If (as was usually the case) the surface of the water did not exactly coincide with either of the points, the true fractional number was taken by estimation. The adjacent points differed in height about 0.001 feet, and a fourth part of this quantity was sufficiently distinct not to be doubtful.

As an example of the precision attainable by the use of this instrument, the following results are given of the comparison of the north hook gauge with the weir, made during the night of October 26, by Mr. John Newell. The results indicate the corrections to be applied to the reading of the hook gauge, to give the true height of the surface of the water in the gauge box, above the top of the weir, each result being a mean of eleven measurements made at equidistant points on the weir.

By the 1st trial, the correction was	—0.03076	feet.
" 2d " " "	—0.03032	"
" 3rd " " "	—0.03076	"
" 4th " " "	—0.03096	"
" 5th " " "	—0.03079	"
Mean	—0.03072	feet.

The extreme variation is between the 2nd and 4th trials, amounting to 0.00064 feet, a quantity scarcely visible to the naked eye; of course, in the mean result of all the trials, the error of observation must be entirely insensible.

It has been remarked that the comparisons made at different times, did not give the same results. Two complete comparisons were made, as follows: —

DATE, 1852.	CORRECTIONS.	
	North hook gauge. Feet.	South hook gauge. Feet.
October 26th.	—0.03072	—0.02786
November 8th.	—0.03250	—0.03069

Considering the care with which these comparisons were made, and the perfection of the method, the differences cannot be attributed to errors of observation, but, rather, to a want of stability in some parts of the apparatus. The corrections determined October 26th, were used in reducing all the experiments made from October 20th, to November 7th, both inclusive; for all subsequent experiments, the corrections found November 8th were used.

The twenty-three experiments numbered from 11 to 33, in table XIII., were made under circumstances as nearly identical as practicable. They were made at different times throughout the course, for the purpose of neutralizing errors of the same class as that just described, the resulting effects of which ought to be shown by the variation in the coefficients deduced from experiments made at different times. These experiments are collected together in the following table: —

DATE, 1852.	Number of experiments.	Mean coefficients.	Differences from the mean deduced from all the experiments, or from 3.3223.
Oct. 20th, P. M., and Oct. 21st, A. M.	6	3.3186	—0.0037
Oct. 21st, P. M., and Oct. 22d, A. M.	8	3.3216	—0.0007
Oct. 29th, P. M.	6	3.3278	+0.0055
Nov. 11th, P. M.	3	3.3207	—0.0016

The extreme variation is between the experiments of October 20th and 29th, in which it amounts to $\frac{1}{361}$. The greatest difference from the mean deduced from all the 23 experiments, is in the coefficient deduced from the experiments of October 29th, in which it amounts to $\frac{1}{661}$. It is fair to presume that similar irregularities, not in any case much exceeding the above, and arising principally from want of stability in the apparatus, exist in other parts of this series of experiments.

144. The capacity of the gauging chamber was obtained by measuring its dimensions. For this purpose, horizontal lines were traced on the sides of the chamber at every foot in height; the widths were then measured at right angles to the sides, at points two feet apart; from these widths, and other necessary measurements, the total area was obtained at each horizontal section. When these measurements were made, the chamber was of course empty, but when filled with water, its dimensions would evidently be somewhat larger, in consequence of the sides and bottom yielding to the pressure. To ascertain what allowance to make for this, a systematic measurement was made in the spaces between the planking and the walls, both when the chamber was empty and when filled to the usual height; similar measurements were made for the bottom, by placing poles vertically, resting upon, and fastened to the bottom; the elevations of the tops of these poles were taken with a levelling instrument, both when the chamber was empty, and when filled. It was thus ascertained that the capacity of the chamber, when filled with water to the usual height, was 11.11 cubic feet greater than when empty.

Two persons made independent measurements of the capacity of the chamber, the results of which differed only about $\frac{1}{4}$ of a cubic foot, a coincidence which must of course be considered as accidental. The capacity finally determined upon for 9.5 feet in height, (which was nearly the depth filled in each experiment,) and including the enlargement resulting from the pressure, was 12138.18 cubic feet.

145. The chamber was not quite water-tight, but the amount of the leakage was determined by noting the rate at which the surface of the water lowered, when none was admitted from the weir, and the waste gate was closed; this was repeated with the water in the chamber at different depths. It was thus found that the mean leakage was 0.035 cubic feet per second; that is, the product of 0.035 multiplied by the number of seconds that the water flowing over the weir continued to enter the chamber during an experiment, must be added to the quantity in the chamber at the moment the water was diverted, in order to give the true quantity that passed over the weir in the same time.

146. It was not convenient to empty the chamber entirely after each experiment, but the heights of the water in the chamber at the beginning and ending of each, were ascertained with great accuracy by means of hook gauges, placed in the boxes *X* and *Y*, figures 1, 2, and 3, plate XII., which were fastened to a post strongly bolted to the wall *A*. A communication was established, at will, between the water in the chamber and either of the boxes, by pipes and cocks. The operation of taking the heights was as follows: the chamber having been sufficiently emptied, the waste gate *K* was closed, the communication of the lower box with the chamber was established, and when the oscillations in the surface had ceased, the height of the water was taken; the cock was then shut, and a signal made for opening the swing gate. When the chamber had been filled, and the flow of water into the chamber diverted by closing the swing gate, the communication with the upper box was opened; when the oscillations had ceased, observations of the water were taken at short and regular intervals, for some minutes, the time and height being noted. In consequence of the leakage of the chamber, the surface lowered slowly, and the continued observations were made for the purpose of being able to infer the exact height at which the water stood in the chamber at the instant that the swing gate was shut, the very slow rate at which the surface of the water in the chamber lowered, permitting this to be done with great precision. For the success, however, of this operation, it was essential that the timekeeper used should agree with the chronometer, by which the times of opening and shutting the swing gate were noted; it was accordingly frequently compared, and any difference noted.

147. Plate XIV. represents the different forms of weir on which experiments were made. All the figures are on the same scale, namely, five feet to an inch, or $\frac{1}{60}$ the full size.

Figure 1 is a longitudinal section, figure 2, a plan, and figure 3, an elevation of what we call the regular weir, that is, a weir in which the contraction is complete, both on the ends and on the bottom.

Figure 4 is an elevation of a weir of precisely the same form as that last described, excepting that it is divided into two equal parts or bays by the partition, which is two feet wide. The upstream side of the partition is in the same vertical plane as the remainder of the weir, having no bolt heads or other projection below the level of the surface of the water.

Figures 5, 6, and 7, represent a weir of precisely the same form as that first above described, excepting that the depth of the canal approaching the weir is diminished.

Figures 8, 9, and 10, represent the same weir as first above described, modified so that the contraction at the ends is suppressed, that is, the canal leading to the weir is of the same width as the weir. These figures also show the apparatus used to ascertain the effect of taking the depths upon the weir at different distances from it, by means of pipes opening near the bottom of the canal.

Figures 11 and 12 represent the upper part of a dam, of the same section as that erected by the Essex Company, in 1846–8, across the Merrimack River at Lawrence, (about nine miles below Lowell). This magnificent work has an overfall 900 feet in length, the perpendicular fall being about 24 feet. This form was experimented upon, in order to obtain a formula for computing the flow of the river over this dam.

DESCRIPTION OF TABLE XIII.

Containing the details of the experiments on the flow of water over weirs, made at the Lower Locks, Lowell, in October and November, 1852.

148. The columns numbered from 1 to 5, require no further explanation than is contained in the respective headings.

149. COLUMN 6. *Duration of the experiment.* This is the interval of time during which the water flowed into the chamber; it is obtained by taking the difference of the corresponding times in column 5.

150. COLUMN 7. *Mean depth upon the weir by observation.* It was found impracticable in many cases, to maintain the canal at a uniform height throughout an experiment, although every endeavor was made. For instance, no experiments were made when the mills were in operation, nor until some hours after the usual time when they ceased drawing water; this rendered it necessary to perform the experiments either during the night, or on Sunday; in consequence of the lateness of the season, advantage was taken of both these opportunities. When any change was made in the level of the water in the canal, for the purpose of varying the depths upon the weir, a considerable time was allowed to elapse before the experiments were resumed, in order that the level of the water might get well established. In spite of all precautions, however, variations frequently occurred in the depths upon the weir, which, with the ordinary mode of taking an arithmetical mean of the several observations of the depth, would have materially affected the accuracy of the results; this difficulty was obviated in a

great degree, by the use of a novel mode of obtaining the mean depth, which will now be explained. Let

$h, h', h'', \text{etc.}\, h^n$ represent the several observed depths upon the weir, the successive values not differing greatly from each other.

$t, t', t'', \text{etc.}\, t^n$, the corresponding intervals of time between the several observations;

T, the sum of all the intervals of time;

Q, the total volume of water actually flowing over the weir in the time T;

H, the mean depth upon the weir that would discharge the volume Q, in the time T;

l, the length of the weir;

C, a constant coefficient:

we shall have, evidently, *very nearly*,

$$Q = \frac{t}{2} C l h^{\frac{3}{2}} + \frac{t+t'}{2} C l h'^{\frac{3}{2}} + \frac{t'+t''}{2} C l h''^{\frac{3}{2}} + \text{etc.} + \frac{t^n}{2} C l h^{n\frac{3}{2}}:$$

$$Q = C l \left(\frac{t}{2} h^{\frac{3}{2}} + \frac{t+t'}{2} h'^{\frac{3}{2}} + \frac{t'+t''}{2} h''^{\frac{3}{2}} + \text{etc.} + \frac{t^n}{2} h^{n\frac{3}{2}} \right):$$

we have also

$$Q = T C l H^{\frac{3}{2}}:$$

whence we derive, by substituting the value of Q previously found,

$$H = \left\{ \frac{\frac{t}{2} h^{\frac{3}{2}} + \frac{t+t'}{2} h'^{\frac{3}{2}} + \frac{t'+t''}{2} h''^{\frac{3}{2}} + \text{etc.} + \frac{t^n}{2} h^{n\frac{3}{2}}}{T} \right\}^{\frac{2}{3}}.$$

As an example of the application of this method, let us take the observations made at the north hook gauge during experiment 74; this is selected, simply because the variations in the depths upon the weir were greater than in any other experiment.

EXTRACT FROM THE NOTES TAKEN AT THE NORTH HOOK GAUGE.

	October 24th, 1852, A. M., 9h, 5', watch 12" fast.		
	TIME.	Reading of the hook gauge.	
	$9^h\ 9'\ 15''$	0.6360	
	10 50	0.6320	
Commencement of the experiment by the time of this watch. $9^h\ 12'\ 12.4''$.	11 45	0.6325	
	12 45	0.6310	1
	13 15	0.6310	1
	14 20	0.6300	1
	14 50	0.6365	3
	15 20	0.6290	1
	16 30	0.6300	1
	17 5	0.6335	2
	17 55	0.6380	3
	18 35	0.6480	5
	19 20	0.6500	6
	20 0	0.6470	5
	20 55	0.6470	5
	21 25	0.6445	4
	22 10	0.6530	7
	22 35	0.6550	7
Ending of the experiment by the time of this watch. $9^h\ 24'\ 49.9''$.	23 5	0.6480	5
	23 45	0.6580	8
	24 35	0.6605	9
	Arithmetical mean reading,	0.6428	

For the purpose of simplifying the operation of finding the mean, it is assumed that we can, without sensible error, use an arithmetical mean of all depths not varying more than 0.002 feet from each other; accordingly an arithmetical mean has been taken of all the readings marked 1 in the margin of the above table, and similar means have been taken of the other readings marked with the same number in the margin. It will be perceived that it was noted at $9^h\ 5'$, that the watch was 12" fast; by another comparison with the chronometer made at $10^h\ 47'$, the watch was 22" fast; from these two comparisons it is inferred that, at the middle of the experiment, the watch was 13.3" fast. Instead of changing the times of all the observations, the time of the commencement and ending of the experiment has been changed to conform to this watch, but for the purpose of this reduction only. By the method adopted, it is assumed that the height of the water did not change until half the interval of time between two consecutive observations had elapsed; accordingly, we find that the time corresponding to the first mean depth, is from the beginning of the experiment to $9^h\ 14'\ 35''$, or 142.6″, and from $9^h\ 15'\ 5''$ to $9^h\ 16'\ 47.5''$, or 102.5″, making 245.1″. The several mean readings and the corresponding times, given in the following table, are obtained in this manner; the depths upon the weir corresponding to the several mean readings, are also given, which are found by subtracting 0.03072 feet from each mean reading, (see art. 143).

Number of the mean reading.	Mean reading of the hook gauge. Feet.	Time corresponding to each mean reading. Seconds.	Mean depths upon the weir, deduced from the several mean readings. Feet.
1	0.63020	245.1	0.59948
2	0.63350	42.5	0.60278
3	0.63725	75.0	0.60653
4	0.64450	37.5	0.61378
5	0.64750	167.5	0.61678
6	0.65000	42.5	0.61928
7	0.65400	62.5	0.62328
8	0.65800	45.0	0.62728
9	0.66050	39.9	0.62978

The quantities in the third column of this table are the values of $\frac{t}{2}$, $\frac{t+t'}{2}$, etc., in the expression given above for H; the quantities in the fourth column are the corresponding values of h, h', etc. The value of T being 757.5, all the quantities in the second member of the equation are known; by substituting these values we find

$$H = 0.6113.$$

The arithmetical mean of the eighteen observations is 0.6428; deducting the correction 0.03072, we find the mean depth to be 0.6121; the difference by the methods is 0.0008.

A similar computation on the observations at the south hook gauge gives

$$H = 0.6099.$$

By taking the arithmetical mean of the observations, we find the depth, by the south hook gauge to be 0.6096.

The mean of the above values of H, or 0.6106, is adopted as the depth on the weir in experiment 74.

A similar reduction has been made of the observations at each hook gauge, in all the experiments; the arithmetical mean of the two results obtained for each experiment, is given in column 7.

Notwithstanding the advantage attending this mode of reduction, it cannot be denied that, for the most perfect experiments, the depth on the weir should be invariable throughout, and that, *cæteris paribus*, the experiments will be the less valuable, the greater the variation. To enable the reader to judge of the relative value of the experiments, as far as it depends upon this variation, the small figures to the left and above the several depths in column 7 are given; they indicate the highest number of values of h, h', h'', etc. used in the reduction of the observations, at either of the hook gauges, in the corresponding experiments.

151. COLUMN 8. *Mean velocity of the water approaching the weir.* This is obtained by dividing the corresponding quantity of water flowing over the weir, given in column 14, by the area of the section of the canal, at the hook gauge boxes. In the weir having contraction at the ends, this would strictly include all the space under the gauge boxes, although, from the form of the walls, it is evident that the current could flow only in a small part of this space; consequently, the portion in which the current could not flow is not included in the areas used.

152. COLUMN 9. *Head due to the velocity in column* 8. This is sufficiently explained in the heading.

153. COLUMN 10. *Depth upon the weir, corrected for the velocity of the water approaching the weir.* In the common formula for the discharge of water over weirs,

$$Q = C l H \tfrac{2}{3}\sqrt{2gH}. \qquad (A)$$

The second member may be separated into three factors, namely: C, the coefficient of contraction; l, the length of the weir; and $H\tfrac{2}{3}\sqrt{2gH}$, the theoretical discharge for the unit of length. According to a well-known elementary theorem in hydraulics, the latter factor may be represented by the area of a segment of a parabola, of which the parameter is $2g$; thus, in figure 5, plate XII, if $AB = H$, and $BC = \sqrt{2gH}$, and the curve AMC is a parabola, of which the vertex is A, we shall have the area of the segment $ABC = H\tfrac{2}{3}\sqrt{2gH}$; also, the velocity of the fluid at any point P will be represented by the ordinate PM. The factor $H\tfrac{2}{3}\sqrt{2gH}$ may also be decomposed into two others: $H = AB$, and $\tfrac{2}{3}\sqrt{2gH}$, which equals the mean value of all the ordinates of the parabola between A and C, and represents the mean velocity of the fluid for the whole height of the orifice. In demonstrating this theorem, it is assumed that the water in the reservoir is at rest; we can, however, easily establish an analogous theorem, in which it is assumed that the water in the reservoir has a velocity approaching the weir, in the direction perpendicular to the plane of the weir. Suppose h to be the head due this velocity; and in figure 6, plate XII., let $AB = H$, and $AD = h$, we shall have for the velocity v', at any point P in the height of the orifice,

$$v' = \sqrt{2g(AP + h)}:$$

but this value of v' is the ordinate corresponding to the abscissa, $AP + h = DP$, of a parabola whose parameter is $2g$. We have also

$$BC = \sqrt{2g(H + h)}.$$

We can, consequently, represent the discharge for the unit of length, by the area of the surface $ABCG$, which is a portion of the segment BCD; the area of $ABCG$ is the difference of the areas of the segments BCD and AGD; the area of BCD is

$$\tfrac{2}{3} BD \times BC = \tfrac{2}{3}(H + h)\sqrt{2g(H + h)},$$

and the area of ADG is

$$\tfrac{2}{3} AD \times AC = \tfrac{2}{3} h\sqrt{2gh},$$

consequently, the area of $ABCG$ is

$$\tfrac{2}{3}(H + h)\sqrt{2g(H + h)} - \tfrac{2}{3} h\sqrt{2gh} = \tfrac{2}{3}\sqrt{2g}\left[(H + h)^{\frac{3}{2}} - h^{\frac{3}{2}}\right];$$

and for the total discharge we have

$$Q' = Cl\tfrac{2}{3}\sqrt{2g}\left[(H + h)^{\frac{3}{2}} - h^{\frac{3}{2}}\right]. \qquad (B)$$

The formula (A) may be put under the form

$$Q = Cl\tfrac{2}{3}\sqrt{2g}\, H^{\frac{3}{2}}. \qquad (C)$$

Suppose H' to represent a depth upon the weir that would give the discharge Q' by the formula (C), we shall have

$$Q' = Cl\tfrac{2}{3}\sqrt{2g}\, H'^{\frac{3}{2}}:$$

substituting the value of Q' in (B), and reducing, we find

$$H' = \left[(H + h)^{\frac{3}{2}} - h^{\frac{3}{2}}\right]^{\frac{2}{3}}. \qquad (D)$$

The equation (B), from which this value of H' is derived, does not agree with that given for a similar case by most writers on hydraulics, who seem generally to have followed Du Buat;* it agrees, however, with the expression given by Weisbach,† who appears to have been the first to point out the error.

The formula (D) was communicated to the author, in 1849, by Mr. Boyden, accompanied by a demonstration somewhat resembling the above.

The values of H', given in column 10, have been computed by the formula (D) from the corresponding values of H and h in columns 7 and 9.

154. Columns 11, 12, and 13 are sufficiently explained by their respective headings.

155. Column 14. *Quantity of water passing the weir per second.* The quantities in this column are obtained by dividing the total quantities given in column 13, by the corresponding intervals of time in column 6.

156. Column 15. *Value of C in the formula*

$$Q = C(l - 0.1\,n\,H')\,H'^{\frac{3}{2}},$$

Q having the corresponding values in column 14.

In the formula proposed at art. 124, namely:—

$$Q = C(l - b\,n\,h)\,h^{a},$$

the values of the constants a and b are to be determined by experiment. The values adopted in the formula by which the coefficients in this column have been computed, namely: $a = \frac{3}{2}$, $b = 0.1$, were determined upon after many trials of other values; in consequence of their giving results according the most nearly with all the experiments, and at the same time having a convenient degree of simplicity. It is quite likely that many other values of a and b (probably an unlimited number) might be found that would accord somewhat nearer with the experiments; a closer approximation than is given by the use of the values adopted, could have, however, but little practical value; much less, it was thought, than would be derived from the use of the simple values adopted. The use of a fractional power, such as $a = 1.47$, deduced from the experiments at the Tremont Turbine (art. 135), is very inconvenient, and, to persons not well skilled in the use of logarithms, offers great difficulty.

---

* *Principes d'Hydraulique, etc., by M. Du Buat.* Paris: 1816. Vol. 1, page 201.

† *Allgemeine Maschinen Encyclopädie.* Leipzig: 1841. Vol. 1, page 489.

157. COLUMNS 16, 17, and 18, are, for the purpose of obtaining correct mean results of the experiments, made under circumstances nearly identical. In consequence of the variations in the height of the canal (art. 150), it was impracticable to repeat the experiments with precisely the same depth upon the weir; by the method adopted for obtaining these mean results, all inconvenience from this source is obviated. As the formula by which the values of C, in column 15, are obtained, is such as to give results agreeing very nearly with experiment, even when the depths differ considerably, it is plain that the values of C deduced from experiments having nearly the same depths, cannot be affected by small variations in the depths, and will be subject to no greater irregularities than if, in the several experiments from which they are deduced, the depths had been precisely the same. We can consequently take a mean coefficient with the same confidence that we could take a mean quantity, if the depths had been precisely the same. These mean coefficients are given in column 16. In column 17 are given depths on the weir, nearly a mean of those in the experiments from which the corresponding mean coefficients have been deduced. In column 18, are given what may be called the mean quantities of water actually found by experiment to be discharged with the corresponding depths in column 17. A method similar to the above was used to reduce the quantities discharged in the experiments of Castel, reported in the *Annales de chimie et de Physique*, vol. 62. Paris: 1836; reprinted in the first volume of the *Annales des Ponts et Chaussées* for 1837.

158. COLUMN 19. *Quantity of water passing the weir, calculated by the formula*

$$Q = 3.33\,(l - 0.1\,n\,H'')\,H''^{\frac{3}{2}},$$

H″ having the corresponding values in column 17.

The coefficient 3.33 is derived from the arithmetical mean of all the coefficients in column 15, which is 3.3318, the two final decimals being omitted for the sake of simplicity. The largest coefficient in column 15, is that deduced from experiment 34, which is 3.3617, exceeding the coefficient adopted by $\frac{1}{105}$ part; the smallest coefficient is that deduced from experiment 4, which is 3.3002, being less than the coefficient adopted, by $\frac{1}{112}$ part; that is, the formula by which the quantities in column 19 are computed, will represent every experiment in the table, within one per cent.

159. COLUMN 20. *Proportional difference, or the absolute difference of the quantities in columns* 18 *and* 19, *divided by the quantity in column* 18. The greatest proportional difference is that deduced from experiments 34 and 35, which is —0.0090, or a little less than one per cent. In these experiments there were two weirs,

about four feet long each, separated by a partition two feet wide; the near neighborhood of the two orifices appears to have affected the discharge. The next largest proportional difference is that deduced from experiments 36 to 43, which is —0.0068, or about $\frac{2}{3}$ of one per cent.; in these experiments, the depth of the water in the canal leading to the weir, was only about three times the depth upon the weir. The experiments with the diminished depth in the canal were made for the purpose of testing the method of correcting the depths, upon the weir, for the velocity of the water approaching the weir (art. 153). They indicate that the method is not strictly accurate, as might have been anticipated, omitting, as it does, all consideration of the effect produced by this velocity, in modifying the contraction. It is well understood that such an effect is produced,* but it is of such a complicated nature, that the investigations hitherto undertaken have thrown but little light upon it.

It will be perceived by referring to column 4, that the experiments 51 to 55 were made under the same circumstances as experiments 44 to 50, excepting that the sheet of water, after passing the weir, was prevented from expanding laterally for a certain distance. This was accomplished by placing boards at the ends of the sheet, as represented by the broken lines at *A*, figures 8 and 9, plate XIV. By referring to column 16, it will be seen that the effect of these boards was to diminish the coefficient from 3.3409 to 3.3270, corresponding to a diminution of the quantity discharged by the weir, with the same depth, of $\frac{1}{240}$, or about four-tenths of one per cent.; in other words, the effect of the boards upon the discharge was the same as would be produced by shortening the weir $\frac{1}{240}$, or $\frac{1}{4}$ inch, at each end. By reference to figure 8, plate XIV., it will be perceived that these boards did not affect the free communication between the atmosphere and the air under the sheet of water; if this communication had been obstructed, so that the pressure of the air under the sheet had been different from that of the atmosphere, it would have affected the discharge.

---

* *Jaugeage des cours d'eau, etc., by M. P. Boileau,* page 40. Paris: 1850.

TABLE

EXPERIMENTS ON THE FLOW OF WATER OVER WEIRS, MADE AT THE

1	2	3		4	5						6	7	8
Number of the experiment.	Date of the experiment. 1852.	Temperatures by Fahrenheit's thermometer. Of the air in the shade.	Of the water.	Reference to the figures on plate XIV., and particular description of the weir.	Time of the commencement and conclusion of the experiment, as indicated by the telegraphic signals. Commencement. H.	min.	sec.	Conclusion. H.	min.	sec.	Duration of the experiment. sec.	Mean depth upon the weir by observation. H. Feet.	Mean velocity of the water approaching the weir at the transverse sect. thro' the holes in the hook gauge boxes, or six feet from the weir. V. Feet.
1	Oct. 27, P.M.			*Figures 1, 2, and 3.* Width of the canal on the upstream side of the weir, 13.96 feet. Mean depth of the canal opposite the hook gauge boxes 5.048 feet below the top of the weir.	10	15	0.8	10	18	13.6	192.8	1.52490	0.7682
2	" " "				11	20	1.3	11	23	14.2	192.9	1.55045	0.7813
3	" " "				11	54	1.3	11	57	10.6	189.3	1.55930	0.7882
4	" 28, A.M.	43.75°	46.5°		0	26	17.9	0	29	20.1	182.2	1.56910	0.7889
5	Oct. 24, P.M.	52°	48.5°	*Figures 1, 2, and 3.* Same as the preceding.	9	4	0.5	9	8	20.5	260.0	1.23690	0.5904
6	" " "				9	33	2.3	9	37	21.5	259.2	1.24195	0.5933
7	" " "				10	0	1.7	10	4	18.4	256.7	1.24795	0.5971
8	" " "				10	31	2.1	10	35	20.6	258.5	1.25085	0.5944
9	" " "				11	0	2.3	11	4	16.1	253.8	1.25290	0.6000
10	" " "				11	30	1.8	11	34	21.4	259.6	1.25490	0.5987
11	Oct. 20, P.M.			*Figures 1, 2, and 3.* Same as the preceding.	10	1	0.8	10	7	22.0	381.2	0.96711	0.4256
12	" " "				10	30	0.9	10	35	44.0	343.1	1.02755	0.4594
13	" " "				11	12	0.7	11	17	47.4	346.7	1.03395	0.4629
14	" " "				11	48	0.6	11	53	43.3	342.7	1.03315	0.4634
15	" 21, A.M.				0	24	59.5	0	30	37.5	338.0	1.04060	0.4680
16	" " "	43°	49°		1	0	0.0	1	5	37.4	337.4	1.03735	0.4666
17	" " P.M.				9	48	8.0	9	54	36.4	388.4	0.96325	0.4233
18	" " "				10	23	1.2	10	29	9.9	368.7	0.97590	0.4304
19	" " "	42°	48.75°		10	52	0.4	10	58	8.7	368.3	0.97950	0.4318
20	" " "				11	23	1.4	11	29	4.2	362.8	0.98885	0.4377
21	" " "				11	53	1.5	11	59	3.5	362.0	0.99460	0.4418
22	" 22, A.M.				0	43	0.0	0	49	48.8	408.8	0.91570	0.3951
23	" " "	42°			1	12	0.7	1	18	40.9	400.2	0.92800	0.4015
24	" " "				1	42	0.3	1	48	26.8	386.5	0.94625	0.4126
25	" 29, P.M.				9	2	3.5	9	7	54.2	350.7	1.01275	0.4517
26	" " "	51.5°	48.25°		9	35	2.7	9	40	51.6	348.9	1.01160	0.4520
27	" " "				10	5	1.2	10	10	59.7	358.5	0.99495	0.4429
28	" " "				10	34	59.8	10	40	39.7	339.9	1.03360	0.4653
29	" " "				11	3	0.2	11	8	29.6	329.4	1.05565	0.4779
30	" " "				11	32	1.8	11	37	27.0	325.2	1.06920	0.4863
31	Nov. 11, P.M.	34°	41.25°		8	56	59.5	9	3	6.0	366.5	0.98370	0.4352
32	" " "				9	30	0.6	9	36	7.8	367.2	0.97820	0.4320
33	" " "				10	0	0.3	10	6	19.8	379.5	0.96700	0.4236
34	Nov. 3, P.M.			*Figure 4.* Width of the canal on the upstream side of the weir, 13.96 feet. Mean depth of the canal opposite the hook gauge boxes, 5.048 feet below top of the weir. Two equal bays separated by a partition 2 feet wide.	9	12	1.3	9	19	37.8	456.5	1.01025	0.3527
35	" " "	45°	48°		9	59	59.6	10	7	18.4	438.8	1.02625	0.3596
36	Oct. 31, A.M.			*Figures 5, 6, and 7.* Width of the canal on the upstream side of the weir, 13.96 feet. Mean depth of the canal opposite the hook gauge boxes, 2.084 feet below the top of the weir. Bottom of the canal horizontal for 25 feet on the upstream side of the weir.	7	17	15.8	7	22	52.9	337.1	1.02805	0.9496
37	" " "				7	47	59.6	7	53	29.9	330.3	1.03720	0.9589
38	" " "	46°	48.75°		9	46	0.5	9	51	34.9	334.6	1.04455	0.9684
39	" " "				10	14	1.4	10	19	23.3	321.9	1.04495	0.9693
40	" " "				10	41	0.7	10	46	28.2	327.5	1.04600	0.9691
41	" " "				11	10	7.8	11	15	32.5	324.7	1.05130	0.9756
42	" " "				11	39	59.4	11	45	8.3	308.9	1.07945	1.0049
43	" " P.M.				0	15	1.4	0	20	15.7	314.3	1.07115	0.9958

XIII.

LOWER LOCKS, LOWELL, IN OCTOBER AND NOVEMBER, 1852.

Number of the experiment.	9 Head due to the velocity in column 8, or the values of h by the formula $h=\frac{v^2}{64.3236}$	10 Depth upon the weir, corrected for the velocity of the water approaching the weir, or the values of H' by the formula $H'=[(H+h)^{\frac{3}{2}}-h^{\frac{3}{2}}]^{\frac{2}{3}}$	11 Length of the weir. L.	12 No. of end contractions. n.	13 Total quantity of water that passed the weir during each experiment, as measured in the lock chamber.	14 Quantity of water passing the weir per second.	15 Value of C in the formula $Q=C(L-0.1nH')H'^{\frac{3}{2}}$ Q having the corresponding values in column 14.	16 Mean value of C for each particular description of weir.	17 Approximate mean depth upon the weir, for each particular description of weir. H''.	18 Quantity of water that would have passed the weir with the depth in column 17, calculated by the formula $Q=C(L-0.1nH'')H''^{\frac{3}{2}}$ C having the corresponding value in column 16.	19 Quantity of water passing the weir, calculated by the formula $Q=3.33(L-0.1nH'')H''^{\frac{3}{2}}$	20 Proportional difference, or the absolute difference of the quantities in columns 18 and 19, divided by the quantity in column 18.
	Feet.	Feet.	Feet.		Cubic feet.	Cubic feet per second.			Feet.	Cubic feet per second.	Cubic feet per second.	
1	0,00917	1.53300	9.997	2	11815.19	61.2821	3.3318					
2	0,00949	1.55945	"	"	12059.49	62.5686	3.3174					
3	0,00966	1.56845	"	"	11964.90	63.2060	3.3230	3.3181	1.56	62.6147	62.8392	+0.0036
4	0,00968	1.57828	"	"	11542.52	63.3508	3.3002					
5	0,00542	1.24208	9.997	2	11723.21	45.0893	3.3412					
6	0,00547	1.24718	"	"	11753.09	45.3437	3.3398					
7	0,00551	1.25325	"	"	11725.57	45.6781	3.3405					
8	0,00549	1.25610	"	"	11760.23	45.4941	3.3159	3.3338	1.25	45.4125	45.3608	—0.0011
9	0,00560	1.25825	"	"	11658.13	45.9343	3.3396					
10	0,00557	1.26022	"	"	11903.41	45.8529	3.3260					
11	0,00282	0.96983	9.997	2	11872.75	31.1457	3.3265					
12	0,00328	1.03071	"	"	11645.35	33.9416	3.3129					
13	0,00333	1.03716	"	"	11869.83	34.2366	3.3110					
14	0,00334	1.03636	"	"	11745.20	34.2725	3.3182					
15	0,00341	1.04388	"	"	11713.35	34.6549	3.3196					
16	0,00339	1.04061	"	"	11651.45	34.5330	3.3233					
17	0,00279	0.96593	"	"	12023.62	30.9568	3.3261					
18	0,00288	0.97868	"	"	11627.93	31.5377	3.3234					
19	0,00290	0.98229	"	"	11659.61	31.6579	3.3179					
20	0,00298	0.99172	"	"	11661.78	32.1438	3.3216					
21	0,00303	0.99752	"	"	11754.36	32.4706	3.3265	3.3223	1.00	32.5486	32.6240	+0.0023
22	0,00243	0.91804	"	"	11721.88	28.6739	3.3218					
23	0,00251	0.93012	"	"	11682.99	29.1929	3.3155					
24	0,00265	0.94881	"	"	11629.93	30.0904	3.3198					
25	0,00317	1.01580	"	"	11678.40	33.3003	3.3211					
26	0,00318	1.01466	"	"	11623.48	33.3147	3.3281					
27	0,00305	0.99789	"	"	11670.68	32.5542	3.3333					
28	0,00337	1.03684	"	"	11697.19	34.4136	3.3297					
29	0,00355	1.05906	"	"	11685.17	35.4741	3.3263					
30	0,00368	1.07274	"	"	11764.18	36.1752	3.3283					
31	0,00294	0.98633	"	"	11702.07	31.9293	3.3251					
32	0,00290	0.98099	"	"	11629.66	31.6712	3.3259					
33	0,00273	0.96969	"	"	11762.21	30.9940	3.3111					
34	0,00198	1.01212	7.997	4	11863.64	25.9883	3.3617					
35	0,00201	1.02820	"	"	11655.85	26.5630	3.3586	3.3601	1.02	26.2686	26.0333	—0.0090
36	0,01402	1.04008	9.997	2	11747.38	34.8484	3.3519					
37	0,01430	1.05039	"	"	11657.37	35.2933	3.3498					
38	0,01458	1.05799	"	"	11953.56	35.7249	3.3548					
39	0,01461	1.05842	"	"	11513.09	35.7660	3.3567					
40	0,01460	1.05946	"	"	11715.10	35.7713	3.3523	3.3527	1.06	35.8026	35.5602	—0.0068
41	0,01480	1.06494	"	"	11712.43	36.0716	3.3548					
42	0,01570	1.09390	"	"	11579.82	37.4873	3.3509					
43	0,01542	1.08535	"	"	11645.21	37.0513	3.3505					

TABLE

EXPERIMENTS ON THE FLOW OF WATER OVER WEIRS, MADE AT THE

1	2	3		4	5						6	7	8
Number of the experiment.	Date of the experiment. 1852.	Temperatures by Fahrenheit's thermometer.		Reference to the figures on plate XIV., and particular description of the weir.	Time of the commencement and conclusion of the experiment, as indicated by the telegraphic signals.						Duration of the experiment.	Mean depth upon the weir by observation. *H.*	Mean velocity of the water approaching the weir at the transverse sect. thro' the holes in the hook gauge boxes, or six feet from the weir. *V.*
		Of the air in the shade	Of the water.		Commencement.			Conclusion.					
					H.	min.	sec.	H.	min.	sec.	sec.		Feet.
44	Nov. 7, A. M.	44°	44°	*Figures* 8, 9, and 10. Mean width of the canal for 20 feet on the upstream side of the weir, 9.992 feet. Mean depth of the canal opposite the hook gauge boxes, 5.048 feet below the top of the weir.	7	50	1.0	7	55	51.2	350.2	0.98675	0.5455
45	" " "				9	38	0.0	9	43	53.7	353.7	0.98490	0.5446
46	" " "				10	11	59.2	10	17	57.4	358.2	0.97450	0.5376
47	" " "				10	43	59.6	10	49	59.0	359.4	0.97620	0.5385
48	" " "				11	15	0.0	11	21	0.4	360.4	0.97600	0.5387
49	" " "				11	48	4.4	11	54	0.7	356.3	0.97775	0.5394
50	" " P. M.				0	21	0.2	0	26	58.3	358.1	0.97690	0.5390
51	Nov. 7, P. M.	42.25°	43.75°	*Figures* 8, 9, and 10. Width and depth same as the preceding. The sheet of water after passing the weir, was prevented from expanding in width, for a certain distance, by boards at each end of the weir, placed in the same planes as the sides of the canal leading to the weir.	8	23	7.6	8	28	52.0	344.4	1.00505	0.5589
52	" " "				8	55	59.7	9	1	46.9	347.2	1.00600	0.5581
53	" " "				9	28	3.1	9	33	51.3	348.2	1.00520	0.5574
54	" " "				9	59	59.8	10	5	52.9	353.1	0.99265	0.5480
55	" " "				10	31	1.5	10	36	54.0	352.5	0.99240	0.5477
56	Oct. 24, P. M.			*Figures* 1, 2, and 3. Width of the canal on the upstream side of the weir, 13.96 feet. Mean depth of the canal opposite the hook gauge boxes, 5.048 feet below the top of the weir.	2	24	59.3	2	32	53.8	474.5	0.81860	0.3405
57	" " "				3	3	0.4	3	11	11.1	490.7	0.80755	0.3338
58	" " "				3	40	0.3	3	48	19.9	499.6	0.79565	0.3272
59	" " "				4	17	4.7	4	25	44.9	520.2	0.77690	0.3170
60	" " "				4	55	1.7	5	3	18.6	496.9	0.80125	0.3306
61	" " "				5	29	1.8	5	37	27.6	505.8	0.79400	0.3258
62	Oct. 31, P. M.			*Figures* 5, 6, and 7. Width of the canal on the upstream side of the weir, 13.96 feet. Mean depth of the canal opposite the hook gauge boxes, 2.014 feet below the top of the weir. Bottom of the canal horizontal, for 18 feet on the upstream side of the weir.	2	20	0.3	2	28	40.5	520.2	0.77115	0.6694
63	" " "				3	0	1.3	3	8	32.4	511.1	0.78725	0.6872
64	" " "	47°	48.75°		3	38	4.4	3	46	7.9	483.5	0.80455	0.7052
65	" " "				4	14	0.2	4	21	5.5	425.3	0.87960	0.7870
66	" " "				4	47	58.0	4	54	56.2	418.2	0.88865	0.7963
67	Nov. 7, P. M.			*Figures* 8, 9, and 10. Mean width of the canal for 20 feet on the upstream side of the weir, 9.992 feet. Mean depth of the canal opposite the hook gauge boxes, 5.048 feet below the top of the weir.	2	7	2.7	2	16	14.7	552.0	0.73620	0.3659
68	" " "				2	43	1.0	2	51	6.8	485.8	0.80195	0.4122
69	" " "				3	17	59.7	3	25	56.0	476.3	0.80950	0.4176
70	" " "				3	51	59.9	3	59	51.7	471.8	0.81495	0.4213
71	" " "				4	25	0.0	4	32	53.9	473.9	0.81325	0.4192
72	Oct. 24, A. M.			*Figures* 1, 2, and 3. Width of the canal on the upstream side of the weir, 13.96 feet. Mean depth of the canal opposite the hook gauge boxes 5.048 feet below the top of the weir.	7	12	2.5	7	25	9.4	786.9	0.59190	0.2182
73	" " "	46.5°	47.75°		7	49	59.8	8	2	52.7	772.9	0.59240	0.2186
74	" " "				9	11	59.1	9	24	36.6	757.5	0.61060	0.2279
75	" " "				10	33	59.7	10	45	14.4	674.7	0.65525	0.2509
76	" " "	59.5°	48°		11	8	1.4	11	19	27.9	686.5	0.64305	0.2449
77	" " "				11	50	0.3	12	1	35.3	695.0	0.63795	0.2419
78	" " P. M.	64.5°	48.5°		0	24	58.5	0	36	42.9	704.4	0.63370	0.2396
79	Oct. 31, P. M.			*Figures* 5, 6, and 7. Width of the canal on the upstream side of the weir, 13.96 feet. Mean depth of the canal opposite the hook gauge boxes, 2.014 feet below the top of the weir. Bottom of the canal horizontal for 18 feet on the upstream side of the weir.	7	6	59.6	7	18	8.6	669.0	0.65150	0.5405
80	" " "				7	46	0.3	7	57	9.3	669.0	0.65590	0.5455
81	" " "	45.25°	48.75°		8	24	0.0	8	34	56.9	656.9	0.65985	0.5496
82	" " "				9	0	59.4	9	12	42.4	703.0	0.63135	0.5193
83	" " "				9	40	1.0	9	51	27.8	686.8	0.64250	0.5309
84	" " "				10	23	1.7	10	34	11.0	669.3	0.65460	0.5439
85	Oct. 31, P. M.			*Figures* 5, 6, and 4. Width, depth, and bottom of the canal same as the preceding. Two equal bays separated by a partition 2 feet wide.	11	43	0.7	11	56	42.8	822.1	0.66940	0.4382
86	Nov. 1, A. M.				0	21	59.9	0	35	26.0	806.1	0.67900	0.4459
87	" " "				1	0	3.8	1	13	17.9	794.1	0.68360	0.4496
88	" " "				1	39	59.8	1	53	9.5	789.7	0.68815	0.4526

XIII — CONTINUED.

LOWER LOCKS, LOWELL, IN OCTOBER AND NOVEMBER, 1852.

	9	10	11	12	13	14	15	16	17	18	19	20
Number of the experiment.	Head due to the velocity in column 8, or the values of h by the formula $h=\frac{v^2}{64.3236}$	Depth upon the weir corrected for the velocity of the water approaching the weir, or the values of H' by the formula $H'=[(H+h)^{\frac{3}{2}}-h^{\frac{3}{2}}]^{\frac{2}{3}}$	Length of the weir. l.	No. of end contractions. n.	Total quantity of water that passed the weir during each experiment, as measured in the lock chamber.	Quantity of water passing the weir per second.	Value of C in the formula $Q=C(l-0.1nH')H'^{\frac{3}{2}}$ Q having the corresponding values in column 14.	Mean value of C for each particular description of weir.	Approximate mean depth upon the weir, for each particular description of weir. H''.	Quantity of water that would have passed the weir with the depth in column 17, calculated by the formula $Q=C(l-0.1nH'')H''^{\frac{3}{2}}$ C having the corresponding value in column 16.	Quantity of water passing the weir, calculated by the formula $Q=3.33(l-0.1nH'')H''^{\frac{3}{2}}$	Proportional difference, or the absolute difference of the quantities in columns 18 and 19, divided by the quantity in column 18.
	Feet.	Feet.	Feet.		Cubic feet.	Cubic feet.			Feet.	Cubic feet per second.	Cubic feet per second.	
44	0.00463	0.99117	9.995	0	11524.62	32.9087	3.3366					
45	0.00461	0.98930	"	"	11616.54	32.8429	3.3394					
46	0.00449	0.97879	"	"	11592.18	32.3623	3.3437					
47	0.00451	0.98051	"	"	11635.28	32.4299	3.3418	3.3409	0.98	32.3956	32.2899	—0.0033
48	0.00451	0.98031	"	"	11680.79	32.4356	3.3434					
49	0.00452	0.98207	"	"	11576.77	32.4916	3.3402					
50	0.00452	0.98122	"	"	11623.93	32.4600	3.3413					
51	0.00486	1.00968	9.995	0	11646.88	33.8179	3.3349					
52	0.00484	1.01061	"	"	11725.23	33.7708	3.3257					
53	0.00483	1.00980	"	"	11743.85	33.7273	3.3254	3.3270	1.00	33.2534	33.2833	+0.0009
54	0.00467	0.99710	"	"	11683.48	33.0883	3.3249					
55	0.00466	0.99684	"	"	11656.76	33.0688	3.3243					
56	0.00180	0.82034	9.997	2	11539.45	24.3192	3.3287					
57	0.00173	0.80923	"	"	11675.86	23.7943	3.3234					
58	0.00166	0.79726	"	"	11628.76	23.2761	3.3237	3.3246	0.80	23.4011	23.4391	+0.0016
59	0.00156	0.77842	"	"	11694.31	22.4804	3.3261					
60	0.00170	0.80230	"	"	11698.38	23.5427	3.3268					
61	0.00165	0.79560	"	"	11719.40	23.1700	3.3188					
62	0.00697	0.77768	9.997	2	11718.53	22.5270	3.3376					
63	0.00734	0.79412	"	"	11887.02	23.2577	3.3406					
64	0.00773	0.81178	"	"	11610.17	24.0128	3.3383	3.3403	0.83	24.8313	24.7548	—0.0031
65	0.00963	0.88855	"	"	11695.06	27.4984	3.3435					
66	0.00986	0.89782	"	"	11671.63	27.9092	3.3417					
67	0.00208	0.73821	9.995	0	11676.57	21.1532	3.3368					
68	0.00264	0.80149	"	"	11709.76	24.1041	3.3422					
69	0.00271	0.81211	"	"	11645.16	24.4492	3.3424	3.3393	0.80	23.8821	23.8156	—0.0028
70	0.00276	0.81760	"	"	11647.58	24.6876	3.3410					
71	0.00273	0.81588	"	"	11658.23	24.5584	3.3341					
72	0.00074	0.59262	9.997	2	11803.57	15.0001	3.3284					
73	0.00074	0.59812	"	"	11614.60	15.0273	3.3303					
74	0.00081	0.61139	"	"	11902.13	15.7124	3.3284					
75	0.00098	0.65621	"	"	11760.38	17.4305	3.3237	3.3275	0.62	16.0382	16.0502	+0.0008
76	0.00093	0.64395	"	"	11659.42	16.9889	3.3306					
77	0.00091	0.63883	"	"	11643.02	16.7597	3.3259					
78	0.00089	0.63456	"	"	11685.54	16.5894	3.3250					
79	0.00454	0.65579	9.997	2	11657.19	17.4248	3.3258					
80	0.00463	0.66027	"	"	11783.50	17.6133	3.3278					
81	0.00470	0.66429	"	"	11674.77	17.7725	3.3278	3.3262	0.65	17.1990	17.2187	+0.0011
82	0.00419	0.63532	"	"	11682.49	16.6181	3.3249					
83	0.00438	0.64664	"	"	11715.28	17.0578	3.3244					
84	0.00460	0.65894	"	"	11748.86	17.5540	3.3266					
85	0.00299	0.67226	7.997	4	11690.02	14.2197	3.3382					
86	0.00309	0.68195	"	"	11703.92	14.5192	3.3378					
87	0.00314	0.68669	"	"	11644.82	14.6642	3.3378	3.3368	0.68	14.4541	14.4247	—0.0020
88	0.00318	0.69118	"	"	11678.11	14.7880	3.3333					

COMPARISON OF THE PROPOSED FORMULA WITH THE RESULTS OBTAINED BY PREVIOUS EXPERIMENTERS.

160. We find on record a great number of experiments on the discharge of water over weirs; in the present state of the science of hydraulics, however, a large proportion of them can be considered only in the light of first approximations; of great value undoubtedly, at the respective epochs at which they were made; but it could serve no useful purpose to compare them with the results obtained with the more perfect apparatus used of late years. Three sets of experiments have been made in France within the last thirty years, on a comparatively minute scale, it must be admitted, but with complete apparatus, and conducted with great care. They were made by Poncelet and Lesbros at Metz, in 1827 and 1828; by Castel at Toulouse, in 1835; and by Boileau at Metz, in 1846. It will be recollected that the application of the proposed formula to the discharge over weirs in which the contraction at the ends is complete, is limited to depths on the weir, not exceeding one third of the length of the sheet; this limitation permits the comparison to be made with only a portion of the results obtained by Poncelet and Lesbros, and by Castel. Boileau operated on weirs in which the end contraction was suppressed, and to which form the limitation does not apply.

161. *Comparison of the proposed formula, with the results obtained by Poncelet and Lesbros.* These experiments are to be found among the magnificent series made at the expense of the French Government, and recorded at length in *Expériences hydrauliques sur les lois de l'écoulement de l'eau* by M. M. Poncelet and Lesbros, Paris: 1832; and in the continuation under the same title by M. Lesbros, Paris: 1851. In table XXXIX., of the last mentioned work, are given the coefficients for computing the discharge over weirs of a variety of forms, and of certain lengths, and with certain depths of water, by the formula

$$d = mlh\sqrt{2gh},$$

in which d is the discharge, m the coefficient, l the length, h the depth, and $g = 9.8088$ metres, or 32.1817 feet. The comparison can be usefully made with only one of the forms experimented upon, namely: that in which the orifice was made in a thin plate, in the plane side of a reservoir; the orifice being at a great distance from the bottom and lateral sides, and the discharge made freely into the air.

In table XIV. are given the quantities computed according to Lesbros, for all the depths for which he gives values of m, determined by experiment, and which are within the limitation required by the proposed formula, namely: that the depth shall not exceed one third of the length. The quantities are also given as computed by the proposed formula. It will be perceived by the final column of the table, that the proportional differences are nearly constant, and that the quantities by the proposed formula are too small by a little more than two per cent. If the coefficient of the proposed formula was changed from 3.33 to 3.41, the computed results would agree very nearly. It should be recollected that the constants in the proposed formula have been determined from experiments in which the depths upon the weir were from 0.6 to 1.6 feet, or about eight times the depths in the experiments by Poncelet and Lesbros. It is the general result of all the precise experiments on the discharge through openings of a variety of forms, in a thin plate, that, for very small heads, the coefficients require to be increased; which proves that the law of the discharge varying as the square root of the head, does not hold good for very small heads. The comparison in table XIV. affords the same indications; and the constancy of the proportional differences, indicates that the correction of the length, to compensate for the effect of the end contraction, is practically correct, both for large and small depths upon the weir. It would not be difficult so to determine the values of the constants in the formula

$$Q = C(l - bnh)h^{a},$$

as to represent the experiments both of Poncelet and Lesbros and the Lower Locks experiments with nearly the same degree of exactness that the latter are represented, with the constants that have been adopted. This would undoubtedly be an advantage in some particular cases in practice, but if it was intended to make the formula general, the sacrifice of simplicity would be more than an equivalent disadvantage.

TABLE XIV.

The length of the weir is constant, and equal to 0.6562 feet.

1	2	3	4	5
Depth on the weir. Feet.	Value of the coefficient m according to Lesbros.	Quantity of water discharged by the formula $Q = m\,l\,h\sqrt{2gh}$; m having the corresponding value in the preceding column. Cubic feet per second.	Quantity of water discharged by the formula $Q = 3.33(L - 0.1nH)H^{\frac{3}{2}}$. Cubic feet per second.	Proportional difference, or the absolute difference divided by the quantity in column 3.
0.06562	0.417	0.0369	0.0360	—0.0245
0.08202	0.414	0.0512	0.0500	—0.0225
0.09843	0.412	0.0670	0.0655	—0.0228
0.11483	0.409	0.0838	0.0820	—0.0207
0.13124	0.407	0.1019	0.0997	—0.0209
0.14764	0.405	0.1210	0.1184	—0.0212
0.16404	0.404	0.1413	0.1379	—0.0239
0.18045	0.402	0.1622	0.1583	—0.0243
0.19685	0.401	0.1844	0.1794	—0.0271
0.21326	0.399	0.2069	0.2012	—0.0274

162. *Comparison of the proposed formula with the results obtained by Castel.* An abstract of these experiments may be found in the *Annales de Chimie et de Physique*, vol. 62. Paris: 1836; and in the *Annales des ponts et chaussées*, vol. 1, for 1837. Paris. It appears to have been a leading idea in these experiments, to imitate, as nearly as possible, the forms and proportions of the weirs ordinarily used in practice for gauging streams of water; in fact, to reproduce them on a small scale, anticipating that the rules deduced from precise experiments upon them might be applied, without modification, to gaugings on a large scale. The weir was formed by damming up a wooden canal, 2.4279 feet in width, by a thin plate of copper, in which the weir was formed, the crest being 0.5578 feet above the bottom of the canal; the width of the weir varying from about ¼ of a foot to 2¼ feet. The latter width is so near that of the canal, that the end contraction must have been sensibly modified, so that any comparison of the results obtained from it would be of little use; they have consequently been omitted. In the abstract referred to, a table is given of the coefficients deduced from the experiments, for a variety of widths and depths. In table XV. are given the quantities computed with these coefficients, for all the widths and depths to which the proposed formula is applicable; also the quantities as computed by the proposed formula. In consequence of the small dimensions of the canal, the water approaching the weir had a sensible velocity; in table XV.

the depths on the weir, for which the quantities have been computed by the proposed formula, have been corrected for this velocity. It will be seen by referring to the final column, that the proportional differences are considerably greater, and have less uniformity than in the comparison with the experiments of Poncelet and Lesbros; nevertheless, there is a certain harmony in the results of both comparisons, and they serve to show how unsafe it is, in the present state of the science of hydraulics, to apply rules to gauging streams of water passing over weirs, of which the dimensions differ greatly from those in the experiments from which the rules have been deduced.

TABLE XV.

Width of the canal leading to the weir 2.4279 feet; height of the crest of the weir above the bottom of the canal 0.5578 feet.

1	2	3	4	5	6	7	8
Length of the weir.	Depth on the weir.	Value of the coefficient m, in the formula $Q=m\frac{2}{3}LH\sqrt{2gH}$, according to Castel.	Quantity of water discharged by the formula $Q=m\frac{2}{3}LH\sqrt{2gH}$, m having the corresponding value in the preceding column.	Head due the mean velocity of the water in the canal leading to the weir by the formula $h=\frac{v^2}{64.4}$	Depth on the weir, corrected for the velocity of the water in the canal by the formula $H'=\left((H+h)^{\frac{3}{2}}-h^{\frac{3}{2}}\right)^{\frac{2}{3}}$	Quantity of water discharged by the formula $Q=3.33(L-0.1nH')H'^{\frac{3}{2}}$	Proportional difference, or the absolute difference of the quantities in columns 4 and 7, divided by the quantity in column 4.
Feet.	Feet.		Cubic feet per second.	Feet.	Feet.	Cubic feet per second.	
0.3281	0.09843	0.618	0.0335	0.00001	0.09844	0.0317	—0.0537
0.6562	0.19685	0.604	0.1852	0.00016	0.19701	0.1796	—0.0302
"	0.16404	0.611	0.1425	0.00010	0.16414	0.1380	—0.0311
"	0.13124	0.619	0.1033	0.00006	0.13130	0.0998	—0.0339
"	0.09843	0.624	0.0676	0.00003	0.09846	0.0655	—0.0318
0.9843	0.32809	0.604	0.5976	0.00120	0.32924	0.5778	—0.0331
"	0.26247	0.606	0.4290	0.00072	0.26316	0.4189	—0.0237
"	0.19685	0.610	0.2805	0.00036	0.19720	0.2755	—0.0176
"	0.16404	0.616	0.2155	0.00023	0.16426	0.2109	—0.0211
"	0.13124	0.623	0.1559	0.00014	0.13138	0.1519	—0.0257
"	0.09843	0.631	0.1026	0.00006	0.09849	0.0993	—0.0322
1.3124	0.39371	0.621	1.0769	0.00337	0.39687	1.0266	—0.0468
"	0.32809	0.621	0.8192	0.00225	0.33022	0.7876	—0.0386
"	0.26247	0.620	0.5852	0.00134	0.26375	0.5682	—0.0291
"	0.19685	0.622	0.3813	0.00067	0.19749	0.3720	—0.0245
"	0.16404	0.626	0.2920	0.00043	0.16446	0.2842	—0.0266
"	0.13124	0.632	0.2109	0.00025	0.13148	0.2042	—0.0320
"	0.09843	0.636	0.1379	0.00012	0.09855	0.1332	—0.0341
1.6404	0.32809	0.631	1.0405	0.00363	0.33147	1.0003	—0.0386
"	0.26247	0.632	0.7457	0.00218	0.26452	0.7192	—0.0355
"	0.19685	0.632	0.4843	0.00108	0.19788	0.4692	—0.0312
"	0.16404	0.633	0.3690	0.00069	0.16470	0.3578	—0.0304
"	0.13124	0.636	0.2653	0.00039	0.13161	0.2566	—0.0327
"	0.09843	0.642	0.1740	0.00019	0.09861	0.1671	—0.0393
1.9685	0.32809	0.644	1.2743	0.00545	0.33308	1.2174	—0.0446
"	0.26247	0.644	0.9118	0.00326	0.26549	0.8725	—0.0431
"	0.19685	0.645	0.5931	0.00163	0.19838	0.5675	—0.0432
"	0.16404	0.644	0.4505	0.00103	0.16502	0.4320	—0.0410
"	0.13124	0.645	0.3229	0.00058	0.13179	0.3094	—0.0417
"	0.09843	0.651	0.2117	0.00027	0.09869	0.2012	—0.0495

163. *Comparison of the proposed formula, with that obtained by Boileau.* The experiments from which Boileau deduced his formula, are given at length in *Jaugeage des cours d'eau a faible ou a moyenne section*, by M. P. Boileau. Paris: 1850. Boileau has particularly studied the discharge in the form of weir in which the contraction at the ends is suppressed; that is to say, the form in which the weir occupies the whole width of the canal conducting the water to it. The proposed formula is applicable to this case, by making $n=0$. Boileau experimented on three weirs of this form; one of them was 5.30 feet in length, with the crest 1.54 feet above the bottom of the canal; the other two were 2.94 feet in length, the crest in one being 1.12 feet above the bottom of the canal; and in the other 1.61 feet above the bottom; the depths on the weir varying from 0.19 feet to 0.72 feet. By a train of reasoning combined with the results of his experiments, Boileau has arrived at the following formula for weirs of this form:—

$$Q=\frac{S+H}{\sqrt{(S+H)^2-H^2}}\,0.417\,LH\sqrt{2gH},$$

in which

$Q=$ the discharge.

$S=$ the height of the crest of the weir, above the bottom of the canal, which is supposed to be horizontal for a short distance, upstream from the weir.

$H=$ the depth on the weir, taken before the sheet begins to curve in consequence of the discharge.

$L=$ the width of the canal, and also the length of the weir.

$g=9.8088^{m}$.

The coefficient 0.417 is determined from a mean of 14 experiments.

Adopting the English foot as the unit, and reducing, we have

$$Q=3.3455\,\frac{S+H}{\sqrt{(S+H)^2-H^2}}\,LH^{\frac{3}{2}}. \qquad (A)$$

For this form of weir, the proposed formula becomes

$$Q=3.33\,LH'^{\frac{3}{2}}: \qquad (B)$$

H' being the depth upon the weir, corrected for the velocity of the water approaching the weir.

These formulas differ so essentially that they can be conveniently compared only by applying them to particular cases. In the formula (A), as S increases relatively to H, the factor $\frac{S+H}{\sqrt{(S+H)^2-H^2}}$ approaches unity, which is the limit when S is infinitely greater than H; in the latter case we have also, $H'=H$; the formulas (A) and (B) then become identical, excepting the coefficients, that in (B) being $\frac{1}{216}$ less than in (A). Hence we may conclude that for any length of weir, and for any depth upon it, providing that the depth of the canal leading to the weir, is very great relatively to the depth on the weir, the quantities computed by the formulas (A) and (B) will differ $\frac{1}{216}$ only.

In practice, however, S is seldom very great, relative to H. Let us take an example conforming more nearly to the usual cases that occur in practice. Let $H=1$ foot, $S=3$ feet, $L=10$ feet, by the formula (A), $Q=34.552$ cubic feet per second. In the formula (B), H' is the depth on the weir, corrected for the mean velocity of the water approaching the weir; this velocity is equal to the quotient of the area of the section of the canal, divided by the quantity. But the quantity itself depends on this velocity. The formula (B), if put under a form to give the quantity directly from the measured depth upon the weir, would become very complicated; it will be equally exact and much easier, to find the quantity by successive approximations as follows.

1st approximation.

Assume $H'=1$, then $Q=33.3$.

2nd approximation.

If $Q=33.3$, the mean velocity of the water in the canal leading to the weir is $\frac{33.3}{10(H+S)}=0.8325$; and for the head due this velocity we have

$$h=\frac{(0.8325)^2}{2g}=0.011;$$

$$H'=\left[(H+h)^{\frac{3}{2}}-h^{\frac{3}{2}}\right]^{\frac{2}{3}}=1.0103;$$

$$Q=33.816.$$

A third approximation in a similar manner gives $Q=33.817$.

The proportional difference of the quantities by the two formulas is about $\frac{1}{47}$, or a little over two per cent.

Boileau, in establishing his formula, assumes that the living force in the entire section of the canal is expended in increasing the discharge over the weir; in the method adopted in this work for correcting the depth on the weir for the velocity of the water in the canal, it is assumed that the living force in the part of the section of the canal equal to the area of the orifice of discharge only, is expended in increasing the discharge; as applied to a weir of the form under consideration, it is clear that neither of these assumptions is strictly true; the latter, however, appears to be the most rational, and to agree the best with experiment.

PRECAUTIONS TO BE OBSERVED IN THE APPLICATION OF THE PROPOSED FORMULA.

164. $$Q = 3.33\,(L - 0.1\,n\,H)\,H^{\frac{3}{2}};$$

in which

$Q =$ the discharge, in cubic feet per second;
$L =$ the length of the weir;
$n =$ the number of end contractions;
$H =$ the depth on the weir;

the English foot being the unit of measure.

When the contraction is complete at each end of the weir, $n = 2$; when the weir is of the same width as the canal conducting water to it, the end contraction is suppressed, and $n = 0$.

This formula is only applicable to rectangular weirs, made in the side of a dam, which is vertical on the upstream side, the crest of the weir being horizontal, and the ends vertical; also, the edges of the orifice presented to the current must be sharp; for, if bevelled or rounded off in any perceptible degree, a material effect will be produced on the discharge; it is essential, moreover, that the stream should touch the orifice only at these edges, after passing which it should be discharged through the air, in the same manner as if the orifice was cut in a thin plate. See fig. 3, plate XVIII.

The formula is not applicable to cases in which the depth on the weir exceeds one third of the length; nor to very small depths. In the experiments from which it has been determined, the depths have varied from 7 inches to nearly 19 inches, and there seems no reason why it should not be applied with safety to any depths between 6 inches and 24 inches.

The height of the surface of the water in the canal, above the crest of the weir, is to be taken for the depth upon the weir; this height should be taken at a point far enough from the weir to be unaffected by the curvature caused by the discharge; if more convenient, it may be taken by means of a pipe opening near the bottom of the canal near the upstream side of the weir, which pipe may be made to communicate with a box placed in any convenient situation; and if the box and pipe do not leak, the height may be observed, in this manner, very correctly (art. 175). However the depth may be observed, it may require to be corrected for the velocity of the water approaching the weir.

The end contraction must either be complete, or entirely suppressed; the necessary distance from the side of the canal or reservoir to the end of the weir, in order that the end contraction may be complete, is not definitely determined; in experiments 1 to 4, table XIII., the depth on the weir was about 1.5 feet, and the distance from the side of the canal to the end of the weir, about 2 feet; the proposed formula applies well to all these experiments. In cases where there is end contraction, we may assume a distance from the side of the canal to the end of the weir equal to the depth on the weir, as the least admissible, in order that the proposed formula may apply.

As to the fall below the weir, requisite to give a free discharge to the water, it is not definitely determined; a comparison of experiments 49, 50, and 51, table X., indicates that, when the depth on the weir is 1 foot, and the entire sheet, after passing the weir, strikes a solid body at about 0.5 feet below the crest of the weir, the discharge, with the same depth, is diminished about $\frac{1}{1005}$. By experiments 1 and 2, table XII., it appears that, when the sheet passing the weir, falls into water of considerable depth, the depth on the weir being about 0.85 feet, no difference is perceptible in the discharge, whether the water is 1.05 feet or 0.235 feet below the crest of the weir; it is very essential, however, in all cases, that the air under the sheet should have free communication with the external atmosphere. With this precaution it appears that, if the fall below the crest of the weir is not less than half the depth upon the weir, the discharge over the weir will not be perceptibly obstructed. If the sheet is of very great length, however, more fall will be necessary, unless some special arrangement is made to supply air to the space under the sheet at the places that would otherwise not have a free communication with the atmosphere.

In respect to the depth of the canal leading to the weir, experiments 36 to 43, table XIII., show that, with a depth as small as three times that on the weir, the proposed formula agrees with experiment, within less than one per cent.; this proportion may be taken as the least admissible, when an accurate gauging is required.

It not unfrequently happens that, in consequence of the particular form of the canal leading to the weir, or from other causes, the velocity of the water in the canal is not uniform in all parts of the section; this is a frequent cause of serious error, and is often entirely overlooked. If great irregularities exist, they should be removed by causing the water to pass through one or more gratings, presenting numerous small apertures equally distributed, or otherwise, as the case may require, through which the water may pass under a small head; these gratings should be placed as far from the weir as practicable.

If the canal leading to the weir has a suitable depth, it will be requisite only when great precision is required, to correct the depth upon the weir for the velocity of the water in the canal by the formula (D) (art. 153); thus, in experiment 42, table XIII., the water in the canal had a mean velocity of about 1 foot per second, the effect of which was to increase the discharge about two per cent.; in experiment 82, in which the velocity was about 0.5 feet per second, the discharge was increased about one per cent.; these examples will enable the operator to judge, in each case, of the necessity of going through the troublesome calculation for correcting the depth on the weir.

MISCELLANEOUS EXPERIMENTS ON THE FLOW OF WATER, MADE AT THE LOWER LOCKS, IN NOVEMBER, 1852.

On the discharge of water over a dam of the same section as that erected by the Essex Company, across the Merrimack River at Lawrence, Massachusetts.

165. As these experiments cannot be usefully compared with those on weirs of more regular form, they have not been included in table XIII.; and as they are of less general interest, they will not be given with much detail.

The form of the dam is represented by figures 11 and 12, plate XIV. (art. 147); the other apparatus was the same as that used for the experiments in table XIII.

The end contraction was suppressed by making the canal leading to the overfall of the same width as the overfall itself. The water in the hook gauge boxes communicated only with the water contained in the spaces between the masonry and the wood-work forming the sides and bottom of the canal leading to the overfall; as there was a free communication between the water at A, figures 11 and 12, and that near the hook gauge boxes, and as the water between these places was sensibly at rest, we may consider that the height of the water was taken at A.

166. In table XVI. these experiments are exhibited in sufficient detail to be intelligible.

Columns 1 and 2 require no explanation.

Column 3. The heights contained in this column are above the mean level of the crest of the dam, which was very nearly horizontal for a distance of 2.95 feet from C to D. These heights have not been corrected for the velocity of the water approaching the weir; indeed, from the manner in which they were observed, no correction was necessary.

Column 4. The quantities in this column have been obtained in the manner described in the explanation of table XIII. (art. 155).

Column 5. *Quantity of water passing over the dam, calculated by the formula*

$$Q = 3.01208\, l h^{1.53}.$$

This formula was arrived at by trial of various powers of h, and was adopted as representing, the most nearly, the results of the five experiments in the table; it should be distinctly understood, however, that it is not applicable to depths much greater or less than in the experiments from which it is deduced. In April, 1852, the depth of water flowing over the dam at Lawrence, was 10 feet; if the quantity then passing over the dam was computed by this formula, it is probable that it would be greatly in error.

COLUMN 6. *Proportional difference.* It will be observed that the greatest proportional difference is 0.0085, or less than one per cent.; we may therefore say with confidence, that we can compute the flow of water over the Lawrence dam, when free from ice or other obstruction, for any depth not greater than 20 inches or less than 7 inches, without being liable to an error exceeding one per cent.

TABLE XVI.

Time, from November 10th, 8h, 57', P. M., to November 11th, 0h, 12', A. M.
Temperature of the air at 10h, 50', P. M., 34.50° Fahrenheit.
" " water " " " 41.75° "
The air calm.

1	2	3	4	5	6
Number of the experiment.	Length of the overfall. l.	Mean height of the surface of the water in the hook gauge boxes, above the top of the horizontal crest of the dam. Feet. h.	Quantity of water passing over the dam, as measured in the lock chamber. In cubic feet per second.	Quantity of water passing over the dam calculated by the formula $Q = 3.01208\, l\, h^{1.53}$ In cubic feet per second.	Proportional difference, or the absolute difference of the quantities in columns 4 and 5, divided by the quantity in column 4.
89	9.995	0.58720	13.385	13.332	−0.0040
90	"	0.79035	20.892	21.005	+0.0054
91	"	0.97670	28.914	29.039	+0.0043
92	"	1.32520	46.183	46.317	+0.0029
93	"	1.63380	64.346	63.804	−0.0085

EXPERIMENTS TO ASCERTAIN THE EFFECT OF TAKING THE DEPTHS UPON A WEIR AT DIFFERENT DISTANCES FROM IT, BY MEANS OF PIPES OPENING NEAR THE BOTTOM OF THE CANAL.

167. It is often a matter of great doubt and uncertainty, to know at what distance from the weir the depth of the water upon it should be observed; very often also it becomes a matter of necessity to observe the depth at a distance from the weir so small that, according to some, the quantity of water passing the weir, computed in the usual manner, would be liable to sensible

error. For the purpose of obtaining some light upon this point these experiments were undertaken, and, as they were made with all the precautions for insuring accuracy that could be devised, they will be described with some detail.

168. Figures 8, 9, and 10, plate XIV., represent the form of the weir, and the system of pipes used for these experiments. The canal leading to the weir was of the same width as the weir, so that the end contraction was suppressed. The pipes were of lead, about three fourths of an inch interior diameter, the lower extremities of which, numbered from 1 to 8, were about three inches above the bottom of the canal, and terminated in holes in the board *CC*; the side of the board at which they opened was vertical, and in the axis of the canal; the ends of the pipe did not project through the board; the other extremities of the pipes were fastened by small flanges to the bottoms of the hook gauge boxes; holes were made in the bottoms of the boxes corresponding to each pipe, and communication between the boxes and the pipes could be controlled at pleasure, by plugging up these holes. It will be readily perceived that heights of the water observed by this apparatus are not necessarily the true elevations of the surface of the water immediately over the orifices of the pipes, but that they are the elevations of the surface in the hook gauge boxes; an elevation which is due to the statical pressure on the orifice of the pipe.

169. In order to obtain the heights at different distances from the weir, observations were necessarily made with both hook gauges at the same time, one of which was always in communication with a pipe opening at 6 feet from the weir, the apertures in the bottom of the box, communicating with all the other pipes, being plugged up; at the other hook gauge, either pipe might be in communication with the box, all the other apertures being plugged up; thus, the depth at six feet from the weir was observed in each experiment, to be used as a standard with which the depth observed simultaneously at any other distance might be compared; this mode of proceeding was rendered necessary, in consequence of the impossibility of maintaining the level of the water uniform for any considerable length of time.

170. In considering the sources of error to which the observations with the hook gauges were liable, it appeared that four kinds required to be specially guarded against, namely: *First*, imperfect comparison of the gauges, with the top of the weir. *Second*, defective stability, in consequence of which the relative elevations of the gauges and the weir might not be constant. *Third*, errors in the graduation of the gauges. *Fourth*, the difference in the habit of observers, in making the point of the hook coincide with the surface of the water; or, what we may call, the personal error. In relation to the *first*, we must bear in mind

that the requirement here is not so much that the absolute height above the top of the weir should be exactly determined, as that the difference of the heights at two points, at different distances from the weir, should be determined correctly; if then we know the relative heights of the two gauges, the object can be attained, even if we do not know precisely the height of either of them, relatively to the weir. The heights of the gauges relative to each other, could easily be ascertained at any time, by closing up all the apertures in each box, except those communicating with pipes, numbers 4 and 5, which, it will be seen by reference to figure 9, had a common orifice at their lower extremities; consequently, the surface of the water in both boxes must have been at the same level. The correction to be applied to the reading of one of the hook gauges, was taken as previously determined for the experiments on the discharge over the weir, and the correction for the other gauge, was deduced from simultaneous observations on both gauges, when the boxes communicated with a common orifice, in the manner just described. The *second* source of error was guarded against as much as practicable, by making the observations for the correction just described, at nearly the same time as the experiments to which it was to be applied. The danger of error from the *third* source was much diminished by making the observations for the correction, with nearly the same depth upon the weir as in the experiments to which it was to be applied. The *fourth* source of error was eliminated by determining the correction separately for each pair of observers. In short, these four sources of error were reduced to a minimum by determining for each session of the experiments, and for each pair of observers, the relative corrections to be applied to the readings of the hook gauges, to give the depths upon the weir; the depths, when the observations for these corrections were made, being nearly the same as in the experiments to which they were to be applied.

171. In table XVII. are given the results of the observations made for the purpose of obtaining the relative corrections for the gauges, for each session of the experiments, and for each pair of observers. In computing the depth upon the weir by the north hook gauge, the correction —0.03072 is applied to the mean reading of the gauge, (art. 143); the mean reading of the south hook gauge is given; as the water in both boxes is at the same height, the difference between the depth upon the weir, as determined by the north hook gauge, and the mean reading of the south hook gauge, must give the correction for the last named gauge.

TABLE XVII.

DATE, 1852.	Time of beginning the observation.	North hook gauge, in communication with pipe No. 5 opening near the bottom of the canal at 6 feet from the weir.		South hook gauge, in communication with pipe No. 4, opening near the bottom of the canal at 6 feet from the weir.			
		Observer.	Arithmetical mean depth on the weir. Feet.	Observer.	Arithmetical mean reading of the gauge. Feet.	Correction to be applied to the mean reading to give the depth on the weir. Feet.	Mean correction for each session, and each pair of observers. Feet.
November 3.	9^h 13′ P.M.	Francis	1.01180	Avery	1.03760	—0.02580	
" "	10 0 "	"	1.02617	"	1.05375	—0.02758	—0.02669
November 3.	11^h 16′ P.M.	Haeffely	1.00739	Newell	1.03377	—0.02638	—0.02638
November 3.	9^h 29′ P.M.	Francis	1.01984	Newell	1.04625	—0.02641	
" "	10 45 "	"	1.01073	"	1.03716	—0.02643	—0.02640
" 4.	1 47 A.M.	"	1.04532	"	1.07169	—0.02637	
November 3.	11^h 0′ P.M.	Francis	1.00807	Haeffely	1.03431	—0.02624	
" 4.	1 58 A.M.	"	1.04734	"	1.07350	—0.02616	—0.02620
November 7.	7^h 50′ A.M.	Francis	0.98775	Avery	1.01362	—0.02587	
" "	9 38 "	"	0.98555	"	1.01195	—0.02640	
" "	2 7 P.M.	"	0.73665	"	0.76357	—0.02692	
" "	8 22 "	"	1.00696	"	1.03287	—0.02591	—0.02639
" "	8 56 "	"	1.00677	"	1.03311	—0.02634	
" "	9 28 "	"	1.00580	"	1.03244	—0.02664	
" "	10 0 "	"	0.99338	"	1.01973	—0.02635	
" "	10 31 "	"	0.99294	"	1.01961	—0.02667	
November 7.	8^h 4′ A.M.	Francis	0.98932	Newell	1.01478	—0.02546	
" "	9 49 "	"	0.98019	"	1.00597	—0.02578	—0.02572
" "	2 26 P.M.	"	0.78315	"	0.80906	—0.02591	
November 7.	9^h 20′ A.M.	Haeffely	0.99305	Newell	1.01997	—0.02692	—0.02692

172. It will be perceived, by an examination of table XVII., that there are greater irregularities in the comparisons by some observers, than in those by others; this is to be attributed, principally, to the different degrees of experience and skill in the observers.

173. In table XVIII. are given the details of the experiments, to ascertain the effect of observing the depths upon the weir, at different distances from the weir, by means of pipes opening near the bottom of the canal. In order to obtain the depth upon the weir by the north hook gauge, the correction —0.03072 has been applied to the mean readings of this gauge. The correction for the south hook gauge is taken from the final column of table XVII., for the corresponding session and pair of observers. From want of time, pipes number 6 and 7 were not made use of.

It will be perceived, by referring to the final column of table XVIII., that the differences in the heights, at the different distances tried, are very inconsiderable, and such as could be detected only by the most delicate means of observation.

174. Two comparisons were made in a similar manner, of the heights, when one gauge box communicated with a pipe opening near the bottom of the canal, and the other with a pipe opening through the side, at about 4.2 feet above the bottom, the orifices of both being at 6 feet from the weir, as represented at *B*, figures 8, 9, and 10, plate XIV.; the following are the results.

First comparison, made November 7th, beginning at 3^h, 52′, P.M.

Francis, at north hook gauge, with pipe No. 5, depth on weir	0.81616 feet.
Avery, at south hook gauge, with pipe *B* " "	0.81641 "
Difference	+0.00025 feet.

Second comparison, made November 7th, beginning at 4^h, 5′, P.M.

Francis, at north hook gauge, with pipe No. 5, depth on weir	0.81775 feet.
Newell, at south hook gauge, with pipe *B* " "	0.81776 "
Difference	+0.00001 feet.

These differences are so minute that we may conclude that the depth was the same whether the pipe opened near the bottom of the canal or at 4.2 feet above.

175. These experiments, taken in connection with those of Boileau,* who has arrived at similar results, leave no doubt as to the propriety, whenever convenience requires it, of observing the depths upon the weir by means of a pipe opening into the dead water, near the bottom of the canal on the upstream side of the weir.

* *Jaugeage des cours d'eau, by M. P. Boileau.* Paris: 1850.

TABLE XVIII.

DATE, 1852.	Time of beginning the observation.	North hook gauge. Pipe No. 5 opens at 6 feet from the weir. " " 6 " 8 " " " " " 7 " 10 " " " " " 8 " 12 " " "			South hook gauge. Pipe No. 1 opens at 1 inch from the weir. " " 2 " 2 feet " " " " 3 " 4 " " " " " 4 " 6 " " "			Difference in the depths upon the weir, the pipe opening at 6 feet from the weir being the standard.	Mean difference in the depths upon the weir, the pipe opening at 6 feet from the weir being the standard.
		Number of the pipe.	Observer.	Corrected depth upon the weir.	Number of the pipe.	Observer.	Corrected depth upon the weir.		
November 3	11ʰ 53′ P. M.	5	Francis	1.01267	1	Newell	1.01321	+ 0.00054	
" 4	0 27 A. M.	5	"	1.01439	1	Haeffely	1.01459	+ 0.00020	
" 7	10 33 "	5	Haeffely	0.97530	1	Newell	0.97547	+ 0.00017	+ 0.00033
" "	10 44 "	5	Francis	0.97644	1	Avery	0.97683	+ 0.00039	
" "	10 56 "	5	"	0.97658	1	Newell	0.97695	+ 0.00037	
November 4	0ʰ 37′ A. M.	5	Haeffely	1.02189	2	Newell	1.02286	+ 0.00097	
" "	1 23 "	5	Francis	1.04220	2	Haeffely	1.04263	+ 0.00043	+ 0.00050
" "	1 34 "	5	Haeffely	1.04472	2	Newell	1.04481	+ 0.00009	
November 7	11ʰ 48′ A. M.	5	Francis	0.97829	3	Avery	0.97883	+ 0.00054	
" "	0 21 P. M.	5	"	0.97734	3	"	0.97800	+ 0.00066	+ 0.00060
November 4	1ʰ 5′ A. M.	8	Francis	1.03882	4	Newell	1.03940	− 0.00058	
" "	1 11 "	8	Haeffely	1.03701	4	"	1.03881	− 0.00180	
" 7	11 9 "	8	Francis	0.97501	4	"	0.97677	− 0.00176	− 0.00153
" "	11 15 "	8	"	0.97579	4	Avery	0.97761	− 0.00182	
" "	11 31 "	8	"	0.97530	4	Newell	0.97700	− 0.00170	
November 7	2ʰ 43′ P. M.	5	Francis	0.80266	1	Avery	0.80311	+ 0.00045	
" "	3 0 "	5	"	0.80731	1	Newell	0.80806	+ 0.00075	+ 0.00060
November 7	4ʰ 25′ P. M.	5	Francis	0.81346	2	Avery	0.81432	+ 0.00086	
" "	4 41 "	5	"	0.80972	2	Newell	0.81079	+ 0.00107	+ 0.00096
November 7	3ʰ 18′ P. M.	8	Francis	0.80984	4	Avery	0.81072	− 0.00088	
" "	3 36 "	8	"	0.81362	4	Newell	0.81491	− 0.00129	− 0.00108

176. It has been stated (art. 164) that the formula

$$Q = 3.33\,(L - 0.1\,n\,H)\,H^{\frac{3}{2}} \qquad (1.)$$

is applicable only to a weir in which the crest is horizontal. Professor James Thomson,* of Queen's College, Belfast, has deduced from formula (1.) a formula for the discharge over symmetrical triangular notches or weirs, figure 4, plate XVIII., viz.:—

$$Q_2 = 2.664\,m\,H_2^{\frac{5}{2}}, \qquad (2.)$$

in which

Q_2 = the discharge in cubic feet per second.

m = the cotangent of the inclination of the crest to the horizon, on each side of the vertex D, equal to $\frac{E\,D}{A\,E}$.

H_2 = the depth $B\,D$ on the vertex of the notch; the line $A\,B\,C$ being the level of the surface of the water, far enough from the notch, to be unaffected by the curvature caused by the discharge.

We can easily deduce from (2.) a formula for the case in which the crest of the weir has a uniform inclination from one end to the other.

Formula (2.) gives the discharge for the notch $A\,D\,C$, figure 4, plate XVIII., in which $A\,B = C\,B$. The discharge Q_3 for one half of the notch $A\,B\,D$ is

$$Q_3 = 1.332\,m\,H_2^{\frac{5}{2}}. \qquad (3.)$$

The discharge Q_4 of the portion of the notch $F\,G\,B\,D$ is the difference of the discharge of $A\,B\,D$ and $A\,F\,G$. Calling $F\,B = L$ and $F\,G = H_3$,

$$Q_4 = 1.332\,m\,H_2^{\frac{5}{2}} - 1.332\,m\,H_3^{\frac{5}{2}},$$

from which we deduce

$$Q_4 = 1.332\,m\left(H_2^{\frac{5}{2}} - H_3^{\frac{5}{2}}\right). \qquad (4.)$$

By its definition $m = \frac{L}{H_2 - H_3}$;

substituting this value of m in (4.), we have

$$Q_4 = 1.332\,\frac{L}{H_2 - H_3}\left(H_2^{\frac{5}{2}} - H_3^{\frac{5}{2}}\right). \qquad (5.)$$

* Civil Engineer and Architects' Journal for April, 1863.

Introducing the correction for the end contraction, formula (5.) becomes

$$Q_4 = 1.332 \frac{L - 0.1\, n \frac{H_2 + H_3}{2}}{H_2 - H_3} \left(H_2^{\frac{5}{2}} - H_3^{\frac{5}{2}}\right). \qquad (6.)$$

Formula (6.) is, of course, applicable only to weirs of the same section in the direction of the flow, as formula (1.), from which it is deduced (see figure 3, plate XVIII.), the depth at one end of the weir being H_2 and at the other end H_3. When the difference in the depths is small, relatively to the mean depth, the quantity computed for the mean depth by formula (1.) for horizontal crests will differ but little from the quantity computed by formula (6.), as will be seen by the following examples:—

Let $L = 10$, and the mean depth on the weir $= 1$ foot.

By the formula for a horizontal crest, . . . $Q = 32.6340$ cub. ft. per sec.

If the crest is 0.01 foot higher at one end than at the other, by formula (6.) . . . . $Q_4 = 32.6340$ " "

If the crest is 0.1 foot higher at one end than at the other, by formula (6.) . . . . $Q_4 = 32.6442$ " "

The formula for the discharge of a weir, deduced from the theoretical velocity of water issuing from an orifice, is

$$Q = \tfrac{2}{3} \sqrt{2\, g\, H}\, L\, H. \qquad (1.)$$

Q, L, and H having the same signification as in art. 164, and g being the velocity acquired by a body, at the end of the first second of its fall in a vacuum, which varies with the latitude of the place and its height above the level of the sea.

In this formula $L\,H$ represents the area of the orifice and $\frac{2}{3}\sqrt{2\,g\,H}$ the mean velocity. Applying this formula to a weir in which the contraction is complete, both at the ends and on the crest, two corrections must be introduced; that for the ends amounting, as we have seen (arts. 123, 124), in weirs of considerable length in proportion to the depth of water flowing over, to a diminution of the length, by a quantity depending only on the depth, and which we have found by experiment (art. 156) to be $0.1\, n\, H$, making the effective area of the weir $(L - 0.1\, n\, H)\, H$.

The correction for the contraction on the crest may be applied in the form of a coefficient m of the velocity, which then becomes $\frac{2}{3}\, m \sqrt{2\, g\, H}$.

Introducing these corrections, formula (1.) becomes

$$Q = \tfrac{2}{3}\, m \sqrt{2\, g\, H}\, (L - 0.1\, n\, H)\, H. \qquad (1.)$$

Taking H from under the radical, we have

$$Q = \tfrac{2}{3}\, m \sqrt{2\,g}\, (L - 0.1\, n\, H)\, H^{\frac{3}{2}}. \qquad (2.)$$

Formula (2.) is identical with that determined by experiment and given in art. 164, except that the coefficient 3.33 is replaced by $\tfrac{2}{3}\, m \sqrt{2\,g}$. In order that both formulas may give the same value of Q, we must have

$$\tfrac{2}{3}\, m \sqrt{2\,g} = 3.33.$$

Substituting the value of g for Lowell, where the experiments were made, we find

$$m = 0.6228.$$

Substituting this value of m in (2.), we have

$$Q = 0.4152 \sqrt{2\,g}\, (L - 0.1\, n\, H)\, H^{\frac{3}{2}}, \qquad (3.)$$

by which formula the discharge of a weir, in which the depth flowing over is not greater than one third of the length and the contraction complete, may be computed for any latitude and height above the sea, by introducing the corresponding value of g, which is given for several latitudes and heights above the sea in a table at the end of this volume.

A METHOD OF GAUGING THE FLOW OF WATER IN OPEN CANALS OF UNIFORM RECTANGULAR SECTION, AND OF SHORT LENGTH.

177. The distribution of the Water Power at Lowell among the different manufacturing establishments, in accordance with the rights of the several parties, renders it necessary to make frequent gaugings of the quantities of water drawn by them respectively. In all the leases of Water Power given by the Proprietors of the Locks and Canals on Merrimack River there is the following provision:—

"For the purpose of ascertaining the quantity of water drawn from the said canals or either of them by the said party of the second part or their assigns, the said Proprietors shall have the right, from time to time, as they may desire, by their duly authorized Agent, Engineer, or other officer, and with the necessary workmen and assistants, to enter upon the premises of the said party of the second part, and to do all acts (with as little injury as may be) necessary or proper for the measuring and ascertaining the quantity of water so drawn as aforesaid. And to this end the said party of the second part shall render all needful and proper facilities; and in case they shall suffer any loss or damage by the acts and doings of the said Proprietors in so measuring and ascertaining the quantity of water drawn as aforesaid, they shall be entitled to compensation therefor, to be paid by the said Proprietors, the amount of which shall be ascertained and determined by arbitrators appointed and acting according to the provisions of the Agreement of 1848 before mentioned."

From the nature of the case, it is necessary to make the gaugings when the manufacturing operations are proceeding in the usual manner. The large number of persons employed, varying from five hundred in the smallest establishment to more than two thousand in the largest, renders any interference with the ordinary course of the work very objectionable. The delicacy of many of the operations also, as well as the large pecuniary interests involved, renders such establishments extremely sensitive to any interruption to their normal condition. As the only mode of avoiding frequent and troublesome controversies under the above provision in the leases, the methods adopted for gauging the quantity of water drawn have been limited to such as could be applied without affecting the usual course of operations in the manufacturing establishments.

It will readily be understood, that this limitation is often an embarrassment to the Engineer charged with the duty of making the gaugings. The simple and exact method of the weir is rarely applicable, as it would, in most cases, detract materially from the effective fall operating upon the water-wheels. Seldom less than two feet fall would be required for this purpose, and the cases are exceptional where such an amount of fall could be taken from that usually used, without causing interruption to the manufacturing operations dependent upon the power. The same objection applies to gaugings by means of apertures of any kind, excepting, however, the apertures by which the water is applied directly to the water-wheels; such apertures are, however, constructed more with reference to the requirements of the water-wheel as a motor than to the making of accurate gaugings of the quantities of water passing through them, and if the Engineer attempts to compute the flow through them by the known laws of hydraulics, he generally finds himself beset with such difficulties and uncertainties as to prevent any confidence in the results, except as approximations.

178. In the gaugings at Lowell the weir is sometimes, although rarely, admissible; gaugings by means of the apertures by which the water is applied directly to the water-wheels are more frequent, but, as a rule, only where experiments have been made on the discharge of the particular water-wheel or one of the same form. The water drawn by the Suffolk Manufacturing Company and by the Tremont Mills is now ascertained in this manner, all the water drawn by them being used on turbines of the same form and dimensions as that experimented upon at the Tremont Mills in the year 1851; an account of the experiments on which is given in the first part of this work. A similar course is adopted at two other establishments in which the water is used upon turbines; in both of these cases the discharge of a turbine of each different pattern has been determined by means of weirs, under various circumstances as to height of gate, velocity of rotation, etc., and from the data thus obtained tables have been prepared; by means of which the discharge is at any time readily obtained, from the observed height of gate, velocity of rotation of wheel and the fall. Care must be taken, however, that the wheels are in good running order when the observations are made.

179. Generally, the water used at a manufacturing establishment is all drawn from the same canal or watercourse, at several points on the same bank, through covered penstocks; and from each of these the water is delivered to one or more water-wheels, and in some instances to several smaller apertures, where water is drawn for other purposes than for that of furnishing mechanical power. To gauge the quantity of water drawn simultaneously at all these points would be a work of much difficulty, under the most favorable circumstances, and when hampered with

the limitation that it must be done without interference with the ordinary operations of the establishment, it becomes impracticable. The difficulties mainly disappear, however, if the gauge can be made in gross before or after the water enters the establishment; and it has been a matter of great interest here to devise and perfect methods by which this could be satisfactorily done.

180. In the year 1830, the quantity of water drawn at one of the cotton-mills of the Hamilton Manufacturing Company was measured by means of a gauge-wheel 15 feet in length and 19.25 feet in diameter, which operated in a manner somewhat similar to an ordinary wet gas-meter. The gauge-wheel was placed in the tail-race of the mill, where all the water used in it was discharged. The quantity of water thus gauged was about 90 cubic feet per second.*

181. In the year 1841, Messrs. James F. Baldwin, George W. Whistler, and Charles S. Storrow, three eminent engineers, were appointed Commissioners to determine the quantities of water drawn from the canals of the Proprietors of the Locks and Canals on Merrimack River, by the several manufacturing companies at Lowell. The following extracts are from their reports, which have never been printed.

Extract from first report, dated October 8, 1841: —

"Upon considering how we should best effect the object in view, various methods occurred to us, given in the books on the subject, by which the quantity of water passing through a canal is deduced by calculation from elements easily measured, such as the velocity at the surface, the slope and the several dimensions of the canal. For many purposes these rules would be sufficient, and if applied here would give us an approximation. The experiments on which they depend having been, however, generally conducted on a small scale, and not always consistent with each other, we did not feel willing to trust to their decision interests so important as those involved in the question before us. The application of such rules would occasion, it is true, but little expense, but for that little expense would furnish only very imperfect information.

"It appeared to us, therefore, that the only satisfactory mode of proceeding was to make a direct and positive measurement of the quantity of water flowing through the Merrimack and Western Canals, which afford greater facilities for the purpose than the others, and by that means to obtain not only the true quantity passing there at the present time, but to test a rule of easy application to the other canals, by which the quantity which they convey can be ascertained without the expense of a similar measurement, and by which also the quantities passing in any of the canals may at any future time be very easily determined.

"In pursuance of this plan we selected a convenient spot in the Western Canal, near the Tremont and Suffolk Mills, where it is about twenty-nine feet wide and eight feet deep.

* See Journal of the Franklin Institute of Pennsylvania, Vol. XI. 2d Series, 1833.

We there excavated the earth from the sides and formed a basin about eighty feet across in the widest place, and raised the bottom so as to leave the depth only about four feet six inches. We there placed across the canal seven paddle-wheels, sixteen feet in diameter and ten feet long each, with narrow and solid piers between them, and coupled the shafts, to make them all revolve together as one piece. These wheels were made with great care, and were so accurately fitted as to run within about a quarter of an inch of the apron or floor below them, and the piers at the sides, thus filling, as nearly as possible in practice, the whole of the seven spaces included between the piers. By driving sheet piling across the head of the apron and into the banks, we obliged all the water of the canal to pass between the piers and drive the wheels. The apron was formed of timbers cut to a true sweep, corresponding to the circle described by the bottom of the floats, and was of sufficient length, in the direction of the current, for one float to enter it at the upper side before the preceding float had left it on the lower. If the wheel, therefore, accurately fitted the apron and the piers, it is evident that when two successive floats were over the apron at the same time, the body of water included in the space between them and the apron (which we call a bucket) was cut off from the rest and passed by itself; and as the wheels revolved, all the water of the canal could only pass in this manner by successive buckets full. A clock fixed upon the end of the shaft showed the number of revolutions of the wheel, and consequently the number of buckets passed in any given time. If we, therefore, could tell just the quantity of water contained in a bucket, or between the two floats when over the apron, we had simply to multiply it by the number of buckets passed, and we had at once the whole quantity of water for the given time.

"To ascertain the quantity of water in a bucket, knowing all the dimensions of our wheels, we only needed to get the depth of the water above the apron. This would vary according to the variations in the level of the canal, and was observed and noted every five minutes during the whole period of our experiments, by means of gauges fixed upon some of the piers and upon the floats themselves.

"The foregoing description shows the manner in which we obtained a direct measurement in the Western Canal. To obtain the other object, that is, to test a simpler mode of measuring, to be used in the other canals and in future in this, we placed at some distance above the wheels a flume or wooden trunk of a section nearly equal to that of the canal, to the bottom and sides of which it thus formed a lining. The bottom of the flume very nearly coincided with the bottom of the canal, and was covered, as well as the sides, with plank carefully jointed so as to form a smooth and even surface. The length of the flume was 150 feet; the width, 27.22 feet; the water was generally about eight feet deep. Being of such a size it produced no sensible disturbance in the flow of the water, and gave us means of accurately measuring the dimensions of the stream as it passed through it. Sheet piling was of course driven at its upper end, so as to throw all the water through it. Its lower end was 151 feet distant from the wheels.

"Simultaneously then with the observations which we made on the quantity at the wheels, we carried on another series at the flume, through which, of course, the same quantity was passing. We carefully noted, about once in 2.5 minutes, the depth of water and the number of seconds in which a small float, placed in the centre of the stream, at the surface, passed through a space of 130 feet in length, measured on the flume. The width of the flume and the depth being known, we knew, therefore, at the moment of each observation, the section of the stream and the velocity of the water at the surface in the centre.

"It was long since ascertained by experiments made on a small scale, that a certain ratio exists between the surface velocity thus measured and the *mean* velocity, or that which, multiplied by the section of the stream, gives the true discharge; and as in the present case we knew by the wheels the true quantity passing, we were able to test this simple rule, and see how much it should be altered and corrected, if at all, in order to give accurate results with bodies of water so very much larger than those hitherto experimented upon.

"It is this rule, thus corrected, and further tested by a similar course of experiments with wheels and a flume in the Merrimack Canal, that we propose to use for the measurements in the other canals at Lowell. The expense of erecting the wheels is great, and they could not of course be left permanently in the canals. The flumes cost much less, interfere neither with the navigation nor with the passage of the water, and are intended to remain in place as long as may be desired. At any future time, therefore, it would only be necessary to measure the depth of water in them, and the surface velocity, and deduce at once, by the rule, the quantity passing through them."

Extract from the third and final report of the Commissioners, dated December 17, 1842:—

"In our first report, made in October, 1841, we stated that we hoped to make our experiments serve, not only to give us a measurement of the quantity of water passing at the present moment, but to test and verify a method or rule for finding the quantity at a future time, without the necessity of the heavy expenditure now incurred. As the result of our labor, we recommend for future use the following rule for measuring the quantities passing through the open flumes which we have erected in the various canals leading the water to the mills.

"Multiply the depth in feet by the width in feet of the stream where it passes through the flume, and the product by the velocity at the surface, in feet per second; this velocity being found by noting the time in which a small float, just immersed in the quickest part of the stream, passes through a given distance. Multiply the quantity then found by

0.847 in the Western Canal,
0.814 in the Merrimack Canal,
0.835 in the Hamilton and Appleton Canal,
0.830 in the Eastern Canal,
0.810 in the Lowell Canal,

and the result is the number of cubic feet per second passing through the flume.

"Should there be in future any great change in the velocity with which the water passes through the canals, these constant numbers or multipliers would require some alteration. We may state, in general terms, that these numbers should be increased in case of a greatly increased velocity, and diminished for a velocity greatly diminished. Within the limits, however, of the ordinary variations in the canals, as they are now used, they may be considered as fixed and constant for the same canal.

"To show the application of the rule, and how far it can be relied on for accuracy, we refer to the annexed table marked A. In that table the numbers of column 6 show the depth of the water in the flume, which in the first experiment was 8.03 feet. Column 7 shows the number of seconds in which the float ran 130 feet, which in that case was 41.47 seconds. Column 8 gives the surface velocity per second, 3.135 feet (in experiment 1), found by dividing 130 feet, the distance run, by 41.47 seconds, the time occupied. Multiplying the depth, 8.03, by the width, 27.22, which gives 218.58, and this product by the velocity, 3.135, we find the quantity, 685.25, given in column 9. This quantity, we may observe, would be the true quantity, if the velocity of every portion of the stream was the same as the surface velocity. Multiplying 685.25 by 0.847, which is the constant multiplier for this canal, we obtain 580.41 in column 11 for the number of cubic feet per second, according to calculation. Actual measurement at the gauge-wheels gives us 586.69 in column 12, for the true quantity. Calculation, therefore, gives in this case 6.28 cubic feet less than measurement, as shown in column 13; and 6.28, the amount of error, is only about one per cent of the true quantity 586.69, or, more exactly, 0.0107 of 586.69, as shown in column 14.

"We refer to our report made in October, 1841, for a description of the manner in which our experiments, made simultaneously at the flumes and at the gauge-wheels, were conducted. The experiments then described as made in the Western Canal were repeated this year, in exactly the same manner, in the Merrimack Canal, where the velocity was about two thirds as great as in the other.

"Table A shows the results of all the experiments in both canals. Comparing the calculated quantities in each experiment, given in column 11, with the measured quantities in column 12, it will be seen, that in the Western Canal the greatest difference between the calculation and the measurement is about one per cent, and the mean difference in the experiments in that canal is something less than one half per cent. In the Merrimack Canal there is one experiment, the 19th, in which the proportional difference was between three and four per cent. In all the other experiments it was less than two per cent, and the mean difference of the whole was about one per cent. The experiment, No. 19, in which the greatest difference occurred, was manifestly made under less favorable circumstances than any of the rest, there having been a great irregularity in the depth of the water, as shown by the note in the table. In estimating the degree of accuracy which the flume rule will give us, it would perhaps be proper to throw out this experiment.

"In addition to Table A, which contains all our experiments, and is the one from which the constant coefficients are determined, we have given two other tables, B and C, which contain the same experiments divided into short periods, with two others, on the accuracy of which we could not place quite so much reliance. These tables show, of course, greater variations between calculation and measurement than the other, and could hardly fail to do so; just as a single observation is less accurate than the mean of several made with equal

TABLE *A*.

COMPARISON OF THE FLUMES WITH THE GAUGE-WHEELS.

Description of the Experiments.	1	2	3				4	5	6	7	8	9	10	11	12	13	14	15
	No. of the Exp.	Date.	Time.				Duration of the Exp.	No. of Observations at the Flume.	Depth of water in the Flume.	Time in which the Float ran 130 feet.	Surface Velocity.	Quantity calculated by Surface Velocity.	Coefficient for this case only.	Quantity calculated by the rule for the Flume.	Quantity by actual measurement at the Gauge-Wheels.	Quantity calculated compared with quantity measured.	Proportional differences.	Extreme variations of level in the Flume.
			Beginning.		Ending.													
		1841.	H.	Min.	H.	Min.	Minutes.		Feet.	Seconds.	Feet per Sec.	Cubic feet per Sec.		Cubic feet per Sec.	Cubic feet per Sec.	Cubic feet per Sec.		Inches.
Experiments at the	1	Aug. 28	10	30	11	30	60	19	8.03	41.47	3.135	685.25	0.856	580.41	586.69	— 6.28	0.0107	0.75
flume in the West-	2	" 30	10	25	12	10	105	30	8.16	41.70	3.117	692.33	0.848	586.40	587.02	— 0.62	0.0011	1.50 gradual.
ern Canal.	3	" 31	4	25	5	20	55	17	8.14	42.35	3.069	680.00	0.853	575.96	580.15	— 4.19	0.0072	1.50 "
	4	Sept. 1	9	45	10	10	25	9	8.02	40.44	3.214	701.62	0.849	594.27	595.96	— 1.69	0.0028	0.50
Width of flume,	5	" 1	10	45	11	35	50	22	7.52	39.73	3.272	669.76	0.843	567.29	564.68	+ 2.61	0.0046	2 75 gradual.
27.22 feet.	6	" 6	9	35	10	30	55	19	8.14	41.31	3.147	697.28	0.838	590.60	584.35	+ 6.25	0.0107	1.00
	7	" 6	10	50	11	10	20	7	8.10	41.71	3.117	687.24	0.845	582.09	580.84	+ 1.25	0.0022	0.25
Coefficient of surface	8	" 7	9	20	9	35	15	8	7.66	39.25	3.312	690.57	0.848	584.91	585.90	— 0.99	0.0017	0.50
velocity, 0.847.	9	" 7	9	35	10	20	45	15	7.74	39.06	3.328	701.15	0.845	593.87	592.62	+ 1.25	0.0021	0.75
	10	" 7	10	25	12	15	110	41	7.63	38.95	3.338	693.27	0.846	587.20	586.82	+ 0.38	0.0006	1.75 gradual.
Means.											3.205		0.847				0.0044	
		1842.																
Experiments at the	11	July 27	9	50	10	40	50	17	8.56	63.352	2.052	525.90	0.814	428.08	427.87	+ 0.21	0.0005	2.25 gradual.
flume in the Mer-	12	" 29	2	30	3	30	60	20	8.57	65.500	1.985	509.32	0.825	414.59	420.41	— 5.82	0.0138	0.75
rimack Canal.	13	Aug. 2	9	20	9	45	25	10	8.73	60.700	2.142	559.87	0.805	455.73	450.95	+ 4.78	0.0106	1.50 gradual.
	14	" 2	9	50	10	10	20	7	8.85	63.286	2.054	544.25	0.812	443.02	441.75	+ 1.27	0.0029	0.25
Width of flume,	15	" 2	10	25	12	0	95	37	8.45	60.216	2.159	546.21	0.817	444.61	446.44	— 1.83	0.0041	2.25 gradual.
29.94 feet.	16	" 2	2	30	4	30	120	44	8.60	62.273	2.088	537.63	0.799	437.63	429.82	+ 7.81	0.0181	3.75 "
	17	" 8	11	0	12	0	60	24	8.70	64.042	2.030	528.77	0.817	430.42	432.17	— 1.75	0.0040	2.00 "
Coefficient of surface	18	" 10	3	45	4	30	45	20	8.70	46.950	2.769	721.26	0.805	587.11	580.52	+ 6.59	0.0114	1.50
velocity, 0.814.	19	" 18	2	35	4	15	100	34	7.78	61.676	2.108	491.02	0.846	399.69	415.42	— 15.73	0.0379	3.25 irregular*
	20	" 19	9	25	10	40	75	35	7.78	59.143	2.198	511.99	0.797	415.76	407.99	+ 7.77	0.0190	3.25
	21	" 22	2	45	3	30	45	18	7.67	60.111	2.162	496.48	0.817	404.13	405.83	— 1.70	0.0042	0.50
	22	" 22	4	0	4	30	30	13	8.72	68.154	1.907	497.87	0.810	405.27	403.10	+ 2.17	0.0054	1.75
Means.											2.138		0.814				†0.0110	

* In 18 minutes water fell 3.25 inches; in 24 minutes rose 3 inches; in 21 minutes *fell* 2.25 inches; in 25 minutes rose 2.25 inches; in 12 minutes fell 2.25 inches.

† Mean, omitting experiment No. 19, 0.0085

TABLES *B* AND *C*.

COMPARISON OF THE FLUMES WITH THE GAUGE-WHEELS.

Description of the Experiments.	1 No. of the Exp.	2 Date.	3 Time.				4 Duration of the Exp.	5 Depth of Water in the Flume.	6 Time in which the Float ran 130 feet.	7 Surface Velocity.	8 Quantity calculated by Surface Velocity.	9 Coefficient for this case only.	10 Quantity calculated by the rule for the Flume.	11 Quantity by actual measurement at the Gauge-Wheels.	12 Quantity calculated compared with quantity measured.	13 Proportional differences.
			Beginning.		Ending.											
		1841.	Hours.	Min.	Hours.	Min.	Minutes.	Feet.	Seconds.	Feet per Second.	Cubic feet per Second.		Cubic feet per Second.	Cubic feet per Second.	Cubic feet per Second.	
TABLE *B*.	1	Aug. 28	10	30	11	0	30	8.04	41.900	3.103	679.09	0.866	575.19	588.16	—12.97	0.0220
			11	0	11	30	30	8.02	41.000	3.171	692.23	0.845	586.32	585.22	+ 1.10	0.0019
Experiments at the flume in the Western Canal.	2	Aug. 30	10	25	11	0	35	8.12	41.700	3.117	688.94	0.843	583.53	580.88	+ 2.65	0.0046
			11	0	11	35	35	8.18	41.700	3.117	694.03	0.854	587.84	593.01	— 5.17	0.0087
			11	35	12	10	35	8.19	41.700	3.117	694.88	0.844	588.56	586.85	+ 1.71	0.0029
Width of flume, 27.22 feet.	3	Aug. 31	4	25	4	50	25	8.10	42.870	3.032	668.50	0.864	566.22	577.52	—11.30	0.0196
			4	50	5	20	30	8.18	41.890	3.103	690.91	0.842	585.20	581.78	+ 3.42	0.0059
	6	Sept. 6	9	35	10	5	30	8.11	41.000	3.171	700.01	0.839	592.91	587.60	+ 5.31	0.0090
			10	5	10	30	25	8.17	41.600	3.125	694.97	0.837	588.64	581.66	+ 6.98	0.0120
Coefficient of surface velocity, 0.847.	10	Sept. 7	10	25	11	5	40	7.65	39.810	3.265	679.88	0.854	575.86	580.57	— 4.71	0.0081
			11	5	11	45	40	7.64	38.670	3.362	699.16	0.833	592.19	582.69	+ 9.50	0.0163
			11	45	12	15	30	7.61	38.080	3.414	707.19	0.849	598.99	600.06	— 1.07	0.0018
		Sept. 1	10	10	10	45	35	7.83	40.530	3.207	683.52	0.851	578.94	581.60	— 2.66	0.0046
			11	35	12	0	25	7.70	39.680	3.276	686.62	0.833	581.57	571.87	+ 9.70	0.0169
Mean.																0.0098
TABLE *C*.	11	1842. July 27	9	50	10	15	25	8.50	64.000	2.031	516.87	0.822	420.73	425.12	— 4.39	0.0103
			10	15	10	40	25	8.58	61.667	2.108	541.51	0.795	440.79	430.58	+10.21	0.0237
Experiments at the flume in the Merrimack Canal.	12	July 29	2	30	3	0	30	8.56	64.200	2.025	518.98	0.809	422.45	419.70	+ 2.75	0.0066
			3	0	3	30	30	8.58	66.800	1.946	499.90	0.848	406.92	423.75	—16.83	0.0397
	15	Aug. 2	10	25	10	50	25	8.48	61.300	2.121	538.50	0.822	438.34	443.22	— 4.88	0.0110
Width of flume, 29.94 feet.			10	50	11	15	25	8.40	59.889	2.171	546.00	0.820	444.44	447.82	— 3.38	0.0075
			11	15	11	40	25	8.46	59.800	2.174	550.66	0.813	448.24	447.78	+ 0.46	0.0010
			11	40	12	0	20	8.50	59.750	2.176	553.77	0.806	450.77	446.62	+ 4.15	0.0093
Coefficient of surface velocity, 0.814.	16	Aug. 2	2	30	3	0	30	8.45	61.333	2.120	536.34	0.808	436.58	433.26	+ 3.32	0.0077
			3	0	3	30	30	8.61	60.167	2.161	557.07	0.790	453.45	440.34	+13.11	0.0298
			3	30	4	0	30	8.69	64.300	2.022	526.08	0.803	428.23	422.48	+ 5.75	0.0136
			4	0	4	30	30	8.68	64.182	2.025	526.26	0.805	428.38	423.79	+ 4.59	0.0108
	17	Aug. 8	11	0	11	30	30	8.74	62.769	2.071	541.93	0.801	441.13	434.09	+ 7.04	0.0162
			11	30	12	0	30	8.65	65.545	1.983	513.56	0.837	418.04	430.22	—12.18	0.0283
	18	Aug. 10	3	45	4	10	25	8.67	46.545	2.793	725.01	0.795	590.16	576.36	+13.80	0.0239
			4	10	4	30	20	8.73	47.444	2.740	716.17	0.817	582.96	585.27	— 2.31	0.0039
	19	Aug. 18	2	35	3	5	30	7.75	60.500	2.149	498.63	0.831	405.88	414.27	— 8.39	0.0203
			3	5	3	35	30	7.81	60.909	2.134	499.00	0.852	406.19	424.97	—18.78	0.0442
			3	35	4	15	40	7.78	63.143	2.059	479.61	0.857	390.40	410.90	—20.50	0.0499
	20	Aug. 19	9	25	10	0	35	7.71	57.875	2.246	518.46	0.787	422.03	407.90	+14.13	0.0346
			10	0	10	40	40	7.84	60.150	2.161	507.25	0.802	412.90	406.87	+ 6.03	0.0148
Mean.																0.0194

care. Little inaccuracies are unavoidable in measuring the time and depth; and an occasional eddy or cross current, or other accidental cause, may vary the observed velocity, and consequently the calculated discharge; and such inaccuracies have less influence on the result of observations long continued than of a smaller number, taken in a shorter time. Still, the comparison of calculation and measurement given in these two tables shows that the mean difference, on the whole, is but from one to two per cent, sometimes in excess and sometimes in deficiency.

"It may be of some interest to compare the accuracy shown by our tables with that shown in the table of M. De Prony, a French engineer of the highest reputation, whose rules have generally been adopted in France. To determine the relation between the surface and the mean velocity, he used seventeen experiments made in France by Du Buat, on a small scale, in little wooden troughs about eighteen inches wide, with depths of from two to ten inches, and velocities varying from six inches per second to four feet and three inches per second. He gives 0.816 as the decimal by which to multiply the surface velocity in order to reduce it to the mean velocity. Comparing his quantities so calculated for these seventeen experiments with the quantities actually measured, he finds proportional differences amounting, for the mean of the whole, to a little less than five per cent, the greatest difference being about fourteen per cent. As his velocities varied very much, he found that he could calculate the discharge more accurately by varying the number 0.816, taking a smaller number for low velocities and a larger number for high velocities. Calculating the quantities by his most exact rule, in which due influence is given to the variation of velocities, he found results differing from measurement about three per cent on an average, after throwing out two of the seventeen experiments which showed a much greater difference, and which he considered less satisfactory than the rest. He remarks, that, as his most correct rule gives a result which is within about one thirtieth of the truth, it ought to be considered as more than sufficiently correct for practical purposes. Our own table *A* shows a much closer correspondence between calculation and measurement, as it naturally should do, because the variations of velocity and change of circumstances in our experiments in each canal were much less than they were in the experiments made in France. It is a remarkable circumstance that the rules of M. De Prony should apply as closely as they do to our case, where the section of the stream is 400 or 500 times as large as it was in his experiments." *

* Prony's more correct formula, reduced to the English foot as the unit, is

$$v = \frac{V(V + 7.78188)}{V + 10.34508}. \qquad (1.)$$

in which $V =$ the surface velocity in the middle of the stream, and $v =$ the mean velocity. (*Storrow on Water-Works*, p. 96.)

Putting the constants in (1.) equal to A and B, we can determine their values from the experiments in table *A*.

Taking the mean values, we have at the Western Canal $V = 3.205$ and $v = 0.847 \times 3.205$, and at the Merrimack Canal $V = 2.138$ and $v = 0.814 \times 2.138$.

182. In connection with the measurement of the quantities of water drawn by the several manufacturing corporations at Lowell, undertaken in the year 1852, the method of gauging in measuring flumes placed in the feeding canals naturally received much attention. The flumes constructed in the years 1841 and 1842 were generally in good order, and had been used at intervals as originally intended. Serious doubts, however, arose, as to whether it was safe to apply the rules deduced from the experiments at the flumes in the Western and Merrimack canals, to those in some of the other canals. In both the Western and Merrimack canals the water at its arrival at the flumes had passed through more than a thousand feet of canal, of nearly uniform section, without having any part of its volume abstracted. In the Western canal, the nearest bend on the up-stream side of the flume was about six hundred feet distant, and in the Merrimack canal about two thousand feet. It was thought that, in the passage of the water to the flumes, under these circumstances, the velocities in different parts of the section would become adjusted, according to the natural laws governing the flow of water in regular channels of great length, and that while it might be safe to compute the flow in other flumes, similarly situated as to the approaches, from the observed surface velocity, it would not be so in other cases, where the length of canal, immediately above the flume, in which the direction, section, and velocity were nearly the same as in the flume, was too short to allow of such an adjustment of velocities in different parts of the section. All the other measuring flumes were much less favorably situated in this respect than those in the Western and Merrimack canals; one of them designed to gauge the largest quantity, being immediately below the entrance to the canal, and the others were liable to be affected by bends, and other irregularities, at short distances above them. The difficulty lay in the uncertainty as to whether the velocities at the surface and at other parts of the section would

---

Substituting these values in (1.), we have

$$0.847 \times 3.205 = \frac{3.205\,(3.205 + A)}{3.205 + B},$$

and

$$0.814 \times 2.138 = \frac{2.138\,(2.138 + A)}{2.138 + B},$$

from which two equations we find $A = 1.889$ and $B = 2.809$, and the formula becomes

$$v = \frac{V\,(V + 1.889)}{V + 2.809}. \qquad (2.)$$

Formulas (1.) and (2.) appear to differ very much, but it will be found that at ordinary velocities they give values of v which differ but little. In figure 1, plate XVIII., the line $A\,B\,C$ represents the values of v by formula (1.), and the line $A\,B\,D$ the values by formula (2.). Both formulas give the same value of v when $V = 1.41$, corresponding to the intersection of the two lines at B.

bear the same relations to each other under different circumstances as to the approaches; there were strong reasons for believing that they would not, and that consequently, the quantity computed by the rules for deducing it from the surface velocity would be liable to errors of such magnitude as to render the results valueless.

183. No substitute for the method of gauging in the flumes could be devised, and the proper course appeared to be to adopt some method of arriving at the mean velocity which should not be open to the objections urged to that of deducing it from the surface velocity. What appeared to be required was a correct and convenient method of taking into account the velocities in every part of the section. There are several well-known methods designed to accomplish this result. Woltman's mill, or tachometer, has been much used for this purpose, but, to insure correct results, its application is one of much delicacy, and in our large channels would require much time. Submerged floats, Pitot's tube, the hydrometric pendulum, and many other contrivances are described in the books on hydraulics. The most promising appeared to be that of obtaining the mean velocity by means of light rods or staves loaded at one end so that they would float vertically, or nearly so, and extend nearly to the bottom of the channel. The advantages of this method were suggested long since. The following extract is from a paper on rivers and canals, by T. H. Mann, read before the Royal Society of London, and printed in their Transactions for the year 1779: —

"The best and most simple method of measuring the velocity of the current of a river or open canal, that I know of, is the following: —

"Take a cylindrical piece of dry, light wood, and of a length something less than the depth of the water in the river; round one end of it let there be suspended as many small weights as may be necessary to keep up the cylinder in a perpendicular situation in the water, and in such a manner that the other end of it may just appear above the surface of the water. Fix to the centre of that end which appears above water a small and straight rod, precisely in the direction of the cylinder's axis; to the end, that when the instrument is suspended in the water, the deviations of the rod from a perpendicularity to the surface of it may indicate which end of the cylinder advances the fastest, whereby may be discovered the different velocities of the water at different depths; for if the rod inclines forwards according to the direction of the current, it is a proof that the surface of the water has the greatest velocity; but if it inclines back, it shows that the swiftest current is at the bottom; if it remains perpendicular, it is a sign that the velocities at the surface and bottom are equal.

"This instrument being placed in the current of a river or canal receives all the percussions of the water throughout the whole depth, and will have an equal velocity with that of the whole current *from the surface to the bottom* at the place where it is put in, and by that means may be found, both with ease and exactness, the mean velocity of that part of the river for any determinate distance and time.

"But to obtain the mean velocity of the whole section of the river, the instrument must be put successively both in the middle and towards the sides, because the velocities at those places are often very different from each other. Having by this means found the *difference of time required for the currents to run over an equal space; or, the different distances run over in equal times, the mean proportional* of all these trials, which is found by dividing the common sum of them all by the number of trials, *will be the mean velocity of the river or canal.*"

Mann does not claim to have been the first to propose this method, and it is probably to be found in the works of some of the older hydraulicians. It is frequently mentioned by more modern writers; generally, however, as one of the modes which have been proposed, but without much stress being laid upon it, as being a convenient and accurate method. Buffon gauged the Tiber by this method, using for floats small bundles of rods, so loaded at one end as to float almost vertically, and extending from the surface nearly to the bottom. Krayenhoff * made some use of it in gauging rivers in Holland, previous to the year 1813; but in applying it to natural watercourses the irregularities in the depth must often present difficulties, not met with in rectangular channels of uniform section.

184. This method of obtaining the mean velocity of water flowing in open channels, not being that commonly used by engineers, or given by writers of authority on the subject as an accurate and established method, it was necessary, at its first introduction here, — large pecuniary interests being involved, — to prove its accuracy, or at least to ascertain within what limits of error it could be applied. Accordingly, in the year 1852, some direct comparisons were made between the results obtained by gauging the flow through rectangular channels, in which the mean velocity was measured by means of loaded tubes, and by gauging the same volume of water by means of weirs; the formula for computing the flow over weirs having been determined by experiments on a suitable scale. These comparisons are described in the first edition of this work. They indicate a close correspondence in the results arrived at by the two methods; as might be expected, however, the quantity deduced from the mean velocity of the tubes was, generally, a little in excess of the mean velocity as deduced from the gauge of the same volume of water at the weirs; the greatest difference being in the comparisons in

* *Recueil des observations Hydrographiques et Topographiques faites en Hollande, par* C. R. T. Krayenhoff. Amsterdam, 1813.

The floats used by Krayenhoff were wooden poles, loaded with lead at the bottom, and buoyed up by copper floats at the surface of the water.

which the shortest tubes were used, the excess in this case, however, being only about four per cent. These comparisons furnished the means of making corrections of the *flume measurements*, (by which term is to be understood the product of the mean velocity of the tubes into the section,) in order that the results might be substantially the same as would be given by weir measurements; and also established, beyond question, that the method could be relied upon, when applied under favorable circumstances, to give results sufficiently near the truth to meet the practical requirements of all the parties in interest.

The experiments of 1852 were not sufficiently numerous and varied to afford the data for a formula of correction of general application, and arrangements having been subsequently made between the lessors and the lessees, which involved more frequent and more accurate gaugings, it was deemed expedient to perfect the method as far as practicable; and also to ascertain the extent to which we were liable to err in applying the method in some peculiar circumstances, such as high winds, or with irregular currents and eddies in the measuring flumes. Accordingly, in the year 1856, an extensive series of experiments was made for these purposes, an account of which is given below.

185. In long straight channels, in which the section occupied by the water is uniform, and the quantity of water flowing is constant, at a distance, greater or less, from the place where the water is admitted, a certain relation is established between the slope of the surface and the mean velocity; and also between the velocities of the water in different parts of the section; that is, the *regime* is established, and the stream is said to be in a state of *uniform motion*. The comparative velocities at different depths, in any vertical plane which is parallel with the direction of the current, are called the *scale of velocities*. Most of the rules given by writers on hydraulics for the motion of water in open channels are for the case of uniform motion.

It is generally assumed that the resistance to the motion of water all proceeds from the bed, by which is meant the bottom and sides of the channel, and that the maximum velocity in symmetrical channels of the usual forms is at the surface, and in the middle of the stream.

186. When the air in contact with the surface of water, flowing in an open channel, is moving in the same direction, and with the same velocity, as the surface of the water, it is clear that it can have no effect on the motion of the water; but such exact conformity in the motion of the air and water is uncommon; ordinarily, the air has some motion relatively to that of the water, and either retards or accelerates the velocity of the surface. That the air may produce a material effect on the scale of velocities is apparent from the following considerations.

Let us suppose the surface of the water to move, relatively to the air, with the same velocity as the water at the bottom moves relatively to the bed; also, that the inequalities of the surface of the water caused by the action of the air and those in the bed of the stream are alike; and suppose, also, that a sheet of water of uniform thickness, in contact with the bed, is at rest; we shall then have the water near the bottom moving over a bed of water, and the water at the surface moving under a bed of air, and as both beds have the same inequalities, they will cause the same retardation in the velocity of the water, except as these beds, from the nature of the substances of which they are composed, offer more or less resistance. These resistances will be of the same nature as is experienced by a body moving in a resisting medium. According to well-known principles, the retardation in this case is as the square of the velocity of the moving body, relatively to that of the medium,* and as the density of the medium. The density of the air is about $\frac{1}{840}$ of that of water; a body moving through the air, with the same velocity, will therefore be retarded $\frac{1}{840}$ as much as if it moved through water. Consequently, in the case supposed, if the relative velocity of the air and the surface of the water is the same as that of the bed and bottom of the stream, the retardation at the surface will be $\frac{1}{840}$ of that at the bottom. The retardation being as the square of the relative velocity, if the air is moving in the opposite direction to the motion of the water, with a relative velocity equal to $\sqrt{840} = 29$ times the velocity of the water at the bottom, the retardation at the surface and at the bottom will be the same, and the maximum velocity will be found at half the depth.

This supposed case is designed merely to show the mode in which the air acts in modifying the scale of velocities, and to afford some idea of the extent of its influence.

187. It follows, from what is said in the preceding section, that in all cases, except when there is a wind blowing in the direction of the current, of equal or greater velocity than the water at the surface of the stream, the air will retard the surface velocity.

Many attempts have been made to determine, experimentally, *the scale of velocities* at different depths. Du Buat, who experimented in very small wooden

---

* The retardation will be as the square of the velocity, only, when the inequalities of the surfaces in contact remain constant. But the inequalities of the surface of the water will increase and diminish with the velocity; consequently, the retardation at the surface of the water will be in a higher ratio than the square of the velocity. It is, however, sufficient for the present purpose to assume it to be as the square of the velocity. The relative thickness of the beds of water and air will also have an important effect; but it need not be considered here.

canals, reports that he found the maximum velocity at the surface. Defontaine, who experimented on the Rhine, thought, allowance being made for the wind, that the maximum velocity was at the surface. Hennocque experimented on an arm of the Rhine near Strasburg; according to Boileau, he found the maximum velocity as follows: —

In a calm or very slight breeze blowing up stream, at about one fifth of the depth below the surface.

In a strong wind blowing up stream, at about half the depth.

In a strong wind blowing down stream, at the surface of the current.

Baumgarten, who experimented on the canal from the Rhone to the Rhine, reports that he found the maximum velocity between one fifth and one third of the depth from the surface.

Boileau, who experimented in small wooden canals, reports that he found the maximum velocity at one fifth of the depth below the surface.

Messrs. Humphreys and Abbot,* of the United States corps of Topographical Engineers, in connection with their operations for gauging the flow of the Mississippi, made an elaborate series of experiments with submerged floats, to determine the scale of velocities. They report, that, as a mean result, they found the maximum velocity, when there was little or no wind, at about three tenths of the depth from the surface.

Messrs. Darcy and Bazin,† in their extensive series of experiments on the flow of water in open channels, made at the expense of the French government, report that they found the maximum velocity below the surface.

188. In their work, previously cited, Humphreys and Abbot give what they term the grand-mean curve, determined from very numerous observations on the Mississippi at Carrolton and Baton Rouge, in Louisiana, in the year 1851; the mean depth of the river being 82 feet, and the mean velocity 3.3814 feet per second. The curve thus determined is a parabola, of which the equation is

$$V = -0.79222\, d_{\prime\prime}^2 + 3.2611,$$

in which $V =$ the velocity in feet per second at any depth $d_{\prime\prime}$ above or below the axis of the curve; $d_{\prime\prime}$ being taken in fractional parts of the whole depth of the river, which is taken as unity; and the axis being 0.297 of the whole depth below the surface.

* Report upon the Physics and Hydraulics of the Mississippi River, by Captain A. A. Humphreys and Lieutenant H. L. Abbot. Philadelphia, 1861.

† *Recherches Hydrauliques entreprises par* M. H. Darcy, *continuées par* M. H. Bazin. Paris, 1865.

If we put $d_{\prime\prime\prime}$ = the depth, in feet, above or below the axis, and substitute for $d_{\prime\prime}$ in the preceding equation its value $\frac{d_{\prime\prime\prime}}{82}$, we shall have

$$V = -0.00011782\, d_{\prime\prime\prime}^2 + 3.2611.$$

The axis being $0.297 \times 82 = 24.354$ feet below the surface. When $d_{\prime\prime\prime} = 0$, then $V = 3.2611$ feet per second, which is the velocity at the axis of the curve and the maximum. For the velocity at the surface we have $d_{\prime\prime\prime} = -24.354$ and $V = 3.1912$. For the velocity at the bottom we have $d_{\prime\prime\prime} = 57.646$, and $V = 2.8696$.

189. In the experiments of Humphreys and Abbot, the direction of the wind was noted and its force estimated, a calm being called 0 and a hurricane 10; they made no experiments, however, when the force exceeded 4. In the experiments from which the grand-mean curve was determined, the mean estimated down force of the wind is stated to have been 0.2. They found that the direction and force of the wind produced a marked effect upon the position of the axis, or, in other words, upon the depth below the surface at which the velocity was a maximum. Their grand-mean curve indicates that when the wind was blowing down stream with the force 0.2, the maximum velocity was about 0.3 of the whole depth; and they state that they always found it below the surface in a calm, and they infer, from their elaborate experiments, that even when the wind was blowing down stream with a velocity equal to that of the current, that the maximum velocity is generally, if not always, below the surface. It is difficult to understand how this can be the case in a long, straight, uniform channel. The Mississippi, as is well known, is very crooked, and the disturbing effects of bends in a large stream are felt at great distances down stream; and probably no point could be found below the mouth of the Ohio at which the velocities in different parts of the section would be free from considerable irregularities from this cause. What effect this may have on the scale of velocities does not appear, but it will scarcely be safe to infer that it would be found to be the same in straight as in crooked channels.

190. Humphreys and Abbot give the following general formulas for the curve representing the scale of velocities, in any vertical plane which is parallel with the direction of the current.

Let V = the velocity at any depth.

v = the mean velocity for the whole stream.

V_m = the mean velocity in the vertical plane under consideration.

V_x = the maximum " " " " " "

V_0 = the surface " " " " " "

V_b = the bottom " " " " " "

f = the number denoting the force of the wind; 0 being a calm or a wind blowing at right angles with the current, and 10 a hurricane; the sign to be — when it blows down stream, and + when it blows up stream.

d = the depth below the surface, at which the velocity is V.

d_a = the depth of the axis of the curve below the surface.

D = the whole depth.

R = the mean radius.

Equation of the curve representing the scale of velocities,

$$V = V_a - \sqrt{\frac{1.69\, v}{\sqrt{D} + 1.5}} \left(\frac{d - d_a}{D}\right)^2, \tag{1.}$$

Depth of the axis of the curve below the surface,

$$d_a = (0.317 + 0.06\, f)\, R. \tag{2.}$$

Mean velocity,

$$V_m = \tfrac{2}{3}\, V_a + \tfrac{1}{3}\, V_D + \tfrac{1}{3}\, \frac{d_a}{D} (V_a - V_D). \tag{3.}$$

Formulas (1.) and (2.) are empirical, and founded, mainly, on the experiments of Humphreys and Abbot. Formula (3.) is purely geometrical, assuming that the scale of velocities is represented by a parabola.

191. In gauging the quantity of water flowing through our measuring flumes, numerous observations are made of the velocity of the tubes in different parts of the width of the flume, and their mean velocity is computed. The tubes cannot extend quite to the bottom of the channel, and the layer of water between the bottom of the tubes and the bottom of the channel, which has usually a less velocity than any other part of the section, will not have its due weight in determining the velocities of the tubes, which will therefore, usually, assume velocities a little greater than they would if they extended to the bottom. Also, if the scale of velocities at different depths is represented by a parabolic curve, as is indicated above, in consequence of the pressures on different parts of the tube being as the squares of the relative velocities of the water and tube, the tubes will assume velocities generally, a little different from the mean velocity of the water above the bottom of the tubes. It is also known that floating bodies do not generally have the same velocity as the water in which they are immersed.

192. Knowing the mean velocity in the whole section, and assuming that the formulas of Humphreys and Abbot apply to our short channels, we can compute the velocity of the tubes in the following manner.

Applying formulas (1.), (2.), and (3.) to the plane in the direction of the current, in which the mean velocity is a mean of the whole section, we have

$$v = V_m.$$

The equation of the parabola representing the scale of velocities is of the form

$$V = A - B\ (d - d_v)^2.$$

The values of A and B can be determined by equations (1.), (2.), and (3.).

In (2.) the values of f and R are given by observation, hence the value of d_v is known, which can be substituted in (1.), in which, besides the co-ordinates V and d, all the quantities will be known excepting V_v, which can be determined by (3.), as follows:—

In (1.) when $d = D$

$$V_D = V_v - \sqrt{\frac{1.69\ v}{\sqrt{D + 1.5}}} \left(\frac{D - d_v}{D}\right)^2, \qquad (4.)$$

and when $d = o$

$$V_o = V_v - \sqrt{\frac{1.69\ v}{\sqrt{D + 1.5}}} \left(-\frac{d_v}{D}\right)^2. \qquad (5.)$$

Substituting these values in (3.), also V_m for v, and reducing, we have

$$V_v = V_m + \tfrac{1}{3} \sqrt{\frac{1.69\ V_m}{\sqrt{D + 1.5}}} \left(1 - 3\frac{d_v}{D} + 3\left(\frac{d_v}{D}\right)^2\right). \qquad (6.)$$

In which all the quantities in the second member are known, and consequently the value of V_v. Substituting the value of V_v in (4.), we have all the quantities known in the second member, and consequently the value of V_D. The values of V_v and V_D being known, determine two points in the curve, which is sufficient to determine the two constants A and B in its equation.

In experiment 1, table XXII., we have $R = \frac{26.746 \times 9.533}{26.746 + 2 \times 9.533} = 5.5656$; and for a *moderate wind down stream* f is assumed at -0.5. Substituting these values in (2.), we have $d_v = 1.5973$ feet.

In experiment 1, we can determine the mean velocity from the weir measurement and the section of the stream.

$$v = V_m = \frac{681.25}{26.746 \times 9.533} = 2.6719 \text{ feet per second.}$$

We have also $\qquad D = 9.533.$

Substituting these values in (6.), we have

$$V_v = 2.8979 \text{ feet per second.}$$

The quantities in the second members of (4.) and (5.) are now all known, and their values being substituted, give

$$V_D = 2.0899 \text{ and } V_o = 2.8652.$$

In the equation of the curve $V = A - B\,(d - d_o)^2$,

when $d = o$, then $V = 2.8652$ and $2.8652 = A - B\,(-1.5973)^2$

and when $d = 9.533$, then $V = 2.0899$, and $2.0899 = A - B\,(9.533 - 1.5973)^2$.

From these two equations we find

$$A = 2.8979 \text{ and } B = 0.01283,$$

and the equation of the curve representing the scale of velocities in experiment 1 is

$$V = 2.8979 - 0.01283\,(d - 1.5973)^2. \qquad (7.)$$

The curve of which (7.) is the equation is represented by the line $E\,X\,D\,F$, figure 7, plate XV., $O\,X$, $Y\,O\,Y$ being the axes of co-ordinates. Figures 8, 9, and 10 represent the curves deduced in a similar manner from experiments 7, 43, and 47.

193. $O\,A$, figure 7, represents the velocity of the tube. As stated above, this will generally vary slightly from the velocity of the water, the tube having such a velocity that the pressure on its up-stream side will equal the pressure on its down-stream side. These pressures are due to the relative velocities of the water and tube; at the depths where the tube is moving slower than the water there will be a pressure on the up-stream side, and where it moves faster than the water there will be a pressure on the down-stream side. In figure 7 the portion of the tube $B\,D$ will have a pressure on the up-stream side, and the portion $C\,D$ will have a pressure on the down-stream side; the pressure at any point will be proportional to the square of the difference of the velocities of the tube and the water. Adopting this principle, and assuming that the scale of velocities is represented by an equation of the form

$$V = A - Bd^2$$

Putting $V_{,} = A\,X =$ the difference between the velocity of the tube and the maximum velocity of the water; $d_{,} =$ the depth $A\,C$ to which the tube is immersed below the axis of the curve, and retaining the preceding notation, my assistant, Mr. Joseph P. Frizell, finds for the case represented in figure 7, plate XV.

$$\frac{16}{15\sqrt{B}}\,V_{,}^{\frac{5}{2}} - (d_{,} - d_o)\,V_{,}^2 + \tfrac{2}{3}\,B\,(d_{,}^3 - d_o^3)\,V_{,} = \tfrac{1}{5}\,B^2\,(d_{,}^5 - d_o^5). \qquad (8.)$$

In experiment 1, table XXII., we have $d_{,} = 9.482 - 1.5973 = 7.8847$; $d_o = 1.5973$, and $B = 0.01283$. Substituting these values in (8.), and reducing, we have

$$V_{,}^{\frac{5}{2}} - 0.66766\,V_{,}^2 + 0.44152\,V_{,} = 0.10650,$$

from which we find $V_{,} = 0.2655$.

Putting V_t = the velocity of the tube, we have

$$V_t = V_o - V_r = 2.8979 - 0.2655 = 2.6324 \text{ feet per second.}$$

Putting V_{mt} = the mean velocity of the water for the depth to which the tube is immersed, and V_{bt} = the velocity of the water at the bottom of the tube, we have by (7.)

$$V_{bt} = 2.8979 - 0.01283\,(9.482 - 1.5973)^2 = 2.1003 \text{ feet per second,}$$

and by (3.),

$$V_{mt} = \tfrac{2}{3}\,V_o + \tfrac{1}{3}\,V_{bt} + \tfrac{1}{3}\,\frac{d_o}{d_t}\,(V_o - V_{bt}) = 2.6750 \text{ feet per second.}$$

In experiment 1 the tube will, therefore, have a velocity, less than the mean velocity of the water for the whole depth to which the tube is immersed, equal to

$$2.6750 - 2.6324 = 0.0426 \text{ feet per second,}$$

which is about $\frac{1}{63}$ of the velocity of the water.

194. The tube does not at once take the velocity of the water, but after floating a short distance the difference is inappreciable, as will be seen by the following investigation.

The tube when first placed in the water is supposed to be perpendicular and at rest, and to retain its perpendicularity during its motion; the water striking it with the full velocity of the current creates a pressure on its up-stream side; yielding to the pressure, it gradually assumes the velocity of the current. It will, however, be simpler to consider the converse proposition, which will lead to the same result, namely, to assume that the water is at rest and that the tube is impelled against it with a velocity equal to the velocity of the current; it can then be treated as a body moving in a resisting medium of large extent in proportion to the size of the body.

Let V'' = the initial velocity of the tube.

v = the velocity of the tube after traversing any distance, s, in the fluid.

k = a coefficient, depending on the form of the body, but which for a cylinder moving with its axis at right angles to the direction of the motion is about 0.77. (See *Rankine's Applied Mechanics*, London and Glasgow, 1858.)

w = the weight of the body.

A = the area of its greatest transverse section opposed to the motion.

N = the specific gravity of the body.

n = the specific gravity of the fluid.

f = the retarding force.

According to well-known principles, the resistance of the water to the motion of the tube is $k\,A\,n\,\frac{v^2}{2g}$, and $$f = \frac{k A n v^2}{2 g w}. \quad (1.)$$

If the body is a cylinder, put D for its diameter; then for one foot in length of the cylinder we have $A = D$, we have also $w = \frac{1}{4} \pi D^2 N$. Substituting these values in (1.) and reducing, we have

$$f = \frac{2 k n v^2}{\pi g D N}.$$

In the case of a floating body we have $N = n$, and consequently

$$f = \frac{2 k v^2}{\pi g D}.$$

Then (see *Hutton's Mathematics*, "On the Motion of Fluids"), giving dv the negative sign, because v diminishes as s increases,

$$- v\, d v = g f\, d s = \frac{2 k v^2}{\pi D} d s,$$

and hence $$- \frac{d v}{v} = \frac{2 k}{\pi D} d s, \quad (2.)$$

which, by integration, gives

$$s = \frac{\pi D}{2 k} \log. \frac{V'}{v}. \quad (3.)$$

Equation (2.) may be put under the form

$$- \frac{\frac{d^2 s}{d t}}{\frac{d s}{d t}} = \frac{2 k}{\pi D} d s.$$

Multiplying both sides by $\frac{d t}{d s}$ and reducing, we have

$$- \frac{d^2 s}{d s^2} d t = \frac{2 k}{\pi D} d t.$$

Integrating, remembering that $\frac{d t}{d s}$ or $\frac{1}{v}$ is equal to $\frac{1}{V'}$ when $t = o$, we have

$$t = \frac{\pi D}{2 k} \left(\frac{V' - v}{V' v}\right). \quad (4.)$$

Returning to the real case and denoting by s' the distance traversed by the tube in the time t, V' being the velocity of the current, and $v_{,}$ that of the tube at the expiration of the time t, we shall have

$$s' = V' t - s.$$

Substituting the values of s and t by (3.) and (4.) and also $V' - v_{,}$ for v, we have

$$s' = \frac{\pi D}{2 k} \left(\frac{v_{,}}{V' - v_{,}} + \log. \frac{V' - v_{,}}{V'}\right). \quad (5.)$$

195. By equation (5.) we see that, theoretically, the tube never quite attains the velocity of the current; and that the distance it must float in order to attain the velocity of the current, within a given fractional part, is proportional to the diameter of the tube and is independent of the velocity of the current.

In the following experiments, s' was about 20 feet, and $D = 2$ inches $= \frac{1}{6}$ foot. Substituting these values in (5.) we find

$$\frac{V' - v'}{V'} = \frac{1}{64} \text{ nearly,}$$

That is to say:—a tube 2 inches in diameter, after floating 20 feet from the point where it is put into the current, acquires a velocity equal to about $\frac{63}{64}$ that of the current.

196. Observation teaches us that floating bodies move faster than the stream in which they are floating; this is undoubtedly the reason why vessels moving with the current in a calm can be steered; they not only partake of the motion of the water, but they have an independent motion due to the inclination of the surface of the water; the constant intermingling of the upper and lower parts of a stream prevents the water at and near the surface from attaining a velocity as great as it otherwise would. Navier* has investigated this subject; assuming that the velocity of the water is uniform to the depth to which the body is immersed, he finds, adopting our own notation,

$$V_{s} = \sqrt{\frac{2\,g\,Q\,I}{k\,A}}, \qquad (6.)$$

in which

$V_{s} =$ the excess of the velocity of the floating body over that of the water.
$g =$ the velocity imparted by gravity in one second.
$Q =$ the volume of water displaced by the floating body.
$I =$ the slope of the surface.
$k =$ a coefficient depending on the form of the body.
$A =$ the area of the greatest transverse section of the body.

In these experiments the floating bodies are cylinders with the axes vertical, for which case k is nearly 0.77 (art. 194). Put L for the length of the immersed part of such a cylinder and D for the diameter; then

$$Q = \tfrac{1}{4}\,\pi\,D^2\,L, \text{ and } A = D\,L.$$

Substituting these values, and also the values of g and k, in the above equation, and reducing, we have

$$V_{s} = 8.1\,\sqrt{D\,I}. \qquad (7.)$$

* *Architecture Hydraulique, par* BELIDOR. Paris, 1819, page 358.

The value of I can be determined from Eytelwein's formula for the motion of water in open channels, which when the English foot is the unit is *

$$R\,I = 0.000\,024\,265\,1\,v + 0.000\,111\,415\,5\,v^2. \qquad (8.)$$

In which R is the mean radius, I the descent in the unit of length, and v the mean velocity.

Formula (7.) indicates that the excess of velocity is proportional to the square root of the diameter of the tube, and also to the square root of the slope. Except in very small velocities, the velocity of the current is nearly proportional to the square root of the slope; consequently, the excess of the velocity of the floating body over that of the fluid in which it is floating is nearly proportional to the velocity of the current, except when the latter velocity is very small.

In experiment 1 we have $R = 5.5656$ and $v = 2.6719$ (art. 192); substituting these values in (8.) we find $I = 0.000\,154\,56$, we have also $D = \frac{1}{6}$; substituting these values in (7.) we find $V_e = 0.0411$ feet per second, which is about $\frac{1}{65}$ of the mean velocity of the water. Neglecting the small effect this excess of velocity would have on the velocity deduced from the equality of the pressures on the up-stream and down-stream sides of the tube, we find for the computed velocity of the tube in experiment 1, $2.6750 - 0.0426 + 0.0411 = 2.6735$ feet per second; which differs 0.0015 feet per second, or $\frac{1}{1783}$, from the mean velocity of the water for a depth equal to the length of the immersed part of the tube, determined by the formulas of Humphreys and Abbot. The mean velocity of the tube by experiment was 2.6830 feet per second, which exceeds the computed velocity by 0.0095 feet per second, or $\frac{1}{281}$. Similar computations have been made for experiments 7, 43, and 47, table XXII., which are selected as giving a wide range of conditions. The data and results are given in the following table.

TABLE XIX.

1	2	3	4	5	6	7	8	9	10	11
No. of the Exp.	Depth of water in the flume. D Feet.	Mean Radius. R	Length of the immersed part of the tube. d_1 Feet.	Mean velocity of the water deduced from the weir measurement. v Feet per Second.	Assumed value of f	Depth of the axis of the parabola, representing the scale of velocities, below the surface of the water. d_m Feet.	Parameter of the parabola, representing the scale of velocities. B	Maximum velocity of the water. V_m or A Feet per Second.	Velocity of the water at the surface. V_o Feet per Second.	Velocity of the water at the bottom. V_b Feet per Second.
1	9.533	5.5656	9.482	2.6719	—0.5	1.5973	0.01283	2.8979	2.8652	2.0899
7	9.530	5.5645	8.530	2.6539	—0.1	1.7396	0.01280	2.8686	2.8302	2.0902
43	8.172	5.0723	7.120	0.4961	0.0	1.6079	0.00777	0.5871	0.5670	0.2521
47	8.165	5.0696	8.122	0.4842	—0.3	1.5158	0.00770	0.5777	0.5600	0.2374

* A Treatise on Water-Works, by CHARLES S. STORROW. Boston, 1835.

TABLE XIX. — CONTINUED.

No. of the Exp.	12	13	14	15	16	17	18	19	
	Mean velocity of the water for a depth equal to the length of the immersed part of the tube.	Velocity of the tube deduced from the formula founded on the equality of the pressures on the up-stream and down-stream sides of the tube.	Difference between the velocities in column 12 and column 13.	Slope of the surface of the water in the flume, deduced from Eytelwein's formula for the motion of water in open channels.	Excess of the velocity of the tube over that of the water in which it is floating, deduced from Navier's formula.	Computed velocity of the tube.	Mean velocity of the tube by experiment.	Difference between the velocity of the tube by computation and by experiment.	
								Absolute difference.	Proportional difference.
	V_{mt} Feet per Sec.	V_t Feet per Second.	Feet per Second.	I	V_e Feet per Sec.	Feet per Sec.	Feet per Sec.		
1	2.6750	2.6324	0.0426	0.000 154 56	0.0411	2.6735	2.6830	+ 0.0095	+ 0.0036
7	2.7088	2.6752	0.0336	0.000 152 60	0.0409	2.7161	2.7260	+ 0.0099	+ 0.0036
43	0.5247	0.5108	0.0139	0.000 007 78	0.0092	0.5200	0.5190	— 0.0010	— 0.0019
47	0.5111	0.4669	0.0442	0.000 007 47	0.0090	0.4759	0.4950	+ 0.0191	+ 0.0401

197. It will be seen, by column 19 in the preceding table, that the differences between the computed and observed velocities are not very regular; perhaps as much so, however, as could be anticipated, considering the wide difference in the conditions in the experiments of Humphreys and Abbot and in the experiments at the Tremont measuring flume, and that their data for determining formulas (1.) and (2.) are not of a character to afford much confidence in their application to cases where the conditions are so different.

198. From the preceding investigation we infer, that in rectangular channels, in which the natural scale of velocities at different depths is established, and the surface velocity not very much retarded by the wind, the tube is retarded on account of the pressures on the tube being as the squares of the relative velocities of the water and tube at different parts of its length, and is accelerated by the independent motion of the tube due to the slope of the surface of the water, and that the retardations and accelerations compensate each other to a greater or less degree under different circumstances.

Taking a mean of the four experiments in table XIX., the computed velocity of the tube is about $\frac{1}{80}$ less than the observed velocity; and assuming this relation to be of general application, we might, evidently, by a process the reverse of that by which table XIX. is computed, from the observed velocity of the tube, arrive at the mean velocity of the water in the flume. It would, however, involve lengthy computations, and the result would not be free from uncertainty, on account of the doubtful applicability of the formulas of Humphreys and Abbot; and however interesting such an investigation might be as a scientific matter, it will be safer, in practice, to rely upon rules deduced from suitable experiments, even if such rules are empirical.

199. In arranging the programme of these experiments, it was designed to make them under the various circumstances which occur in the gaugings in the

several measuring flumes at Lowell, and as nearly as practicable on the same scale; the only material deviation from what was desired in the latter respect, was in the width of the channel; this was necessarily limited to the width of the canal in which the experimental flume was placed. A series of experiments with tubes of seven different lengths, and with velocities varying from 2.7 to 0.5 feet per second, was made with a flume of as great a width (26.745 feet) as could conveniently be made in the canal, and another series, similar in respect to length of tubes, but with velocities varying from 5.0 to 1.4 feet per second, was made with a flume of half the width of the preceding.

200. The experiments consisted in making a gauge of the quantity of water passing the measuring flume, by observing the velocity of loaded tubes floating down different parts of the section of the flume, and from these observations deducing the mean velocity of the tubes for the whole section; this mean velocity is provisionally assumed to be the mean velocity of the water in the flume, and when multiplied into the area of the section gives the quantity of water passing the flume according to the *flume measurement*. After leaving the flume, the same volume of water is made to pass over a weir, and the depth on the crest being observed, the quantity is computed by means of a formula determined from the experiments made at Lowell, in 1852, and previously described in this work. The quantity thus computed (with a minute correction for leakage in the experiments on the narrow flume) is taken as the true quantity passing through the measuring flume, and the comparison of this quantity with that obtained by the flume measurement determines the correction in that particular experiment.

201. Figures 1 and 2, plate XVI., are a general plan and longitudinal section of the entire apparatus used in the experiments with the wide flume. *A* is the Northern Canal, through which the principal supply of water is primarily conducted from the Merrimack River to the manufacturing establishments. *B*, the Tremont Gates, through which water is at times drawn, to make up any deficiency in the supply in the lower level of the Western Canal. *C* is a grating put across the canal for the purpose of equalizing the flow of the water in different parts of the section of the canal. *D* is a raft or float for the purpose of destroying the oscillations of the surface, caused by the admission of the water at the gates *B*, and which oscillations were partially propagated through the grating *C*. Without the float the oscillations of the surface extended into the measuring flume, and imparted corresponding vertical oscillations to the tubes, causing those extending nearly to the bottom to touch occasionally, which would of course tend to retard them. *E* is the measuring flume. *F* the Tremont Wasteway, over which the occasional supply from the Tremont Gates passes into the lower level of the

Western Canal, *W*; on this wasteway is erected the weir for gauging the water after it has passed through the measuring flume.

Figures 1 and 2, plate XV., are a plan and transverse section of the wide flume. The original section of the canal is lined, from *A* to *B*, with planks about 2.25 inches in thickness, planed on the surface in contact with the current, and fastened to timbers which are securely bolted to the side walls and to stones sunk in the bottom of the canal for the purpose. The lining plank is connected with an old piling, *C D*, put in for another purpose, which extends through the side walls of the canal and into the earth on each side, effectually preventing any flow of water outside of the plank lining. *E F* represents an obstruction in the canal, used in a portion of the experiments for the purpose of creating irregularities in the flow through the measuring flume. *G* is a float of timber and plank for the purpose of destroying the oscillations of the surface of the water caused by the obstruction *E F*. The obstruction and float were used only in experiments 123 to 140, which do not form any part of the series from which the formula of correction is deduced.

202. Figures 3 and 4, plate XV., represent the same measuring flume as figures 1 and 2, with the changes made for the purpose of narrowing the flume. The partition *A B* was placed near the middle of the flume; the dam *C* prevented any flow of water through the part of the flume shut off by the partition. In order to make the flow through the narrow flume more nearly like that through a long canal of uniform section, and in this respect, more like the flow through the wide flume, the partition was extended above the flume from *A* to *D*, a distance of about 100 feet. This extension of the partition was constructed of planks, the lower ends of which were set in the earth forming the bottom of the canal, and the upper ends were secured to timbers and stayed as represented in figure 3. The part of the partition from *A* to *D* was intended to be as nearly impervious to the passage of water through it as it could be conveniently made without jointing the planks; the partition from *A* to *B* was made with more care and was intended to be water tight; the lining of the flume was also intended to be water tight; neither lining nor partition were, however, quite tight. In the experiments with the wide flume, no difficulty was experienced from this cause; in the experiments with the narrow flume it was necessary to ascertain the correction to be applied on account of the leakage. It would occupy much space to give an intelligible description of the operations performed to arrive at the correction to be made on this account, and as it was found to be very small, less than $\frac{1}{1000}$ part of the quantity passing the flume in any experiment, further mention of it is unnecessary.

203. The whole length of the measuring flume was about 100 feet, only 70 feet, however, was included between the upper and lower transit stations *H* and *I*; the principal part of the remainder, *A H*, being about 28.5 feet, was used as an entrance or mouth-piece to the part used for ascertaining the velocity, in order that the eddies and other irregularities incident to the small change in the form and dimensions of the canal, might be, to some extent, obliterated, before reaching the part of the flume used for ascertaining the velocity. This space was also serviceable by giving opportunity for the tubes to become free from considerable oscillations and to attain, sensibly, the velocity of the current.

204. Figures 5 and 6, plate XV., represent two of the loaded tubes, used for ascertaining the velocity of the water in the flume. Figure 5 represents the tube used in experiment 1, in which it extended as nearly to the bottom, *E E*, as appeared to be safe and not touch during its passage. Figure 6 represents the tube used in experiment 7, in which the space between the bottom of the tube and the bottom of the canal was about one foot. The tubes are cylinders, two inches in diameter, made of tinned plates, soldered together, with a piece of lead, *C B*, of the same diameter, soldered to the lower end, and of sufficient weight to sink the tube nearly to the required depth, which was such as to leave about four inches of its length above the surface of the water. The required depth of immersion was marked with red paint at *A*. In order to adjust it precisely, the tube was placed in a tank made for the purpose, and small pieces of lead were dropped into the top of the tube; these rested on the mass of lead, *C B*, and were added until the tube was sunk to the required depth; the orifice *D* was then closed with a cork. The tubes were allowed to remain floating in the tank for some time after they were adjusted, in order to ascertain whether they leaked or not; if they did they were taken out of the tank and filled with water, in order to ascertain the position of the leak, which was then stopped with solder and the operation of adjustment repeated. The centres of gravity of the tubes thus adjusted were at *G*, *G B* in figure 5 being about 1.90 feet, and in figure 6 about 1.78 feet. The centres of gravity being so low, the tubes had a strong tendency to maintain a vertical position. The velocity of the current being, however, generally more rapid near the surface than near the bottom, the upper parts of the tubes must of course, generally, have had an inclination down stream; no special observations were made of the amount of inclination; in the small part projecting above the surface of the water none was apparent, and as it was evidently very small, it has been assumed in all these experiments that the tubes constantly maintained a vertical position.

Tubes of thirty-three different lengths, from six feet to ten feet, six of each length, had been previously provided for the ordinary measurements of the water used by the manufacturing companies. From this stock three or four of each length required for these experiments were selected and specially adjusted for each experiment.

The tubes were put into the water by an assistant standing upon the bridge *K*, figure 1, plate XV.; it is done by a manœuvre requiring a little practice to perform it satisfactorily. The assistant stands with his face up stream, with the tube in hand, the loaded end directed downwards, but up stream, at an angle with the horizon, greater or less, depending on the velocity of the current. At a signal, he pushes the tube rapidly into the water at the angle at which he previously held it, until the painted mark near the upper end of the tube reaches the surface of the water, he retains his hold of the upper end of the tube until the current has brought it to a vertical position, when he abandons it to the current; he then turns round and observes, at its passage under the transit timber *H*, how far the tube is from the left side of the flume, the up-stream face of the timber being, for this purpose, graduated in feet, and distinctly marked and numbered. He also observes its passage under the middle timber *L*, and the lower transit timber *I* in a similar manner. As he makes the observations he calls the distances, which are recorded by another assistant. The mean obtained by adding together the observed distances at the upper and lower transit timbers, and twice the observed distance at the middle timber, and dividing the sum by four, is taken as the mean distance of the tube from the left side of the flume during its passage.

205. The up-stream sides of the timbers *H* and *I* are vertical, and 70 feet apart, and form the upper and lower transit stations. The times when the tube passes the transit stations are noted by an observer at *N*, who has a marine chronometer on a table before him. The passage of the tube at the transit stations is observed by assistants who are seated at *M* and *O*. The signals of the transits are communicated to the observer of the times by means of an electric telegraph erected for the purpose; connected with the telegraph are two break-circuit keys which are conveniently placed within reach of the assistants at *M* and *O*, and a telegraphic call is placed on the table at *N*, near the chronometer. When the tube has been abandoned to the current by the assistant on the bridge *K*, the assistant at *M* puts one of his eyes in the vertical plane forming the upper transit station, and at the instant when the tube passes this plane he depresses the key of the break-circuit, which causes a signal to be made at the call near the chronometer, the observer at *N* noting the time when the signal is made. The chronometer marks half seconds only, but the times are noted, by

estimation, to tenths of seconds. (Art. 142.) The difference of the observed times of the transits at the two stations gives the time during which the tube passes the 70 feet; dividing the distance by the time, the quotient is the velocity in feet per second. Another assistant observed the depth of the water in the flume; this was done during the passage of each tube; the height of the water was observed in the box *P*, figure 1, plate XV., placed between the lining planks and the wall of the canal; there was a communication between this box and the flume by means of a pipe, which opened into the flume near the timber *L*, and about four feet above the bottom of the flume. The box *P* contained a scale graduated to hundredths of feet, the zero point of which was at the mean elevation of the bottom of the part of the flume between the transit stations *H* and *I*. The bottom of the flume was very nearly horizontal, the elevations to obtain the mean were taken at 32 points, the extreme difference observed was 0.027 feet.

206. Printed forms, bound up in books, were prepared, in which the observations were entered. Table XX. compiled from three of these books, contains the observations made in experiment No. 1, together with some of the steps towards obtaining the quantity of water. The distances given in column 1 were arranged and entered previous to commencing the experiment, and were called in order, for the information of the assistant who put in the tubes, by the assistant who observed the times of the transits, as he became ready to make the observations. The intervals of time, given in column 4, are the differences of the times of the transits given in column 3. The velocities of the tubes given in column 5, are taken from table XXVIII., which has been computed, for the purpose of facilitating the ordinary measurements of the water used by the manufacturing companies at Lowell.

207. To find the mean velocity of the tubes, all the observed velocities are plotted on section paper, engraved for the purpose; reduced copies of several of these diagrams are given in plate XVII. The ordinates of the irregularly curved line are intended to represent the mean velocities of the tubes at the corresponding points in the width of the flume; this line is drawn on the original diagram by the eye, which it is plain cannot lead us much astray. The area of the figure *A B C D*, experiment 1, divided by the width of the flume, will evidently give the mean velocity of the tubes. The areas in experiment 1 for each foot in width, excepting the last, are given in column *A*, table XX.; the sum of these areas is 71.768, which being divided by 26.746, the width of the flume, gives 2.6833 feet per second for the mean velocity of the tubes. This last quantity, (assuming it to be the same as the mean velocity of the water,) multiplied by the area of the transverse section of the stream, which in this experiment

is 26.746 × 9.533 = 254.97 square feet, gives 681.16 cubic feet per second, as the quantity of water passing, according to the flume measurement.

208. It will be perceived, by reference to the diagrams in plate XVII., that the observed velocity at the same part of the section is constantly varying; this is not due, in any sensible degree, to errors of observation, but to actual changes in the velocity, due to the unstable condition of the current. In all these experiments, the area of the section, and the quantity of water flowing, were sensibly constant throughout an experiment; the mean velocity must, consequently, have been nearly constant, and the only explanation of the observed variations in the velocity is, that there was a constant interchange of place of currents of different velocities.

209. The water after leaving the measuring flume passed to the weir erected on the Tremont Wasteway, *F*, figures 1 and 2, plate XVI. This weir was in two divisions, each having about 40 feet in length of water-way; the Westerly division, and a part of the Easterly division, are represented on an enlarged scale by figures 3 and 4. Figure 5 is a sectional elevation of the weir and some of the apparatus connected therewith. *A* is a grating for the purpose of equalizing the flow towards the weir, and for obliterating the irregularities in the direction of the currents approaching the weir, which it is obvious, from an inspection of the form of the approaches, would have otherwise existed. The whole length of the grating was 88 feet; the vertical slats were 4 inches wide, in the direction of the current, and one inch thick, the spaces between the slats, for the passage of the water, were about 1.125 inches wide. To equalize the flow still further, horizontal slats 1.5 inches wide were placed on the up-stream side of the grating; they were placed principally at the Westerly part of the grating, on which the current from the measuring flume impinged most directly. The whole length of the grating being divided into five nearly equal parts, the Westerly part had eight horizontal slats, the next part had six slats, the next four, the next two, and the next, or most Easterly part, had none. The effect of this grating was to obliterate all sensible lateral currents; it did not, however, entirely equalize the flow, except in a small portion of the experiments. In experiment 1, in which the discharge over the weirs was 681.25 cubic feet per second, the mean depth on the Easterly division of the weir was 0.0387 feet *less* than on the Westerly division; in experiments 43 to 49, in which the mean discharge was 106.05 cubic feet per second, the mean depth on the Easterly division of the weir was 0.00026 feet *greater* than on the Westerly division. In computing the discharge the mean of the observed depths on the two divisions of the weir is taken, the small inequalities in the depths on the two divisions produce inappreciable effects on the results.

TABLE XX.

1	2	3						4	5	6			7	8
Intended distance of the tube from the left side of the flume, looking down stream.	Length of the immersed part of the tube.	Time of the transits by chronometer No. 320, by Hutton of London. October 7th, 1856. A. M.						Time during which the tube passed from the up-stream station to the down-stream station, a distance of 70 feet.	Mean velocity of the tube.	Observed distance of the tube from the left side of the flume, looking down stream.			Mean distance of the tube from the left side of the flume during its passage.	Depth of water in the flume.
		At the up-stream station.			At the down-stream station.					At the up-stream station.	At the middle station.	At the down-stream station.		
Feet.	Feet.	H.	Min.	Sec.	H.	Min.	Sec.	Seconds.	Feet per Sec.	Feet.	Feet.	Feet.	Feet.	Feet.
0	9.482	7	52	3.3	7	52	33.6	30.3	2.310	0.2	0.8	0.9	0.7	9.550
0	"	"	52	58.8	"	53	29.4	30.6	2.288	0.4	0.5	0.4	0.4	9.550
1	"	"	53	47.3	"	54	15.1	27.8	2.518	1.3	1.3	1.4	1.3	9.552
2	"	"	55	30.0	"	55	58.2	28.2	2.482	2.6	3.5	4.6	3.5	9.546
3	"	"	56	21.2	"	56	51.3	30.1	2.326	2.5	2.2	2.0	2.2	9.549
4	"	"	57	18.8	"	57	45.3	26.5	2.642	4.0	3.9	4.0	3.9	9.541
5	"	"	58	5.6	"	58	30.6	25.0	2.800	5.0	6.0	6.0	5.7	9.545
6	"	"	58	53.8	"	59	21.9	28.1	2.491	6.9	6.9	6.5	6.8	9.551
7	"	8	1	14.0	8	1	38.3	24.3	2.881	6.0	5.3	5.6	5.5	9.550
8	"	"	2	2.4	"	2	28.5	26.1	2.682	7.9	8.5	9.0	8.5	9.550
9	"	"	12	16.7	"	12	41.9	25.2	2.778	9.8	10.4	11.0	10.4	9.530
10	"	"	13	5.3	"	13	30.6	25.3	2.767	10.0	10.9	10.5	10.6	9.525
11	"	"	13	52.5	"	14	20.4	27.9	2.509	11.2	11.5	11.1	11.3	9.518
12	"	"	14	39.8	"	15	3.3	23.5	2.979	12.2	12.3	12.9	12.4	9.519
13	"	"	15	25.4	"	15	49.8	24.4	2.869	14.5	16.8	18.0	16.5	9.526
14	"	"	16	16.4	"	16	40.7	24.3	2.881	14.0	13.8	14.0	13.9	9.529
15	"	"	17	1.4	"	17	27.3	25.9	2.703	14.6	15.0	15.0	14.9	9.524
16	"	"	17	46.9	"	18	12.0	25.1	2.789	16.3	16.0	15.2	15.9	9.516
17	"	"	18	32.4	"	18	56.7	24.3	2.881	16.8	16.0	15.5	16.1	9.525
18	"	"	19	20.9	"	19	45.2	24.3	2.881	17.3	17.0	16.5	16.9	9.530
19	"	"	20	9.0	"	20	33.5	24.5	2.857	19.0	18.5	18.2	18.5	9.528
20	"	"	20	59.7	"	21	24.4	24.7	2.834	19.0	18.9	18.0	18.7	9.530
21	"	"	21	48.4	"	22	13.4	25.0	2.800	20.5	20.9	21.8	21.0	9.523
22	"	"	22	36.6	"	23	2.9	26.3	2.662	22.0	20.8	20.3	21.0	9.519
23	"	"	23	25.3	"	23	50.2	24.9	2.811	22.0	21.7	22.0	21.8	9.528
24	"	"	24	39.7	"	25	6.0	26.3	2.662	24.5	24.6	24.9	24.6	9.523
25	"	"	25	25.3	"	25	52.2	26.9	2.602	24.0	23.1	23.5	23.4	9.529
26	"	"	26	15.0	"	26	45.7	30.7	2.280	26.3	26.2	26.0	26.2	9.528
26.75	"	"	27	21.0	"	27	51.9	30.9	2.265	26.3	26.3	26.3	26.3	9.532
26.75	"	"	28	15.8	"	28	44.2	28.4	2.465	26.2	26.1	26.0	26.1	9.523
										Mean depth of water in the flume = 9.533				

A
Products of the velocity into the width, commencing at the left side of the flume, for each foot in width of the flume, excepting the last, which is for a width 1.[illegible] feet; obtained from a diagram of which the diagram marked Exp. 1 on plate XVII. is a reduced copy.
2.268
2.409
2.493
2.554
2.601
2.633
2.662
2.691
2.722
2.750
2.776
2.799
2.818
2.837
2.850
2.857
2.859
2.853
2.845
2.829
2.802
2.765
2.715
2.653
2.565
4.162
Sum = 71.768
This sum divided by the width of the flume gives the mean velocity of the tubes, viz.: $\frac{71.768}{26.7[illegible]}$ = [illegible] ft. per sec.

210. The up-stream face of the weir *F P*, figure 5, was a vertical plane, 6 feet in height and 88 feet long; the crest of the weir was of the form represented by figure 3, plate XVIII., and was horizontal for a width of 0.5 inches; the up-stream edge presented to the current was as sharp as could be conveniently maintained in wood; the down-stream side of the crest was chamfered off at an angle of 45° with the vertical. The two divisions of the weir were separated by a space *B* four feet wide, and at each of the ends *C* there was a space of two feet; the up-stream faces of these spaces were in the same vertical plane as the up-stream face of the weir, and were deemed to be ample to insure complete contraction at the ends of the sheets of water. The dam or wasteway on which the weir was erected was of a form adapted to the convenient discharge of water over its crest, and for the regulation of the flow over the same; this was, however, not the form to which the ordinary formula for computing the flow over a weir applies, and it was therefore necessary to make such changes in the form of the crest as would permit of such application. It was not deemed admissible to take down the top of the existing dam, and to reconstruct it of suitable form; all that could be done was to make additions which could be removed when the experiments were completed.

211. In order to preserve a sufficient depth of flow over the weir, the crest could not be raised more than one foot above the wasteway. The standards *D*, figures 3 and 5, which formed part of the wasteway and were required to support the flash-boards used in regulating the flow over the wasteway, it was necessary to leave undisturbed; in order that they should not obstruct the flow over the weir, the crest of the latter was placed at a certain distance up stream; this was accomplished by fastening the large timber *E*, figure 5, to the up-stream face of the wasteway, the plank *F*, figures 3 and 5, forming the crest of the weir, was fastened to this timber. As thus arranged, the sheet of water passing over the weir fell vertically, and with very slight obstructions, to the cap of the wasteway, and passed horizontally, a distance of about 1.4 feet from the up-stream face of the weir plank *F*, before it struck the standards *D*.

212. The weir was made in two divisions for the purpose of facilitating the passage of air under the sheet, former observations having shown that air thus situated is rapidly carried away by the water, and unless sufficient means are provided for renewing it, its place will be speedily taken by water, which will materially affect the flow over the weir and prevent the correct application of the formula for computing the discharge. This precaution proved, however, to be insufficient to prevent the space under the sheet from becoming filled with water; it was evident that a portion of the water striking the top of the wasteway flowed back towards the weir and filled the space which ought to be kept free; to prevent

this, the board *G*, figures 3 and 5, was put on; its width was sufficient to reach from the top of the timber *E* very nearly to the underside of the sheet; this remedied the difficulty in a great degree, but, unless the width of the board was properly adjusted to the sheet, it failed to operate satisfactorily; if too low, the water flowed back over the top, if too high, the sheet of water struck the board, in either case very soon filling up the space between the board and the weir plank; at first the only escape of the water from the trough formed by the board *G* and the weir plank was at the ends, and the trough being forty feet long, the escape from the central parts was very slow. This difficulty, however, was remedied by attaching leaden pipes, two inches in diameter, to the board *G;* these pipes were about sixteen feet long and were laid on the inclined surface of the apron of the wasteway, the lower ends of the pipes being about five feet below the upper ends. The Easterly division was first fitted up with twenty-six of such pipes; upon trial this proved to be a much greater number than was necessary to afford escape for the water flowing back over the top of the board *G*, and the Westerly division, which is that shown on figure 3, was provided with only half the number, which proved to be amply sufficient.

It was necessary to readjust the height of the board *G*, whenever a material change was made in the depth of water on the weir. It is represented in figure 5, as it was in experiments 1 to 7, in which the depth on the weir was near the maximum. The top of the board *G*, in these seven experiments was about 0.105 feet below the top of the weir.

213. The depth on the weir was observed at each division separately, by means of hook gauges, similar to that represented by figures 2, 3, and 4, plate XIII. A gauge acting on the same principles is described in article 45. The gauge for the Westerly division was placed in the box *H*, figures 3, 4, and 5, plate XVI.; this box was carefully made so that no water passed into or out of it, except through the pipes in the bottom, and it was strongly fastened to the post *I*, which was firmly set in the earth at the bottom and supported by the braces *K* at the top. When observations were being made with the hook gauge for the depth on the weir, the three pipes *L L L* formed the only communication between the water in the box and the water in the basin between the grating and the weir; the surface of the water in the box was assumed to be at the height giving the mean depth on this division of the weir; subject, however, to a small correction to be described hereafter.

214. The small box *O* was firmly secured to the planking forming the interval between the two divisions of the weir; it had no communication with the water outside of it, except by means of the pipes *N* and *Q*, which furnished the means

of connecting it with either of the hook gauge boxes when desired. The box *O* contained a stationary hook, the point of which was formed by a portion of a sphere of about half an inch in diameter; the coincidence of the level of the surface of the water with the highest part of the spherical surface could be as definitely ascertained, as if the hook had terminated in a sharp point, as in the hook gauges, whilst the spherical surface permitted a levelling-rod to be placed upon it for the purpose described presently. For convenience in using the hook gauges, their zero points were placed several inches above the top of the weir. In order to ascertain the precise elevation of the zero point of one of these gauges relatively to the mean height of the top of the corresponding division of the weir, the water was adjusted to a depth of about one foot on the weir, the three pipes *L* were closed, and the pipe *N* opened. The pipe *N* then furnished a free communication between the boxes *H* and *O*, neither of which at this time had any other orifice for the passage of water in or out. Water was then put into or taken out of these boxes until its surface coincided with the highest part of the spherical surface which formed the point of the stationary hook in the box *O*; when this was done and the water in the boxes free from oscillations, the height of the surface of the water in the box *H* was observed by means of the hook gauge, which evidently gave the height of the point of the stationary hook in the box *O*, by the scale of the hook gauge in the box *M*. The height of the point of the stationary hook in the box *O* above the mean height of the top of the weir was obtained by levelling with a Troughton and Simms dumpy level; this was done with great care and with all the precautions necessary for insuring accuracy; it was done three times during the course of the experiments, with the results given in the following table.

TABLE XXI.

Date.	Height of the point of the Hook in the Box O above the mean height of the top of the Westerly division of the Weir.	Height of the point of the Hook in the Box O above the mean height of the top of the Easterly division of the Weir
1856.	Feet.	Feet.
October 7.	1.0087	1.0112
" 17.	1.0090	1.0111
November 19.	1.0089	1.0127

215. From the observations in the preceding table it is evident that the relative elevations of the weir and the point of the stationary hook were not subject to sensible change. Comparisons between the hook gauges and the stationary hook were made every day, with a depth of about one foot on the weir, and the cor-

rection determined and used in all the experiments of that day. The relative heights of the hook gauges and stationary hook were subject to greater changes than were observed between the stationary hook and the top of the weir. The experiments extended from October 7 to November 13; the difference of height of the stationary hook and the zero of the Westerly hook gauge was greatest on October 8, when it was 0.4402 feet, and least on October 23, when it was 0.4352 feet, the change, which was not abrupt, being 0.0050 feet. The corresponding change at the Easterly hook gauge was 0.0066 feet, the sign and dates being the same as at the Westerly hook gauge. These differences are not very great, and as the corrections were determined daily, no appreciable errors can result therefrom.

216. The experiments of 1852, described in a former part of this work, from which the formula for computing the quantity of water flowing over the weir in these experiments is deduced, were made upon a weir of great simplicity of form, in which the sheet of water passing over the weir had an unobstructed fall of not less than three feet; see figure 1, plate XIII. Other experiments indicated that the sheet of water may meet with great obstructions soon after passing the weir, without its flow over the weir being sensibly affected thereby (see *ante*, page 134), and it was thought, that in these experiments the obstructions to the flow of the water after passing the weir, would affect the discharge over the weir to so small an extent as to be inappreciable. It was highly important, however, to avoid all question on this point; and to determine the matter, a special series of experiments was undertaken.

For this purpose two weirs were erected in the upper chamber of the Lower Locks in Lowell, *K*, figure 1, plate XI. The upper weir was constructed of a form to which the formula for computing the discharge could be applied without objection. The lower weir in a portion of the experiments was of the same form as the upper weir, and in the other portion the form was the same as the weir at the Tremont Wasteway. The experiments consisted in causing the same volume of water to flow over both weirs, and observing the depth assumed by the water on each weir, when the flow had become permanent, the differences in the depths, if any, being due to differences in the forms and conditions of the two weirs.

The Lock chamber is twelve feet wide, and the weirs were each eight feet long, leaving a space of two feet at each end to insure complete contraction. The up-stream faces of the weirs were vertical planes, and the crests and ends were of the same form as the weir at the Tremont Wasteway. The bottoms of the channels on the up-stream sides of both weirs were six feet below the tops of the weirs. The water entered the Lock chamber through the head gates, and under a head of several feet, which caused a great commotion in the water at the upper end of the chamber. The upper weir was placed about sixty feet from the upper end of the chamber, and to obliterate the disturbance in the water before it reached the weir, three gratings, at right angles to the sides, were placed across the chamber at intervals of about twelve feet; each grating contained about one half of the aperture per square foot, for the passage of water, as the grating used at the Tremont Wasteway. The lower grating was about fourteen feet from the weir. The surface of the water between the two upper gratings was nearly all covered by a float of planks, for the purpose of obliterating the oscillations of the surface. The second or lower weir was about thirty-five feet from the upper weir, and similar arrangements were made for obliterating disturbances in the water, as were provided for the upper weir, except that there were only two gratings, the disturbances caused by the fall of the water from the upper weir into the basin below it being much less than were caused by the entrance of the water at the upper end of the chamber. The lower grating was about fourteen feet from the lower weir. The depths of the water on the weirs were observed by means of hook gauges similar to that represented on plate XIII. The difference of the leakages into and out of the part of the chamber included between the two weirs was ascertained, and a correction applied for the same; and also for the rise or fall, if any, of the surface of the water in the same space during the time occupied by an experiment.

In arranging the apparatus, it was designed to make the immediate approach of the water to the two weirs precisely alike. It was not certain, however, that the precautions taken to insure uniformity would produce the desired result. To avoid doubts on this point, the lower weir in part of the experiments, as stated above, was made of the same form as the upper weir, in which case any difference in the depths on the two weirs, the quantity of water flowing being the same at both, and there being no obstructions below, must be due to differences in the immediate approach of the water to the weirs. A series of experiments was made under these circumstances, with different quantities of water flowing, from which it was ascertained, that when the depth on the upper weir was about 0.5 feet, the depth on the lower weir was 0.0008 feet greater; when the depth was about a foot on the upper weir, it was the same on the lower weir; when about 1.5 feet on the upper weir, it was 0.0040 feet less on the lower weir; and when about 2 feet in depth on the upper weir it was about 0.0094 feet less on the lower weir. These differences were probably due to small differences in the relative velocities of the water immediately approaching the weirs, at different depths, and might, doubtless, have been partially remedied by suitable modifications of the gratings. It would have required much time, however, and was not essential to our arriving at correct results, the experiments with the two weirs alike having been sufficiently numerous and varied to enable a table of corrections to be made.

217. Another series of experiments was made with the lower weir like that erected at the Tremont Wasteway, the apron, trough, pipes, standards, etc., being reproduced, as nearly as the length of the weir would permit. The height of the board, forming the down-stream side of the trough, was of course varied in the different experiments, to conform to the corresponding changes at the Tremont weir. The upper weir remained unchanged throughout all the experiments. Water being admitted at the upper end of the chamber, and the flow become permanent, or as nearly so as practicable, observations were made of the depth which the water assumed at the two weirs. It would occupy much space to describe all the experiments made; it will perhaps be sufficient to state some of the results arrived at. After correcting the depth on the lower weir for the differences described in the preceding section, which did not depend on the forms of the weirs, the following differences were found. When the depth on the upper weir was about 0.8 feet, the depth on the lower weir was 0.0007 feet less; when the depth on the upper weir was about 1.5 feet, the depth on the lower weir was the same; when the depth on the upper weir was about 2 feet, the depth on the lower weir was 0.0085 feet greater. This last difference corresponds to a diminution of flow over the lower weir, with the same depth on the weir, of $\frac{1}{158}$.

218. The effect of what appear to be obstructions to the flow over a weir is, generally, to increase the depth on the weir over what it would be if the flow was free; sometimes, however, it has the contrary effect. (See article 137.) The experiments at the Lower Locks described in the preceding section furnished the data for a table of corrections of the depths of water on the Tremont weir, due to the obstructions to the flow of the sheet after passing the crest of the weir. In experiment 1, table XXII., in which this correction has nearly its greatest value, it is — 0.0058 feet.

219. Another small correction was also applied. In the experiments of 1852 (art. 173), it was found that there was no sensible difference in the observed depth upon the weir, whether the external orifice of the pipe, forming the communication between the water approaching the weir and the hook gauge box, was close to the plane of the weir or six feet up stream from that plane, the external orifice of the pipe being at a considerable depth below the top of the weir. In arranging the apparatus at the weir at the Tremont Wasteway, it was thought that there would be less liability to errors in the observed depths, from currents acting on the external orifices of the pipes, if they were very near the plane of the weir, and at the bottom of the canal, and they were accordingly so arranged. In the experiments of 1852, however, on which the formula for computing the flow over the weir is founded, the orifice in the hook gauge box was six feet

from the weir, and in order to ascertain whether any difference could be detected in the observed depths on the weir at the Tremont Wasteway, with the external orifice of the pipe at different distances from the weir, some special experiments were made.

For this purpose an apparatus of pipes similar to that represented in figures 8 and 9, plate XIV., was placed at the bottom of the canal, on the up-stream side of the weir at the Tremont Wasteway. The orifices of the pipes were protected from the action of lateral currents, if any existed, by a second board, placed parallel to the board in which the lower ends of the pipes were inserted, and three inches distant; these boards were placed at right angles to the weir, and the space between them was open at the top and the up-stream end, so that the current flowing towards the weir, flowed through the trough formed by the two boards, by the open ends of the pipes, which, to avoid eddies, did not project beyond the plane of the board. With this apparatus, observations were made of the differences in the depths on the weir, when the different pipes were in communication with the hook gauge box; substantially the same precautions being taken to secure precision in the results as are described in article 170.

Taking the observations made with the pipe opening at 0.52 feet from the weir, as represented at *R*, figures 3, 4, and 5, plate XVI., as the standard; when the depth on the weir was about 0.76 feet, the differences in the depths observed by means of the other pipes were as follows:—

By the pipe opening at 2 feet from the plane of the weir, difference = — 0.0003 feet.
" " " 4 " " " " " = — 0.0003 "
" " " 6 " " " " " = — 0.0004 "
" " " 8 " " " " " = — 0.0001 "
" " " 10 " " " " " = — 0.0003 "
" " " 12 " " " " " = — 0.0012 "

When the depth on the weir was about 1.44 feet, the differences observed were as follows:—

By the pipe opening at 2 feet from the plane of the weir, difference = + 0.0020 feet.
" " " 4 " " " " " = — 0.0009 "
" " " 6 " " " " " = — 0.0013 "
" " " 8 " " " " " = — 0.0054 "
" " " 10 " " " " " = — 0.0089 "
" " " 12 " " " " " = — 0.0124 "

Up to six feet from the weir, these differences are very small; it was thought best, however, to take account of them.

By a discussion of the whole of the experiments a table was formed, for correcting the observed depths on the weir, to what they would have been if observed with the pipe opening at 6 feet from the weir.

When the depth on the weir is	0.5	feet,	this correction is	— 0.0002	feet.
" " " "	"	0.8	"	"	— 0.0004	"
" " " "	"	1.0	"	"	— 0.0006	"
" " " "	"	1.5	"	"	— 0.0014	"
" " " "	"	2.0	"	"	— 0.0023	"

220. By table XVIII., containing the results of similar experiments at the Lower Locks, made about four years previously, it will be seen, that the differences between the depths on the weir, observed by means of a pipe opening at six feet from the plane of the weir, and by a pipe opening at one inch from the plane of the weir (changing the signs to conform to the experiments at the Tremont weir), were as follows:—

When the depth on the weir was about	0.80	feet,	difference =	— 0.00060 feet.
" " " "	1.00	"	" =	— 0.00033 "

The small differences in these results from those obtained at the Tremont Wasteway weir may be explained by the different forms of the approaches to the weirs, and the different arrangement of the apparatus.

221. The formula for computing the quantity of water flowing over weirs, deduced from the experiments made at the Lower Locks in 1852, viz.:

$$Q = 3.33\,(L - 0.1\,n\,H)\,H^{\frac{3}{2}}, \qquad (A.)$$

is adapted to weirs of widely differing proportions, including all the forms on which experiments are given in table XIII. By reference to column 16 in that table, it will be seen, however, that the experiments on each particular description of weir generally give a coefficient differing slightly from the mean value deduced from the whole of the experiments. In case any of those particular forms should be reproduced, it is evident, that the quantity of water flowing over the same could be more accurately computed, by using the corresponding coefficient given in column 16, than by using that given in formula ($A.$), which is a mean, deduced from the whole of the experiments. In determining the formula by which to compute the flow over the weir at the Tremont Wasteway, it was apparent that results more exact could be attained by deducing a new formula from a selection of the experiments given in table XIII., in which the circumstances were most nearly like those at the Tremont Wasteway weir. For this purpose 53 experiments were selected, and the formula deduced from them is

$$Q = 3.318\,(L - 0.08\,n\,H)\,H^{\frac{3}{2}}. \qquad (B.)$$

As applied to the weir at the Tremont Wasteway,

When the depth is 1 foot, the discharge by formula $(A.)$	=	265.09	cubic feet per sec.
And by formula $(B.)$	=	264.40	" " " "
Difference $\frac{1}{384}$	=	0.69	" " " "
When the depth is 2 feet, the discharge by formula $(A.)$	=	746.02	" " " "
And by formula $(B.)$	=	744.84	" " " "
Difference $\frac{1}{632}$	=	1.18	" " " "

When the depth is 3.5527 feet, both formulas give the same discharge.

222. In making these experiments, there were several objects in view, which may be classed under two heads, viz.:—

1st. To determine a formula for correcting the quantity passing a measuring flume, as deduced from the mean velocity of the tubes; there being no unusual disturbing causes.

2d. To ascertain the degree of uniformity in measurements made under like circumstances; and to determine the magnitude of the errors to which we are liable, when measurements are made under exceptionable circumstances, such as high winds and great irregularities in the motion of the water.

The experiments adapted to the *first* object were necessarily made under the normal conditions of freedom from high wind, and from great irregularity in the currents. Table XXII. contains 105 experiments selected as being suitable for this purpose, and table XXV. contains 35 experiments made for the purposes included in the second class.

EXPERIMENTS MADE AT THE TREMONT WEIR AND MEASURING FLUME,

1	2			3		Weir Measurement.								Flume			
No. of the Exp.	Date. 1856.			Temperature, in degrees of Fahrenheit's thermometer, of the atmosphere in shade.	of the water.	4 Total length of the weirs. L	5 Observed depth of water on the Westerly weir.	6 Observed depth of water on the Easterly weir.	7 Mean observed depth of water on the weirs.	8 Corrected depth of water on the weirs. H	9 Quantity of water passing over the weirs, computed by the formula $Q = 3.33(L-0.2nH)H^{\frac{3}{2}}$	10 Correction for the leakage into the flume.	11 Corrected quantity passing the flume, deduced from the weir measurement. Q'	12 Mean width of the flume.	13 Mean depth of water in the flume.	14 Length of the immersed part of the tube.	15 Difference between the depth of water in the flume and the length of the immersed part of the tube, divided by the depth of water in the flume. D
						Feet.	Feet.	Feet.	Feet.	Feet.	Cubic ft. per sec.	Cubic ft. per sec.	Cubic ft. per sec.	Feet.	Feet.	Feet.	
1	Oct.	7	A.M.	52.5	57.0	80.007	1.8972	1.8585	1.8778	1.8839	681.25	0	681.25	26.746	9.533	9.482	0.005
2	"	"	"	53.5	57.0	"	1.8750	1.8395	1.8572	1.8634	670.22	0	670.22	"	9.515	9.430	0.009
3	"	"	"	60.5	57.0	"	1.8556	1.8214	1.8385	1.8448	660.26	0	660.26	"	9.496	9.380	0.012
4	"	"	"	64.5	57.5	"	1.8720	1.8362	1.8541	1.8601	668.61	0	668.61	"	9.510	9.330	0.019
5	"	"	P.M.	65.0	58.0	"	1.8906	1.8548	1.8727	1.8787	678.45	0	678.45	"	9.531	9.230	0.032
6	"	"	"	65.0	58.0	"	1.8896	1.8539	1.8717	1.8777	677.91	0	677.91	"	9.532	9.130	0.042
7	"	"	"	64.0	58.0	"	1.8872	1.8507	1.8689	1.8750	676.45	0	676.45	"	9.530	8.530	0.105
8	"	8	A.M.	51.5	57.0	80.008	1.7846	1.7527	1.7686	1.7752	623.43	0	623.43	"	9.422	9.360	0.007
9	"	"	"	60.5	57.0	"	1.7882	1.7570	1.7726	1.7790	625.42	0	625.42	"	9.426	9.320	0.011
10	"	"	"	63.5	57.0	"	1.7825	1.7495	1.7660	1.7726	622.07	0	622.07	"	9.421	9.2[illegible]0	0.015
11	"	"	P.M.	66.6	57.0	"	1.7810	1.7490	1.7650	1.7716	621.54	0	621.54	"	9.421	9.220	0.021
12	"	"	"	66.1	57.0	"	1.7720	1.7416	1.7568	1.7632	617.20	0	617.20	"	9.412	9.120	0.031
13	"	"	"	62.9	57.2	"	1.7712	1.7394	1.7553	1.7618	616.41	0	616.41	"	9.410	9.020	0.041
14	"	"	"	60.0	58.0	"	1.7626	1.7333	1.7479	1.7546	612.66	0	612.66	"	9.402	8.410	0.106
15	"	9	A.M.	53.0	58.0	80.009	1.5061	1.4864	1.4962	1.5027	486.08	0	486.08	"	9.141	9.080	0.007
16	"	"	"	59.5	58.0	"	1.5045	1.4854	1.4949	1.5015	485.50	0	485.50	"	9.140	9.030	0.012
17	"	"	"	70.5	58.0	"	1.5046	1.4849	1.4947	1.5013	485.40	0	485.40	"	9.138	8.980	0.017
18	"	"	P.M.	80.5	58.0	"	1.5038	1.4842	1.4940	1.5006	485.06	0	485.06	"	9.138	8.930	0.023
19	"	"	"	79.5	58.5	"	1.5033	1.4833	1.4933	1.4999	484.72	0	484.72	"	9.137	8.830	0.034
20	"	"	"	79.5	59.0	"	1.4925	1.4737	1.4831	1.4895	479.71	0	479.71	"	9.126	8.730	0.043
21	"	"	"	74.0	59.0	"	1.4839	1.4639	1.4739	1.4804	475.34	0	475.34	"	9.118	8.120	0.109
22	"	11	A.M.	72.5	59.0	80.010	1.2091	1.1995	1.2043	1.2086	351.03	0	351.03	26.745	8.838	7.830	0.114
23	"	"	"	78.5	59.0	"	1.2131	1.2031	1.2081	1.2124	352.68	0	352.68	"	8.842	8.430	0.047
24	"	"	P.M.	78.0	60.0	"	1.1968	1.1899	1.1933	1.1975	346.22	0	346.22	"	8.830	8.530	0.034
25	"	"	"	77.5	60.0	"	1.1942	1.1857	1.1899	1.1941	344.75	0	344.75	"	8.827	8.630	0.022
26	"	13	A.M.	57.5	60.0	80.011	1.1985	1.1886	1.1935	1.1977	346.31	0	346.31	"	8.827	8.780	0.005
27	"	"	"	59.0	60.0	"	1.1854	1.1759	1.1806	1.1847	340.76	0	340.76	"	8.815	8.730	0.010
28	"	"	"	64.5	60.0	"	1.1820	1.1726	1.1773	1.1813	339.24	0	339.24	"	8.810	8.680	0.015
29	"	"	P.M.	69.0	60.0	"	1.3715	1.3555	1.3635	1.3691	422.96	0	422.96	"	8.997	7.980	0.113
30	"	"	"	68.0	60.0	"	1.3533	1.3382	1.3457	1.3512	414.72	0	414.72	"	8.981	8.600	0.042
31	"	"	"	67.0	60.0	"	1.3580	1.3427	1.3503	1.3559	416.87	0	416.87	"	8.985	8.700	0.032
32	"	"	"	65.5	60.0	"	1.3524	1.3387	1.3455	1.3510	414.63	0	414.63	"	8.978	8.800	0.020
33	"	14	A.M.	40.0	59.0	80.012	1.3752	1.3618	1.3685	1.3742	425.32	0	425.32	"	9.006	8.850	0.017
34	"	"	"	40.5	59.0	"	1.3772	1.3635	1.3703	1.3760	426.15	0	426.15	"	9.009	8.900	0.012
35	"	"	"	42.0	59.0	"	1.3681	1.3552	1.3616	1.3673	422.13	0	422.13	"	8.997	8.960	0.004
36	"	"	P.M.	45.0	58.0	"	0.9792	0.9756	0.9774	0.9801	256.58	0	256.58	"	8.609	8.230	0.044
37	"	"	"	43.5	58.0	"	0.9840	0.9818	0.9829	0.9856	258.74	0	258.74	"	8.615	8.330	0.033
38	"	"	"	41.5	58.0	"	0.9746	0.9732	0.9739	0.9766	255.21	0	255.21	"	8.604	8.430	0.020
39	"	15	A.M.	34.5	56.0	"	1.0016	0.9971	0.9993	1.0022	265.29	0	265.29	"	8.631	8.480	0.017
40	"	"	"	39.0	56.0	"	0.9916	0.9872	0.9894	0.9922	261.34	0	261.34	"	8.620	8.530	0.010
41	"	"	"	47.5	56.0	"	0.9841	0.9805	0.9823	0.9850	258.51	0	258.51	"	8.611	8.570	0.005
42	"	"	P.M.	52.5	56.0	"	0.9942	0.9900	0.9921	0.9950	262.44	0	262.44	"	8.626	7.620	0.117
43	"	"	"	50.0	56.0	"	0.5504	0.5502	0.5503	0.5513	108.43	0	108.43	"	8.172	7.120	0.129
44	"	16	A.M.	38.5	54.0	80.014	0.5445	0.5444	0.5444	0.5454	106.70	0	106.70	"	8.167	7.720	0.055
45	"	"	"	47.0	54.0	"	0.5403	0.5406	0.5404	0.5414	105.53	0	105.53	"	8.164	7.920	0.030
46	"	"	"	55.5	54.0	"	0.5394	0.5395	0.5394	0.5404	105.24	0	105.24	"	8.163	8.070	0.011
47	"	"	P.M.	60.5	54.0	"	0.5410	0.5412	0.5411	0.5421	105.74	0	105.74	"	8.165	8.122	0.005
48	"	"	"	58.5	54.0	"	0.5342	0.5350	0.5346	0.5356	103.84	0	103.84	"	8.159	8.020	0.017
49	"	21	"	71.0	53.5	"	0.5447	0.5454	0.5450	0.5460	106.88	0	106.88	"	8.171	8.070	0.012
50	"	27	A.M.	35.0	47.0	"	1.9105	1.8662	1.8883	1.8943	686.93	0	686.93	"	9.540	9.380	0.017
51	"	"	"	36.0	47.0	"	1.9152	1.8714	1.8933	1.8992	689.58	0	689.58	"	9.548	9.330	0.023
52	"	"	"	39.0	47.0	"	1.9106	1.8659	1.8882	1.8941	686.82	0	686.82	"	9.543	9.130	0.043
53	"	"	"	39.0	47.0	"	1.8879	1.8453	1.8666	1.8726	675.22	0	675.22	"	9.521	8.532	0.104
54	"	"	"			"	1.8909	1.8484	1.8696	1.8757	676.89	0	676.89	"	9.525	9.476	0.005
55	"	"	"	57.0	47.0	"	1.8742	1.8320	1.8531	1.8593	668.07	0	668.07	"	9.508	9.230	0.029
56	"	"	P.M.	59.5	48.0	"	1.8758	1.8334	1.8546	1.8609	668.94	0	668.94	"	9.510	9.431	0.008

FROM WHICH THE FORMULA OF CORRECTION $C = 0.116\,(\sqrt{D} - 0.1)$ IS DETERMINED.

Measurement. No. of the Exp.	16 Mean velocity of the tubes throughout the flume, by the diagram. Feet per sec.	17 Quantity of water passing the flume, deduced from the mean velocity of the tubes. Q'' Cubic ft. per sec.	18 Difference between the quantity of water passing the flume deduced from the mean velocity of the tubes, and the quantity deduced from the weir measurement. $Q''-Q'$ Cubic ft. per sec.	19 Proportional difference, or the difference in the preceding column, divided by the quantity deduced from the flume measurement. $\frac{Q''-Q'}{Q''}$	20 Quantity of water passing the flume deduced from the mean velocity of the tubes corrected by the formula. $Q''' = Q''(1-0.116(\sqrt{D}-0.1))$	21 Proportional difference of the corrected quantity as measured in the flume given in the preceding column, and the weir measurement. $\frac{Q'''-Q'}{Q'}$	22 Remarks on the Force and Direction of the Wind at the Flume, during the Experiments. Force.	22 Direction.	23 General Remarks.
1	2.683	684.16	+ 2.31	+0.0013	685.42	+0.0077	Moderate.	Down stream.	Reduced copies of the diagrams, constructed for the purpose of obtaining the mean velocities of the tubes, in experiments 1 and 7, are given in plate XVII.
2	2.616	665.86	− 4.36	−0.0065	666.26	−0.0059	„	„ „	
3	2.636	669.55	+ 9.29	+0.0139	668.81	+0.0129	„	Irregular.	
4	2.667	678.27	+ 9.66	+0.0142	675.29	+0.0100	„	„	
5	2.649	675.23	− 3.22	−0.0048	669.05	−0.0133	Very gentle.	„	
6	2.713	691.72	+13.81	+0.0200	683.30	+0.0080	Gentle.	Down stream.	
7	2.726	694.86	+18.41	+0.0265	676.80	+0.0005	Hardly perceptible.	„ „	
8	2.451	617.69	− 5.74	−0.0093	618.86	−0.0073	„ „	„ „	
9	2.461	620.49	− 4.93	−0.0079	620.14	−0.0084	„ „	„ „	
10	2.473	623.01	+ 0.94	+0.0015	621.38	−0.0011	„ „	„ „	
11	2.485	626.18	+ 4.64	+0.0074	622.92	+0.0022	Gentle.	„ „	
12	2.487	626.14	+ 8.91	+0.0143	620.62	+0.0035	Very gentle.	„ „	
13	2.484	625.10	+ 8.69	+0.0139	617.67	+0.0020	Hardly perceptible.	„ „	
14	2.525	634.95	+22.29	+0.0351	618.33	+0.0093	Calm.		
15	1.984	485.12	− 0.96	−0.0020	486.04	−0.0001	Hardly perceptible.	Down stream.	
16	1.943	474.94	−10.56	−0.0222	474.41	−0.0228	Calm.		
17	1.967	480.68	− 4.72	−0.0098	478.99	−0.0132	„		
18	1.972	481.97	− 3.09	−0.0064	479.08	−0.0123	Hardly perceptible.	Down stream.	
19	2.003	489.56	+ 4.84	+0.0099	481.77	+0.0001	Calm.		
20	1.985	484.54	+ 4.83	+0.0100	478.51	−0.0025	„		
21	2.025	493.77	+18.43	+0.0373	480.59	+0.0110	„		
22	1.536	363.17	+12.14	+0.0331	353.16	+0.0061	Very gentle.	Down stream.	
23	1.518	353.06	+ 6.28	+0.0178	354.20	+0.0043	„ „	„ „	
24	1.483	350.13	+ 3.91	+0.0112	346.70	+0.0014	„ „	„ „	
25	1.462	345.23	+ 0.50	+0.0014	343.31	−0.0042	Moderate.	„ „	
26	1.453	343.03	− 3.28	−0.0096	344.20	−0.0061	Calm.		
27	1.436	338.56	− 2.14	−0.0063	338.56	−0.0063	Very gentle.	Down stream.	
28	1.446	340.68	+ 1.44	+0.0042	339.79	+0.0016	Moderate.	„ „	
29	1.786	429.81	+ 6.85	+0.0159	418.04	−0.0116	Moderate, sometimes calm.	„ „	
30	1.730	415.58	+ 0.86	+0.0021	410.52	−0.0101	Hardly perceptible.	„ „	
31	1.737	417.33	+ 0.46	+0.0011	413.51	−0.0081	Moderate.	„ „	
32	1.723	413.69	− 0.94	−0.0023	411.70	−0.0071	Hardly perceptible.	„ „	
33	1.779	428.00	+ 3.28	+0.0077	427.09	+0.0042	Brisk but variable.	„ „	
34	1.772	426.99	+ 0.84	+0.0020	426.52	+0.0009	Strong but variable.	Generally across.	
35	1.762	424.02	+ 1.96	+0.0046	425.90	+0.0089	„ „ „	„ „	
36	1.138	261.92	+ 5.34	+0.0204	258.59	+0.0078	„ „ „	„ „	
37	1.130	260.38	+ 1.64	+0.0063	257.91	−0.0032	„ „ „	„ „	
38	1.110	255.47	+ 0.26	+0.0010	254.24	−0.0038	„ „ „	„ „	
39	1.155	266.68	+ 1.39	+0.0052	265.74	+0.0017	Very moderate.	Irregular.	
40	1.118	257.81	− 3.55	−0.0137	257.81	−0.0135	„ „	„	
41	1.135	261.59	+ 2.99	+0.0114	262.39	+0.0150	Moderate.	Irreg., but gen. up stream.	
42	1.167	269.18	+ 6.74	+0.0250	261.62	−0.0021	Very moderate.	Irregular.	
43	0.519	113.39	+ 4.96	+0.0437	109.98	+0.0143	Hardly perceptible.		
44	0.498	108.76	+ 2.06	+0.0189	107.06	+0.0034	Calm.		
45	0.497	108.52	+ 2.99	+0.0276	107.60	+0.0196	Very moderate.	Generally across.	
46	0.500	109.18	+ 3.94	+0.0361	109.12	+0.0360	„ „	Irregular.	
47	0.495	108.09	+ 2.35	+0.0217	108.46	+0.0251	Moderate, sometimes calm.	Down stream.	
48	0.486	106.01	+ 2.17	+0.0205	105.64	+0.0173	Hardly perceptible.	„ „	
49	0.497	108.52	+ 1.64	+0.0151	108.40	+0.0142	Calm.		
50	2.701	689.07	+ 2.14	+0.0031	686.64	−0.0004	„		
51	2.707	691.16	+ 1.58	+0.0023	687.02	−0.0037	Hardly perceptible.		
52	2.774	708.05	+21.23	+0.0300	699.23	+0.0181	Calm.		
53	2.737	696.89	+21.67	+0.0311	678.90	+0.0055	„		
54	2.654	676.22	− 0.67	−0.0010	678.52	+0.0024	„		
55	2.658	676.02	+ 7.95	+0.0118	670.51	+0.0087	„		
56	2.596	660.36	− 8.58	−0.0130	661.17	−0.0116	„		

EXPERIMENTS MADE AT THE TREMONT WEIR AND MEASURING FLUME,

1	2	3		4	5	6	7	8	9	10	11	12	13	14	15
				Weir Measurement.								Flume			
No. of the Exp.	Date. 1856.	Temperature, in degrees of Fahrenheit's thermometer, of the atmosphere in shade.	of the water.	Total length of the weirs. L	Observed depth of water on the Westerly weir.	Observed depth of water on the Easterly weir.	Mean observed depth of water on the weirs.	Corrected depth of water on the weirs. H	Quantity of water passing over the weirs, computed by the formula $Q = 3.33(L - 0.2nH)H^{\frac{3}{2}}$	Correction for the leakage into the flume.	Corrected quantity passing the flume, deduced from the weir measurement. Q'	Mean width of the flume.	Mean depth of water in the flume.	Length of the immersed part of the tube.	Difference between the depth of water in the flume and the length of the immersed part of the tube, divided by the depth of water in the flume. D
				Feet.	Feet.	Feet.	Feet.	Feet.	Cubic ft. per sec.	Cubic ft. per sec.	Cubic ft. per sec.	Feet.	Feet.	Feet.	
57	Oct. 27 P.M.	58.0	48.0	30.014	1.5186	1.4955	1.5070	1.5137	491.42	0.00	491.42	26.745	9.151	9.097	0.006
58	" " "			"	1.5011	1.4797	1.4904	1.4969	483.31	0.00	483.31	"	9.135	9.047	0.010
59	" " "	54.0	48.0	"	1.5014	1.4794	1.4904	1.4969	483.31	0.00	483.31	"	9.133	8.850	0.031
60	" 28 A.M.	49.0	48.0	"	1.5294	1.4981	1.5092	1.5157	492.40	0.00	492.40	26.746	9.154	8.747	0.044
61	" " "			"	1.5212	1.5097	1.5103	1.5175	493.28	0.00	493.28	"	9.158	9.000	0.017
62	" " "	51.0	48.0	"	1.5210	1.4995	1.5102	1.5168	492.94	0.00	492.94	"	9.157	9.050	0.012
63	" " "	51.0	48.0	"	1.5073	1.4867	1.4970	1.5035	486.50	0.00	486.50	"	9.145	8.150	0.109
64	Nov. 10 A.M.	29.0	44.0	"	0.6888	0.6872	0.6880	0.6894	151.55	—0.14	151.41	13.372	8.305	8.150	0.019
65	" " "			"	0.6861	0.6858	0.6859	0.6873	150.86	—0.14	150.72	"	8.301	8.000	0.036
66	" " "			"	0.6850	0.6844	0.6847	0.6861	150.46	—0.14	150.32	"	8.298	7.900	0.018
67	" " "	37.0	44.0	"	0.6831	0.6821	0.6826	0.6840	149.77	—0.14	149.63	"	8.299	8.250	0.006
68	" " "			"	0.6806	0.6801	0.6803	0.6817	149.02	—0.14	148.88	"	8.295	8.100	0.024
69	" " "	42.0	44.0	"	0.6781	0.6777	0.6779	0.6793	148.24	—0.14	148.10	"	8.294	7.300	0.120
70	" " P.M.	40.5	44.0	"	0.6840	0.6840	0.6840	0.6854	150.23	—0.14	150.09	"	8.299	8.200	0.012
71	" " "	36.0	44.0	"	0.9022	0.8985	0.9003	0.9025	226.80	—0.19	226.61	"	8.510	8.130	0.045
72	" " "			"	0.9004	0.8965	0.8984	0.9006	226.09	—0.19	225.90	"	8.511	7.530	0.115
73	" " "	34.0	44.0	"	0.9054	0.9002	0.9028	0.9051	227.78	—0.19	227.59	"	8.514	8.410	0.012
74	" 11 A.M.	22.0	42.0	"	0.9069	0.9027	0.9045	0.9069	228.46	—0.19	228.27	"	8.519	8.330	0.022
75	" " "			"	0.9071	0.9025	0.9048	0.9071	228.53	—0.19	228.34	"	8.518	8.230	0.034
76	" " "			"	0.9107	0.9058	0.9082	0.9105	229.82	—0.19	229.63	"	8.522	8.360	0.019
77	" " "	28.0	42.0	"	0.9085	0.9039	0.9062	0.9085	229.06	—0.19	228.87	"	8.517	8.460	0.007
78	" " "	36.0	42.0	"	1.1051	1.0929	1.0990	1.1025	305.98	—0.22	305.76	"	8.707	7.708	0.115
79	" " "			"	1.1057	1.0924	1.0990	1.1025	305.98	—0.22	305.76	"	8.710	8.300	0.047
80	" " "			"	1.1044	1.0913	1.0978	1.1013	305.48	—0.22	305.26	"	8.702	8.400	0.035
81	" " "			"	1.1032	1.0910	1.0971	1.1006	305.19	—0.22	304.97	"	8.709	8.500	0.024
82	" " "			"	1.1022	1.0902	1.0962	1.0997	304.82	—0.22	304.60	"	8.708	8.550	0.018
83	" " P.M.			"	1.1020	1.0884	1.0952	1.0987	304.40	—0.22	304.18	"	8.707	8.600	0.012
84	" " "	46.0	42.0	"	1.0886	1.0769	1.0827	1.0861	299.19	—0.22	298.97	"	8.693	8.650	0.005
85	" " "	46.0	42.0	"	1.3672	1.3399	1.3535	1.3591	418.36	—0.35	418.01	"	8.955	8.800	0.017
86	" " "			"	1.3701	1.3433	1.3567	1.3623	419.83	—0.35	419.48	"	8.958	8.550	0.046
87	" " "			"	1.3702	1.3402	1.3552	1.3608	419.15	—0.35	418.80	"	8.960	8.650	0.035
88	" " "			"	1.3678	1.3419	1.3548	1.3604	418.96	—0.35	418.61	"	8.956	8.900	0.006
89	" " "			"	1.3820	1.3505	1.3662	1.3719	424.26	—0.35	423.91	"	8.971	8.850	0.013
90	" " "			"	1.3794	1.3491	1.3642	1.3698	423.30	—0.35	422.95	"	8.967	7.950	0.113
91	" " "	40.0	42.0	"	1.3788	1.3488	1.3638	1.3694	423.11	—0.35	422.76	"	8.962	8.750	0.024
92	" 12 A.M.	34.0	40.0	"	1.6555	1.5954	1.6254	1.6323	550.05	—0.41	549.64	"	9.213	9.150	0.007
93	" " "			"	1.6472	1.5888	1.6180	1.6249	546.32	—0.41	545.91	"	9.210	9.100	0.012
94	" " "			"	1.6480	1.5830	1.6155	1.6224	545.02	—0.41	544.61	"	9.207	9.050	0.017
95	" " "			"	1.6363	1.5795	1.6079	1.6147	541.21	—0.41	540.80	"	9.201	8.800	0.044
96	" " "			"	1.6519	1.5890	1.6204	1.6273	547.53	—0.41	547.12	"	9.215	8.900	0.034
97	" " "			"	1.6464	1.5862	1.6163	1.6231	545.42	—0.41	545.01	"	9.208	8.200	0.109
98	" " "	36.0	40.0	"	1.6460	1.5900	1.6180	1.6249	546.32	—0.41	545.91	"	9.208	9.000	0.023
99	" " P.M.	42.0	40.0	"	1.8508	1.7655	1.8081	1.8144	644.14	—0.48	643.66	"	9.392	9.350	0.004
100	" " "			"	1.8396	1.7588	1.7992	1.8056	639.48	—0.48	639.00	"	9.383	9.300	0.009
101	" " "			"	1.8486	1.7674	1.8080	1.8144	644.14	—0.48	643.66	"	9.386	9.100	0.030
102	" " "	43.0	40.0	"	1.8582	1.7721	1.8151	1.8214	647.85	—0.48	647.37	"	9.396	9.000	0.042
103	" " "			"	1.8687	1.7819	1.8253	1.8316	653.27	—0.48	652.79	"	9.407	8.400	0.107
104	" " "			"	1.8446	1.7565	1.8005	1.8069	640.17	—0.48	639.69	"	9.382	9.250	0.014
105	" " "	42.0	40.0	"	1.8375	1.7538	1.7956	1.8022	637.68	—0.48	637.20	"	9.381	9.200	0.019

FROM WHICH THE FORMULA OF CORRECTION $C = 0.116\,(\sqrt{D} - 0.1)$ IS DETERMINED.

Measurements.			18	19	20	21	22		23
No. of the Exp.	16 Mean velocity of the tubes throughout the flume, by the diagram.	17 Quantity of water passing the flume, deduced from the mean velocity of the tubes. Q''	Difference between the quantity of water passing the flume deduced from the mean velocity of the tubes, and the quantity deduced from the weir measurement. $Q''-Q'$	Proportional difference, or the difference in the preceding column, divided by the quantity deduced from the flume measurement. $\frac{Q''-Q'}{Q''}$	Quantity of water passing the flume deduced from the mean velocity of the tubes corrected by the formula $Q''' = Q''(1-0.116(\sqrt{D}-0.1))$.	Proportional difference of the corrected quantity as measured in the flume given in the preceding column, and the weir measurement. $\frac{Q'''-Q'}{Q'}$	Remarks on the Force and Direction of the Wind at the Flume, during the Experiments.		General Remarks.
	Feet per sec.	Cubic ft. per sec.	Cubic ft. per sec.				Force.	Direction.	
57	1.991	487.27	− 1.15	−0.0023	485.54	−0.0059	Very gentle, variable.	Down stream.	
58	1.969	481.00	− 2.31	−0.0048	481.06	−0.0048	" "	" "	
59	1.978	483.14	− 0.17	−0.0004	478.88	−0.0092	" "	" "	
60	2.029	496.76	+ 4.36	+0.0088	490.44	−0.0040	Very gentle.	Irregular.	
61	2.013	493.07	− 0.21	−0.0004	491.33	−0.0040	Calm.		
62	2.034	498.21	+ 5.27	+0.0106	497.66	+0.0096	Very gentle, variable.	Up stream.	
63	2.040	498.97	+12.47	+0.0250	485.65	−0.0017	" "	Down stream.	
64	1.360	151.08	− 0.33	−0.0022	150.42	−0.0065	Moderate, but variable.	Irregular.	Reduced copies of the diagrams, constructed for experiments 67, 68, and 69, are given in plate XVII.
65	1.376	152.69	+ 1.97	+0.0129	151.10	+0.0025	" "	"	
66	1.385	153.63	+ 3.33	+0.0217	151.53	+0.0080	" "	"	
67	1.349	149.69	+ 0.06	+0.0004	150.08	+0.0030	More moderate, but variable.	"	
68	1.370	151.99	+ 3.11	+0.0205	151.02	+0.0144	" "	"	
69	1.381	153.12	+ 5.02	+0.0328	148.71	+0.0043	" "	"	
70	1.373	152.32	+ 2.23	+0.0146	152.15	+0.0137	Moderate, but variable.	Down stream.	
71	2.023	230.23	+ 3.62	+0.0157	227.23	+0.0027			
72	2.020	229.87	+ 3.97	+0.0173	223.49	−0.0107			
73	1.999	226.55	− 1.04	−0.0046	226.30	−0.0057	Moderate.	Down stream.	
74	2.029	231.09	+ 2.82	+0.0122	229.79	+0.0067	Calm.		
75	2.031	231.29	+ 2.95	+0.0128	229.01	+0.0030	Very gentle.	Up stream.	
76	2.029	231.24	+ 1.61	+0.0070	230.22	+0.0026	Moderate.	Generally up stream.	
77	2.001	227.94	− 0.93	−0.0041	228.37	−0.0022	Very gentle.	Irregular.	
78	2.691	313.28	+ 7.52	+0.0240	304.59	−0.0038	Gentle.	Down stream.	
79	2.663	310.20	+ 4.44	+0.0143	306.09	+0.0008	Hardly perceptible.	Uncertain.	
80	2.638	306.91	+ 1.65	+0.0054	303.81	−0.0047	" "	"	
81	2.638	307.09	+ 2.12	+0.0069	305.13	+0.0005	" "	"	
82	2.606	303.51	− 1.09	−0.0036	302.31	−0.0075	" "	"	
83	2.607	303.53	− 0.65	−0.0021	303.19	−0.0033	" "	"	
84	2.526	293.61	− 5.33	−0.0182	294.61	−0.0145	" "	"	
85	3.484	417.20	− 0.81	−0.0019	415.73	−0.0055	Calm.		
86	3.534	423.27	+ 3.79	+0.0090	417.65	−0.0044	Very gentle.	Irregular.	
87	3.532	423.21	+ 4.41	+0.0104	418.94	+0.0005	Calm.		
88	3.454	413.70	− 4.91	−0.0119	414.78	−0.0091	"		
89	3.510	421.06	− 2.85	−0.0068	420.37	−0.0084	"		
90	3.623	434.38	+11.45	+0.0263	422.48	−0.0011	"		
91	3.531	423.17	+ 0.41	+0.0010	420.47	−0.0054			
92	4.451	548.35	− 1.29	−0.0024	549.39	−0.0005	Calm.		Reduced copies of the diagrams constructed for experiments 92, 96, and 97, are given in plate XVII.
93	4.491	541.99	− 3.92	−0.0072	541.39	−0.0083	Hardly perceptible.	Down stream.	
94	4.415	543.50	− 1.11	−0.0020	541.59	−0.0055	" "	" "	
95	4.400	541.33	+ 0.53	+0.0010	531.44	−0.0118	Calm.		
96	4.454	548.87	+ 1.75	+0.0032	543.50	−0.0066	"		
97	4.494	553.64	+ 8.63	+0.0156	538.86	−0.0113	"		
98	4.446	547.41	+ 1.50	+0.0027	544.13	−0.0033	"		
99	5.154	647.65	+ 3.89	+0.0060	650.31	+0.0103	"		
100	5.079	637.28	− 1.72	−0.0027	637.66	−0.0021	"		
101	5.209	658.72	+10.07	+0.0154	648.18	+0.0070	"		
102	5.226	656.67	+ 9.30	+0.0142	648.68	+0.0020	Hardly perceptible.	Down stream.	
103	5.316	672.45	+19.66	+0.0292	654.74	+0.0030	Calm.		
104	5.138	644.39	+ 4.90	+0.0076	643.22	+0.0055	"		
105	5.136	644.21	+ 7.01	+0.0109	641.38	+0.0066	"		
Means for the wide flume, taken algebraically				+0.0088	. . .	+0.0013			
Means for the wide flume, disregarding signs				0.0129	. . .	0.0080			
Means for the narrow flume, taken algebraically				+0.0072	. . .	−0.0011			
Means for the narrow flume, disregarding signs				0.0105	. . .	0.0057			
Taking the whole 90 experiments. Means taken algebraically				+0.0082	. . .	+0.0004			
Means disregarding signs				0.0120	. . .	0.0071			

DESCRIPTION OF TABLE XXII.

223. It will be seen that the experiments are divided into groups of seven. In all the experiments in the same group, the quantity of water passing was intended to be the same. Precise uniformity in the quantity was not essential for the attainment of the object in view, and as such uniformity would have required much time to bring about, it was not attempted. The width of the flume remained constant; the depth of water in the flume depended upon the depth on the weir, which was determined by the quantity of water flowing, and which was, as before stated, nearly constant. We have then in each group seven experiments, in which the width of the flume, the depth of the water, the quantity of water passing, and the mean velocity of the water, are very nearly constant. The only material variation is in the length of the immersed part of the tube. For instance, in the first seven experiments, the length of the immersed part of the tube (column 14) varied from 9.482 feet to 8.530 feet, the depth of water in the flume (column 13) in the same experiments remaining nearly constant.

224. Experiments 1 to 63 are all with the wide flume, figures 1 and 2, plate XV.; the minute variations in the width, given in column 12, arise from the measures having been taken several times during the course; and the same remark applies to the length of the weir, given in column 4. Experiments 64 to 105 are all with the narrow flume, figures 3 and 4, plate XV.

225. Table XXII will be understood from the headings of the several columns, together with what has been said previously, without much further explanation. The mean observed depth of water on the weir is given in column 8. As explained above, this observed depth is subject to several corrections, which it has not been thought necessary to give in detail in the table. It may be well, however, to indicate them for one of the experiments, say the first, in which they are as follows:—

Mean observed depth on the weir, ...	1.8778 feet.
Correction for the difference in the observed depth, when the lower orifice of the hook gauge box pipe is at a point 6 feet from the plane of the weir, instead of 0.52 feet, as in the experiment, ...	— 0.0021 "
	1.8757 "
Correction for the velocity of the water approaching the weir. See section 153, ...	+ 0.0140 "
	1.8897 "

Correction for the obstruction to the flow over the weir, by the apron,
trough, &c. . . . . — 0.0058 feet.
Corrected depth on the weir, as given in column 8. . . . . 1.8839 "

The correction for the leakage into the flume is required only in the experiments with the narrow flume, as is previously explained.

FORMULA OF CORRECTION FOR GAUGINGS MADE WITH LOADED POLES OR TUBES.

226. The absolute difference in the quantities deduced from the weir measurement and from the mean velocity of the tubes is given in column 18, Table XXII, and the proportional difference of the same quantities is given in column 19. The quantity deduced from the weir measurement, given in column 11, is taken as the true quantity passing the flume. By reference to columns 15 and 19 it will be seen, that, when the tube extends nearly to the bottom of the flume, the differences are small, generally less than one per cent. In each group there is one experiment in which the tube does not extend to the bottom within about one foot; in these the differences in the quantities obtained by the two methods are greater, as might be expected; in these, however, the differences are, generally, less than three per cent; in one experiment only (43) does it exceed four per cent.

227. It was anticipated, when the programme of the experiments was arranged, that the differences would be found to vary with the velocity of the water in the flume. If any such relation exists, it should be indicated by the mean values of the proportional differences in the several groups.

Table XXIII., arranged according to velocities, and for each width of flume separately, gives these mean values.

TABLE XXIII.

Numbers of the experiments constituting the group.	Width of the flume: in feet.	Mean velocity of the the tubes in feet per second.	Mean proportional difference.
43 to 49	26.75	0.499	+ 0.0262
36 " 42	"	1.136	+ 0.0079
22 " 28	"	1.476	+ 0.0074
29 " 35	"	1.756	+ 0.0044
15 " 21	"	1.983	+ 0.0024
57 " 63	"	2.008	+ 0.0043
8 " 14	"	2.481	+ 0.0079
1 " 7	"	2.670	+ 0.0097
50 " 56	"	2.690	+ 0.0092
	Means,	1.855	+ 0.0088
64 to 70	13.37	1.371	+ 0.0144
71 " 77	"	2.018	+ 0.0080
78 " 84	"	2.627	+ 0.0038
85 " 91	"	3.524	+ 0.0036
92 " 98	"	4.438	+ 0.0016
99 " 105	"	5.184	+ 0.0115
	Means,	3.194	+ 0.0071
Mean proportional difference for all the experiments.			+ 0.0082

228. By the preceding table it does not appear that the difference depends on the velocity. In both the wide and narrow flume, however, the difference is greatest when the velocity is least, although the velocities in the two cases are very different. Whether this is accidental or depends on some principle is a question I have no means of answering.

229. In the wide flume, the mean proportional difference is 0.0088, or about $\frac{7}{8}$ of one per cent. In the narrow flume, the mean proportional difference is 0.0071, or a little less than $\frac{3}{4}$ of one per cent. Thus, on comparing the whole of the experiments in the two flumes, given in table XXIII., it appears that the proportional differences vary only 0.0017, or about $\frac{1}{6}$ of one per cent.

230. The proportional differences given in column 19 are very irregular, and of the nature of residual quantities, depending upon errors of observation, the instability of the currents and the numerous causes tending to produce differences in the results, derived from the mean velocity of the tubes and the weir measurement. I am unable to assign to each cause its legitimate effect; all I can do is to find an empirical formula that will represent, with sufficient accuracy for practical purposes, the difference in the usual cases which occur in practice. In arranging the programme of experiments, it was designed to cover the usual range

of velocities and proportional depths of immersion of the tubes, and any application of the empirical formula founded on them will generally be free from the objection of being outside the range of the experiments on which it is founded.

231. We have to seek for an expression or formula which will enable us to deduce the real quantity from that deduced from the velocity of the tubes, by assuming that they indicate the mean velocity of the water for the whole depth of the part of the stream in which they float.

In the absence of experimental data it would be rational to assume that the formula of correction is a function of three quantities, viz.:—

1. The width of the flume relatively to its depth.

2. The mean velocity of the current.

3. The depth to which the tube is immersed, relatively to the whole depth of the stream.

1. The sides of the flume must, of course, cause a retardation of the current similar to that produced by the bottom; by reference to the several diagrams on plate XVII. it will be seen that the velocity of the tubes is diminished near the sides. It is not practicable to measure the velocity, by means of the tubes, quite close to the sides, but in drawing the curves, representing the mean velocities of the tubes, it will be seen that the retarding effects of the sides are attempted to be allowed for.

We have experiments only on flumes of two widths, one being twice the width of the other; the depths being nearly the same, the relative width in one will be about twice that in the other. By reference to table XXIII. it will be seen that in the wide flume the mean proportional difference is + 0.0088, the mean velocity being 1.855 feet per second. In the narrow flume, if we take the whole of the experiments, the mean velocity is much greater than in the experiments in the wide flume. If, however, we take the three first groups, which include experiments No. 64 to 84, we have for the mean velocity 2.005 feet per second, and a mean proportional difference of + 0.0087. Comparing the results from the two flumes, it appears that by doubling the relative width, other circumstances remaining nearly the same, the proportional difference has not been sensibly affected. We may, therefore, conclude that the relative width of the flume need not enter into the formula of correction, care being taken, in drawing the curves, representing the mean velocities in different parts of the width of the flume, to inflect the curve downwards at the sides, as has been done in reducing these experiments.

2. As depending on the mean velocity of the current. It results, from Navier's

investigation, that, so far as it depends on the excess of the velocity of the tube above that of the water in which it is floating, the *absolute* difference is proportional to the velocity (art. 196); the *proportional* difference, which we are considering, must therefore be constant, or independent of the velocity. By reference to table XXIII. it will be seen that the mean proportional differences in the several groups of experiments in each flume appear to have two maxima and one minimum; the experiments in which the velocities are least and greatest having the greatest proportional difference, and some intermediate velocity having the least proportional difference. Comparing the whole of the experiments in both flumes, we find in the group having the least velocity the largest proportional difference; but this result having, apparently, no connection with the results deduced from the great mass of the experiments, must, until explained, be considered anomalous. Comparing the results deduced from all the experiments, excepting those comprised in the first group, no connection can be traced between the velocities and the mean proportional differences. We must therefore conclude, that the correction is independent of the velocity.

3. As depending on the depth to which the tube is immersed, relatively to the whole depth of the stream. It is evident that, in the cases in which the natural scale of velocities at different depths has become established, the difference in question must depend mainly upon this circumstance, and its magnitude may be computed by the formulas of Humphrey and Abbot together with those of Navier and Frizell, as has been previously shown (arts. 193, 196); but in these experiments, and in the cases which usually occur in practice, this natural relation is not established, and consequently these formulas do not apply; and there appears to be no alternative but to determine an empirical formula from the experiments, which will serve for practical purposes.

232. In determining the formula of correction, it is assumed that the proportional difference depends only upon the relative depth to which the tube is immersed. Instead of using this relative depth, it has been found more convenient to use a quantity depending directly upon it, viz. the difference between the depth of the water in the flume and the depth to which the tube is immersed, divided by the depth of the water in the flume; this we call D, and its value in each experiment in table XXII. is given in column 15.

For the purpose of more convenient graphic representation, the data given in table XXII. are reduced, by taking means of the values of D within certain limits, and also of the corresponding values of the proportional differences $\frac{Q'-Q}{Q'}$ given in column 19. These means, arranged according to the values of D, are given separately for each width of flume, in table XXIV.

TABLE XXIV.

Width of the flume.	Number of experiments from which the means are deduced.	Greatest and least values of D in the experiments from which the means are deduced.		Mean value of D.	Mean value of the proportional difference. $\frac{Q'-Q}{Q'}$
Feet.		Greatest.	Least.		
26.746	9	0.007	0.004	0.0054	+ 0.00129
"	12	0.012	0.008	0.0107	+ 0.00027
"	8	0.017	0.015	0.0165	+ 0.00400
"	7	0.023	0.019	0.0211	+ 0.00251
"	9	0.034	0.029	0.0318	+ 0.00856
"	9	0.055	0.041	0.0446	+ 0.01577
"	9	0.129	0.104	0.1118	+ 0.03033
13.372	6	0.007	0.004	0.0058	— 0.00503
"	5	0.012	0.009	0.0114	— 0.00040
"	4	0.017	0.013	0.0152	— 0.00080
"	9	0.024	0.018	0.0213	+ 0.00616
"	5	0.035	0.030	0.0336	+ 0.00944
"	7	0.048	0.036	0.0440	+ 0.01269
"	6	0.120	0.107	0.1132	+ 0.02420

233. In the diagram figure 2, plate XVIII. the abscissas represent the mean values of D in the preceding table and the ordinates the corresponding mean values of the proportional differences $\frac{Q'-Q}{Q'}$; the double circles representing the experiments with the wide flume, and the single circles the experiments with the narrow flume. As will be seen, the parabolic curve $A\ B$ represents, nearly, the mean result of all the experiments. Calling the ordinates of the curve C, and the abscissas D, its equation is

$$C = 0.116\,(\sqrt{D} - 0.1) \qquad (1.)$$

C is the proportional difference to be deducted from the quantity directly deduced from the mean velocity of the tubes; Q' being the quantity thus deduced and Q''' being the corrected quantity, we have

$$Q''' = Q' - C\,Q' = (1 - C)\,Q'$$

substituting the value of C in (1.), we have

$$Q''' = \left(1 - 0.116\,(\sqrt{D} - 0.1)\right) Q'. \qquad (2.)$$

Table XXIX. gives the values of the coefficient

$$1 - 0.116\,(\sqrt{D} - 0.1)$$

for the values of D from 0.000 to 0.100, together with the logarithms of the same.

234. Column 20, table XXII., gives the values of Q''' by formula (2.), and column 21 the proportional differences between the values of Q''' and the quantities as measured at the weir. Taking the whole of the experiments together, it will be seen that the mean proportional difference, taken algebraically, is $+$ 0.0004, or, disregarding the signs, 0.0071; the latter quantity is about $\frac{3}{4}$ of one per cent, and is the mean error or discrepancy between the measurement by the weir and the corrected measurement in the flume. It will be observed that the largest discrepancies are in the group of experiments numbered from 43 to 49, in which the velocity was very slow; in one of these experiments, viz. No. 46, the corrected flume measurement is about $\frac{1}{27}$ greater than the weir measurement. In experiment No. 47 the corrected flume measurement is about $\frac{1}{36}$ greater than the weir measurement. In experiment No. 16 the corrected flume measurement is about $\frac{1}{44}$ less than the weir measurement. In all the other experiments, the difference is less than $\frac{1}{50}$, or two per cent.

MISCELLANEOUS EXPERIMENTS AT THE

1	2	3		Weir Measurement								Flume			
		Temperature, in degrees of Fahrenheit's thermometer,		4	5	6	7	8	9	10	11	12	13	14	15
No. of the Exp.	Date. 1856	of the atmosphere in shade.	of the water.	Total length of the weirs. L	Observed depth of water on the Westerly weir.	Observed depth of water on the Easterly weir.	Mean observed depth of water on the weirs.	Corrected depth of water on the weirs. H	Quantity of water passing over the weirs, computed by the formula $Q = 3.33(L-0.2H)H^{\frac{3}{2}}$	Correction for the leakage into the flume.	Corrected quantity passing the flume, deduced from the weir measurement. Q'	Mean width of the flume.	Mean depth of water in the flume.	Length of the immersed part of the tube.	Difference between the depth of water in the flume and the length of the immersed part of the tube, divided by the depth of water in the flume. D
				Feet.	Feet.	Feet.	Feet.	Feet.	Cubic ft. per sec.	Cubic ft. per sec.	Cubic ft. per sec.	Feet.	Feet.	Feet.	
106	Oct. 22 A.M.	53.0	53.0	80.014	0.5509	0.5508	0.5508	0.5518	108.58	0.00	108.58	26.745	8.177	8.070	0.013
107	" " "	57.5	53.0	"	0.5501	0.5501	0.5501	0.5511	108.38	0.00	108.38	"	8.173	8.070	0.013
108	" " "	60.5	53.0	"	0.5504	0.5505	0.5504	0.5514	108.46	0.00	108.46	"	8.175	8.070	0.013
109	Nov. 13 A.M.	32.0	40.0	80.012	1.1583	1.1358	1.1470	1.1508	326.23	—0.27	325.96	13.372	8.760	8.650	0.013
110	" " "			"	1.1661	1.1435	1.1548	1.1587	329.59	—0.27	329.32	"	8.766	8.650	0.013
111	" " "			"	1.1634	1.1409	1.1521	1.1560	328.44	—0.27	328.17	"	8.765	8.650	0.013
112	" " "	36.0	40.0	"	1.1604	1.1373	1.1488	1.1527	327.04	—0.27	326.77	"	8.760	8.650	0.013
113	Oct. 23 P.M.	50.5	53.0	80.014	1.8580	1.8362	1.8471	1.8534	664.91	0.00	664.91	26.745	9.529	9.430	0.010
114	" 30 A.M.			"	1.3612	1.3509	1.3560	1.3616	419.51	0.00	419.51	26.746	9.003	8.600	0.045
115	" " "	67.0	47.0	"	1.3685	1.3588	1.3636	1.3692	423.02	0.00	423.02	"	9.016	8.900	0.013
116	" " P.M.	67.0	47.0	"	1.3391	1.3350	1.3370	1.3424	410.70	0.00	410.70	"	9.015	8.000	0.113
117	" " "	66.0	47.0	"	1.3291	1.3255	1.3273	1.3327	406.28	0.00	406.28	"	9.010	8.850	0.018
118	" " "	64.0	47.0	"	1.2405	1.2221	1.2313	1.2358	362.92	0.00	362.92	"	8.882	7.852	0.116
119	" 31 "				0.9921	0.9857	0.9889	0.9917	261.15	0.00	261.15	"	8.628	8.220	0.047
120	Oct. 22 P.M.	54.0	53.0	80.014	1.8994	1.8539	1.8766	1.8826	680.61	0.00	680.61	26.745	9.526	9.429	0.010
121	" 23 "	48.0	53.0	"	1.8408	1.8144	1.8276	1.8339	654.50	0.00	654.50	"	9.500	9.430	0.007
122	" " "	45.5	53.0	"	1.8244	1.8001	1.8122	1.8186	646.36	0.00	646.36	"	9.479	9.430	0.005
123	Nov. 4 A.M.	52.0	47.0	80.014	1.8837	1.8450	1.8643	1.8704	674.03	0.00	674.03	26.746	9.527	9.430	0.010
124	" " "	53.0	47.0	"	1.8682	1.8321	1.8501	1.8563	666.47	0.00	666.47	"	9.518	9.430	0.009
125	" " "	53.0	47.0	"	1.8588	1.8230	1.8409	1.8472	661.60	0.00	661.60	"	9.509	9.431	0.008
126	" " "	54.0	47.0	"	1.8550	1.8193	1.8371	1.8433	659.51	0.00	659.51	"	9.506	9.432	0.008
127	" " "	54.0	47.0	"	1.8378	1.8023	1.8200	1.8263	650.46	0.00	650.46	"	9.490	9.434	0.006
128	" " "	56.0	47.0	"	1.8217	1.7888	1.8052	1.8116	642.65	0.00	642.65	"	9.476	9.436	0.004
129	" " "	56.0	47.0	"	1.8692	1.8328	1.8510	1.8572	666.95	0.00	666.95	"	9.529	9.437	0.010
130	" " P.M.	56.0	47.0	"	1.8731	1.8271	1.8501	1.8563	666.47	0.00	666.47	"	9.555	9.439	0.012
131	" " "	58.0	47.0	"	1.8670	1.8233	1.8451	1.8514	663.85	0.00	663.85	"	9.551	9.445	0.011
132	" " "	58.0	47.0	"	1.8603	1.8166	1.8384	1.8447	660.26	0.00	660.26	"	9.549	9.440	0.011
133	" " "	58.0	47.0	"	1.8376	1.7950	1.8163	1.8226	648.49	0.00	648.49	"	9.529	9.430	0.010
134	" " "	58.5	47.0	"	1.8780	1.8435	1.8607	1.8669	672.16	0.00	672.16	"	9.562	9.430	0.014
135	Nov. 5 A.M.	39.0	48.0	80.014	1.4879	1.4695	1.4787	1.4852	477.68	0.00	477.68	26.746	9.164	9.040	0.014
136	" " "	40.0	48.0	"	1.4820	1.4631	1.4725	1.4790	474.70	0.00	474.70	"	9.159	9.040	0.013
137	" " "	40.0	48.0	"	1.4823	1.4659	1.4741	1.4806	475.46	0.00	475.46	"	9.168	9.040	0.014
138	" " "	41.0	48.0	"	1.4832	1.4679	1.4755	1.4820	476.14	0.00	476.14	"	9.175	9.040	0.015
139	" " "			"	1.4855	1.4724	1.4789	1.4854	477.77	0.00	477.77	"	9.176	9.040	0.015
140	" " "			"	1.4794	1.4691	1.4740	1.4805	475.42	0.00	475.42	"	9.174	9.040	0.015

TREMONT WEIR AND MEASURING FLUME.

No. of the Exp.	16. Mean velocity of the tubes throughout the flume, by the diagram. Feet per sec.	17. Quantity of water passing the flume, deduced from the mean velocity of the tubes. Q' Cubic ft. per sec.	18. Difference between the quantity of water passing the flume deduced from the mean velocity of the tubes, and the quantity deduced from the weir measurement. $Q'-Q$ Cubic ft. per sec.	19. Proportional difference, or the difference in the preceding column, divided by the quantity deduced from the flume measurement. $\frac{Q'-Q}{Q'}$	20. Quantity of water passing the flume deduced from the mean velocity of the tubes, correc'd by the formula $Q'' = Q'(1-0.116(\sqrt{D}-0.1))$.	21. Proportional difference of the corrected quantity as measured in the flume given in the preceding column, and the weir measurement. $\frac{Q''-Q}{Q}$	22. Remarks on the Force and Direction of the Wind at the Flume, during the Experiments. Force.	22. Direction.	23. General Remarks.
106	0.503	110.09	+ 1.51	+0.0137	109.91	+0.0122	Calm.		These three experiments were made under circumstances as nearly identical as practicable, for the purpose of testing the degree of uniformity attained in the results. The greatest variation in the proportional differences in column 21, from the mean, is [illegible], or [illegible]. Reduced copies of the diagrams constructed for these experiments are given on plate XVII.
107	0.502	109.62	+ 1.24	+0.0113	109.44	+0.0098	"		
108	0.504	110.10	+ 1.64	+0.0149	109.92	+0.0135	"		
109	2.752	322.34	− 3.62	−0.0112	321.81	−0.0127	Calm.		These four experiments were also made under circumstances as nearly identical as practicable, for the same purpose as the preceding. The greatest variation in the proportional differences in column 21, from the mean, is 0.0097 = [illegible]. Reduced copies of the diagrams constructed for these experiments are given on plate XVII.
110	2.811	329.48	+ 0.16	+0.0005	328.94	−0.0012	Very gentle.	Down stream.	
111	2.820	330.49	+ 2.32	+0.0070	329.95	+0.0054	" "	" "	
112	2.796	327.54	+ 0.77	+0.0024	327.01	+0.0007	Calm.		
113	2.542	647.88	−17.03	−0.0263	647.88	−0.0256	Very strong, variable.	Irreg., gen. down stream.	These seven experiments were made when the wind was blowing with considerable force in the direction of the current. It will be seen that the results are less regular than in the preceding seven experiments, and in the experiments in Table XXII., in none of which was there much wind. The mean proportional difference in column 21, in these seven experiments, is −0.0024, which would indicate that the wind blowing down stream had a small effect in diminishing the velocity of the tubes, but the irregularities are too great to permit this inference to be drawn. All that can be safely inferred is, that the wind had no sensible effect on the velocity of the tubes, except to increase the irregularities.
114	1.771	426.55	+ 7.04	+0.0165	421.00	+0.0036	Brisk.	Down stream.	
115	1.725	416.02	− 7.00	−0.0168	415.34	−0.0182	"	" "	
116	1.759	424.07	+13.37	+0.0315	412.45	+0.0043	"	" "	
117	1.695	408.53	+ 2.25	+0.0055	406.91	+0.0016	Strong.	" "	
118	1.563	371.37	+ 8.45	+0.0228	361.01	−0.0053	"	" "	
119	1.174	270.84	+ 9.69	+0.0358	262.17	+0.0231	Brisk, strong at times.	" "	
120	2.720	693.06	+12.45	+0.0180	693.06	+0.0183	Very moderate.	Up stream.	In these three experiments the wind was blowing generally in the opposite direction to the current. The mean proportional difference in column 21 is [illegible], but the results are too irregular to permit any inference to be drawn, except that the effect of the wind was insensible, except in increasing the irregularities.
121	2.494	633.68	−20.82	−0.0329	634.88	−0.0300	Very strong, variable.	Irreg., gen. up stream.	
122	2.531	641.63	− 4.73	−0.0074	613.81	−0.0039	" "	" "	
123	2.595	661.35	−12.68	−0.0192	661.35	−0.0188	Calm.		These twelve experiments were made with an obstruction placed in the canal about [illegible] feet above the flume (X Y, plate XV, and art. [illegible]), which caused great disturbances in the motion of the water in the flume, every part of it being filled with eddies, both horizontal and vertical. The mean proportional difference in column 21, disregarding the signs, is [illegible]. In table XXII. the corresponding mean is 0.0071. Hence we infer, that the irregularities were greater when the current was disturbed in the manner described than when undisturbed, in the ratio of 17 to 10. The mean proportional difference in column 21, regarding the signs, is [illegible], from which, considering the irregularities, all that can be inferred is, that the disturbance of the current had no sensible effect on the results, except in increasing the irregularities. In experiments 123, 124, 127, 128, 131, and 132, the tubes were put into the water in regular order, from the left to the right side of the flume, at intervals of about one foot, passing once across in the usual manner. In experiments 125, 126, 129, 130, 133, and 134, they were put in in the same order, but at intervals of about four feet, and passing across the flume four times, in each experiment, taking different points at each crossing. It was thought possible that the positions of the quick and slow currents might not remain constant throughout an experiment, which, with the ordinary mode of putting in the tubes, might lead to an erroneous result. Comparing the results obtained by the two methods, and disregarding the signs, the mean proportional difference in column 21, in the six experiments in which the tubes were put in in the usual manner, is [illegible]. In the other six experiments the mean proportional difference is [illegible]. The small difference in the results, considering the irregularities, cannot be attributed to the mode of putting in the tubes. Reduced copies of the diagrams constructed for experiments 123 and 124 are given in plate XVII.
124	2.611	664.76	− 1.71	−0.0026	665.16	−0.0020	Hardly perceptible.	Up stream.	
125	2.599	661.00	− 0.60	−0.0009	661.81	+0.0003	Calm.		
126	2.594	659.61	+ 0.10	+0.0002	660.41	+0.0014	"		
127	2.613	663.27	+12.81	+0.0193	665.06	+0.0224	"		
128	2.488	630.54	−12.11	−0.0192	633.23	−0.0147	"		
129	2.539	647.09	−19.86	−0.0307	647.09	−0.0298	Very gentle.	Up stream.	
130	2.633	672.98	+ 6.51	+0.0097	672.23	+0.0086	" "	" "	
131	2.628	671.36	+ 7.51	+0.0112	670.98	+0.0107	Hardly perceptible.	" "	
132	2.566	655.29	− 4.97	−0.0076	654.92	−0.0081	Calm.		
133	2.515	640.98	− 7.51	−0.0117	640.98	−0.0116	"		
134	2.677	684.64	+12.48	+0.0182	683.18	+0.0164	"		
135	2.045	501.32	+23.64	+0.0472	500.25	+0.0472	Strong.	Down stream.	These six experiments were made under similar circumstances to the twelve experiments next preceding, except that there was a high wind blowing in the direction of the current. The mean proportional difference in column 21 is [illegible], which indicates that the joint effect of the disturbance of the current, and the strong wind blowing in the direction of the current, was to increase the velocity of the tubes about two per cent. In experiments 135, 136, 139, and 140 the tubes were put into the water in the usual manner. In experiments 137 and 138 they were put in as described above in experiments 125, 126, &c. The mean proportional difference in column 21, in the four experiments in which the tubes were put in in the usual manner, disregarding the signs, is [illegible], and in the other two experiments the mean is [illegible]; indicating no sensible difference in the magnitude of the irregularities, depending on the manner in which the tubes were put into the water. Reduced copies of the diagrams constructed for experiments 135, 136, and 140 are given in plate XVII.
136	1.947	477.06	+ 2.26	+0.0049	476.28	+0.0033	"	" "	
137	2.032	498.21	+22.75	+0.0457	497.15	+0.0456	"	" "	
138	1.960	480.85	+ 4.71	+0.0098	479.59	+0.0072	"	" "	
139	1.923	471.86	− 5.91	−0.0125	470.63	−0.0149	Strong, but variable.	" "	
140	2.012	493.77	+18.35	+0.0372	492.48	+0.0359	Violent.	" "	

DESCRIPTION OF TABLE XXV.

235. The experiments in this table were made like those in table XXII., and have been reduced in the same manner. The special purposes for which they were made are described in the final column of the table, headed "General Remarks." By referring to the table, it will be seen that the first seven experiments were made for the purpose of testing the degree of uniformity attainable in the results, when the circumstances under which the measurements were made were the same. This is a fundamental question in all kinds and methods of measuring, and is distinct from the errors of observation to which all methods are liable. In geodesic and astronomical methods the difficulties arise principally from the instability of instruments and from atmospheric changes. In measuring the velocity of streams of water, the instability of the currents, mentioned in article 208, appeared to afford a peculiar liability to this trouble, and it was necessary to make special experiments to ascertain the magnitude of the irregularities due to it. In the three experiments, numbered 106 to 108, in which the circumstances were as nearly alike as practicable, the extreme variation is about $\frac{1}{278}$; in the next group of four experiments, in which the circumstances were also alike, as nearly as practicable, the extreme variation is about $\frac{1}{56}$; so far as is known, there was no want of care in the execution of any of these experiments, and the irregularities must be considered as inseparable from the method. In a greater number of trials the extreme variation would probably be greater. We must infer from these seven experiments, that any single measurement is liable to be erroneous to the amount of one per cent, or perhaps rather more; and in any two experiments the errors may be in opposite directions, in which case they may vary from each other two per cent, or rather more. It is of course very desirable that the method should be free from this liability to error; except by accident, however, the quantity of water used at a manufacturing establishment or flowing in a stream will not be found twice alike. An approximation within one or two per cent of the truth is sufficient for most practical purposes; the errors are as liable to be one way as the other, and by repeating the measurement several times and taking the mean, the probabilities are that the result will be very nearly as correct as if the method was free from this liability to error in a single measurement.

236. The seven experiments numbered from 113 to 119 were made, when the wind was blowing with considerable force down stream. Taking the mean, it would appear that the effect of the wind was to cause the corrected flume measurements to be about one quarter of one per cent less than the weir measurements. In these

experiments the length of the immersed parts of the tubes varied from 7.85 feet to 9.43 feet; the length projecting above the water, in each case, was about 0.33 feet; taking a mean, about $\frac{1}{26}$ part of the length projected above the surface of the water, and was liable to be acted upon by the wind. The effect of the wind blowing down stream, with a velocity greater than that of the current, must be to give the tube a greater velocity than it would have in a calm or with the wind blowing up stream. By the mean result of the seven experiments the contrary effect would appear to have been produced. By comparing the differences in these seven experiments, given in column 21, with the corresponding differences in table XXII., it will be seen that the irregularities in the results of the measurements were much greater when the wind was blowing strongly than when it was calm, or nearly so. The extreme variation in the seven experiments is nearly five per cent; under these circumstances, it is apparent that, in order to detect with certainty so small a difference as one quarter of one per cent, a much larger number of experiments is necessary, and that, with the small number made, the real effects may easily be obscured by the irregularities.

237. In experiments 121 and 122 the wind was very strong, but variable, irregular in direction, but generally up stream; the mean result of the two experiments is, that the velocity of the tubes was retarded about $\frac{1}{85}$; but the number of experiments is evidently insufficient to determine it definitely. We may infer from the ten experiments, numbered from 113 to 122, that, although measurements made when the wind is blowing strongly, either up stream or down stream, are subject to greater irregularities than measurements made when there is little or no wind, by making a considerable number of trials, the mean results will vary but little, whether the wind is blowing strongly or not.

238. In the twelve experiments, numbered 123 to 134, there was a great commotion in the stream caused by an obstruction in the channel above, as is explained in the table. The irregularities are increased, but the mean result is not sensibly affected. In the six experiments numbered 135 to 140 there was a similar agitation in the stream, and also a high wind blowing down stream; the effect was to increase the irregularities in the results, and the mean velocity of the tubes appears to have been increased about two per cent.

APPLICATION OF THE METHOD OF GAUGING STREAMS OF WATER BY MEANS OF LOADED POLES OR TUBES.

239. As previously stated, this method is more generally adopted at Lowell, for gauging large volumes of water, than any other. Six measuring flumes have been

constructed in the canals there; all made in a similar manner to that described in article 201, and represented in figures 1 and 2, plate XV. Their principal dimensions and the quantities of water usually gauged in them are as follows:—

The Merrimack flume, about 100 feet long and 50 feet wide, intended to gauge about 1,500 cubic feet of water per second.

The Appleton flume, about 150 feet long and 50 feet wide, intended to gauge about 1,800 cubic feet of water per second.

The Lowell Manufacturing Company's flume, about 150 feet long and 30 feet wide, intended to gauge about 500 cubic feet of water per second.

The Middlesex flume, about 150 feet long and 20 feet wide, intended to gauge about 260 cubic feet of water per second.

The Prescott flume, about 180 feet long and 66 feet wide, intended to gauge about 2,000 cubic feet of water per second.

The Boott flume, about 100 feet long and 42 feet wide, intended to gauge about 800 cubic feet of water per second.

The depths of the water in these flumes are various, usually, however, between eight and ten feet; sometimes, when the river is low, the depth is diminished to about six feet.

It will be seen that the widths of the flumes are not strictly in proportion to the quantities of water intended to be gauged in them; the widths and depths have usually been determined by the dimensions of the canals in which they are placed.

240. Under the existing arrangements at Lowell, a daily account is usually kept of the excess of water, if any, drawn by each manufacturing company, over and above the quantity to which it is entitled under its lease. In ordinary times this is arrived at with sufficient exactness by means of occasional measurements, but when the flow of water in the river is too small to supply the wants of all, it is necessary to make frequent measurements of the quantity of water drawn by those who habitually draw an excess. In the latter case the usual course of proceeding is this. A gauging party, consisting of one or more engineers and a sufficient number of assistants, is assigned to each flume where measurements are required. Arrangements are made so that the observations for a single gauge occupy about half an hour. Several gauges are made during the day, the intervals between the times when the observations are made being occupied by the same party in working out the results, which, as soon as obtained, are communicated to the proper local authorities at the manufacturing establishments where the water is drawn. This is done to enable them to adjust the amount of machinery they run, so as to draw only the quantity of water to which they are entitled. If

they continue to draw an excess after due notice, they are liable to heavy penalties. It is essential to the proper working of these arrangements that the results of the gaugings should be arrived at and communicated as speedily as possible; with this view, as well as to reduce the expense, engraved diagrams and printed forms and tables have been prepared, and all the apparatus provided and preparations made which can in any way facilitate the operation.

241. The mode of making the observations for a gauge in a measuring flume is substantially the same as that practised in the experimental flume in the Tremont Canal, and fully described in articles 204 and 205. With the view, however, of reducing the number of assistants required, a stop-watch beating quarter seconds is used instead of a marine chronometer, and the electric telegraph is dispensed with. The observer with the stop-watch takes his position at the upper transit station, and starts the watch when the tube passes it; he then walks to the lower transit station and stops the watch when the tube passes it. By this method two observers are dispensed with. Another observer notes the depth of the water in the flume, and also records the distances of the tubes from the left side of the flume, which are observed and called by the assistants who put in and take out the tubes. One other assistant is required to carry back the tubes to the up-stream station, making five in the party.

242. Ordinarily, about an hour is occupied in making the observations for a measurement. The following measurement is given in detail as an example of the whole process. The flume in which it was made is situated a short distance below a bend of about ninety degrees in the canal, which produces a great irregularity in the current, the velocity being much greater on the right-hand side of the flume than on the left-hand side; sometimes there is no sensible current on the left-hand side. It being inconvenient to perform the measurement under such circumstances, the difficulty was remedied by placing an obstruction near the lower end of the bend; the up-stream face of this obstruction was an oblique plane, so placed as to direct a part of the current towards the left-hand side of the flume. Although far from producing a uniform velocity in all parts of the flume, it removed all the trouble in making the measurement due to the original irregularity. The remaining irregularities in the velocity are indicated by the inflections of the curved line *A B* on plate XIX.

The mean width of the part of the flume between the upper and lower transits is 41.76 feet.

TABLE XXVI.

GAUGE OF THE QUANTITY OF WATER PASSING THE BOOTT MEASURING FLUME, MADE BETWEEN 10 HOURS 30 MINUTES AND 11 HOURS 30 MINUTES, A. M., MAY 17TH, 1860.

Distance from the left-hand side of the flume at which the tube was put into the water.	Length of the immersed part of the tube.	Time during which the tube passed from the up-stream transit station to the down-stream transit station, a distance of 70 feet.	Mean velocity of the tube.	Distance of the tube from the left-hand side of the flume during its passage.			Depth of water in the flume.
				At the upper transit station.	At the lower transit station.	Mean.	
Feet.	Feet.	Seconds.	Feet per second.	Feet.	Feet.	Feet.	Feet.
0.0	8.40	33.3	2.102	0.3	0.8	0.55	8.510
1.5	"	31.0	2.258	1.8	1.6	1.70	8.481
3.0	"	30.2	2.318	3.2	2.1	2.65	8.450
4.5	"	28.3	2.473	4.4	4.5	4.45	8.470
6.0	"	29.5	2.373	6.2	5.4	5.80	8.445
7.5	"	27.0	2.593	8.2	10.1	9.15	8.438
9.0	"	26.2	2.672	9.7	10.4	10.05	8.440
10.5	"	25.0	2.800	10.5	8.8	9.65	8.470
12.0	"	25.8	2.713	12.3	10.9	11.60	8.483
13.5	"	25.2	2.778	13.8	15.5	14.65	8.490
15.0	"	25.0	2.800	15.2	18.0	16.60	8.500
16.5	"	29.5	2.373	17.0	20.4	18.70	8.498
18.0	"	27.0	2.593	18.0	17.8	17.90	8.505
19.5	"	28.8	2.431	19.7	19.0	19.35	8.505
21.0	"	30.7	2.280	21.1	20.9	21.00	8.522
22.5	"	31.8	2.201	23.4	29.3	26.35	8.533
24.0	"	33.7	2.077	23.7	22.1	22.90	8.510
25.5	"	33.8	2.071	26.5	29.7	28.10	8.495
27.0	"	31.0	2.258	27.0	25.2	26.10	8.483
28.5	"	31.0	2.258	28.6	26.5	27.55	8.495
30.0	"	29.0	2.414	31.0	34.3	32.65	8.550
31.5	"	28.0	2.500	32.1	30.0	31.05	8.630
33.0	"	31.0	2.258	32.5	28.1	30.30	8.610
34.5	"	26.2	2.672	34.6	36.7	35.65	8.625
36.0	"	28.8	2.431	36.5	35.0	35.75	8.642
37.5	"	28.5	2.456	37.5	35.5	36.50	8.612
39.0	"	28.0	2.500	40.1	40.5	40.30	8.578
40.0	"	28.0	2.500	39.0	39.6	39.30	8.578
41.0	"	29.2	2.397	41.2	40.6	40.90	8.560
0.0	"	34.3	2.047	0.5	0.4	0.45	8.471
10.0	"	26.5	2.642	9.8	8.7	9.25	8.580
20.0	"	32.2	2.174	20.9	19.9	20.40	8.605
30.0	"	30.8	2.273	31.5	33.8	32.65	8.635
41.0	"	30.5	2.295	41.4	40.6	41.00	8.610
						Mean	8.5294

A Products of the mean velocities into the widths, for each foot in width, excepting the last product, which is for a width of 0.76 feet; commencing at the left-hand side of the flume.
2.073
2.193
2.284
2.359
2.422
2.478
2.529
2.577
2.623
2.666
2.705
2.744
2.776
2.801
2.811
2.798
2.747
2.648
2.514
2.363
2.219
2.172
2.120
2.098
2.105
2.130
2.163
2.023
2.246
2.289
2.331
2.373
2.413
2.450
2.483
2.510
2.531
2.544
2.540
2.504
2.417
2.264 × 0.76 = 1.721
Sum, 101.523
Mean velocity of the tubes, $\frac{101.523}{41.76} = 2.4311$ ft. per sec.

243. The mean velocity of the tubes is obtained by means of a diagram, a copy of which, on the same scale as the original, is given in plate XIX. The small circles represent the several observations, the abscissa and ordinate of each being the mean distance from the left-hand side of the flume and the observed velocity of the tube as given in table XXVI. The curved line represents the mean and is drawn by the eye, giving due weight to each observation. The mean velocity is 2.4311 feet per second, and is found by taking a mean of the ordinates of the curve; the process is given in column *A*, table XXVI.

The mean depth of the water in the flume was	8.5294 feet.
The length of the immersed part of the tube was	8.4000 "
Difference	0.1294 "

Then D (art. 232) $= \frac{0.1294}{8.5294} = 0.0152.$

The mean section of the water-way in the flume was

$$41.76 \times 8.5294 = 356.188 \text{ square feet.}$$

And the quantity of water passing, by the tube measurement, was

$$356.188 \times 2.4311 = 865.929 \text{ cubic feet per second} = Q'.$$

This is to be corrected by formula (2.), art. 233.

Substituting for D its value 0.0152, we have for the coefficient of correction $1 - 0.116\,(\sqrt{D} - 0.1) = 0.99730$ (see table XXIX.) and the corrected quantity $Q'' = 0.99730 \times 865.929 = 863.59$ cubic feet per second.

244. In the preceding example the entire volume of water flowing through the canal was gauged. It often happens that only a portion of the entire flow of the stream is to be gauged, namely, the quantity drawn out of the canal at a single orifice or branch canal. In this case a flume of suitable dimensions is constructed and connected with the edges of the orifice or the sides and bottom of the branch canal, so that no water can enter the orifice or branch canal except through the measuring flume. A rough preliminary estimate of the quantity should be made by some other method; this will enable the sectional area of the measuring flume to be determined, so that the velocity in it may be convenient for observation, say between one foot and three feet per second, although it may exceed these limits, in either direction, if the circumstances are such as to require it. It will generally be most convenient to place the flume so that its axis will be parallel, or nearly so, with the axis of the canal. Its

length will usually be limited by local circumstances and economical considerations; a considerable length in which to measure the velocity of the loaded tubes is desirable, although not essential. If the means for observing the transits and the times of the same are good, a less length is necessary than in cases where the means of observing are less perfect. By means of the electric telegraph and a skilled observer of the chronometer, as in the experiments at the Tremont measuring flume (art. 205), an interval of a few seconds between the times of the transits at the upper and lower stations will enable a good gauge to be made. If the observations are made in the less perfect manner practised at the Boott measuring flume, and described in art. 241, a considerably longer interval is necessary in order to attain equally accurate results. There seems to be scarcely any limit to the shortness of the time admissible in the first case, if corresponding care and precautions are adopted in making the observations.* In the second case, it will depend much on the degree of skill of the observer. The method has not been used extensively enough, as yet, to enable a limit to be definitely fixed. A practised observer, with a stop-watch beating quarter seconds, the transit stations being twenty-five feet apart, has been able to observe both transits, when the time between them was ten seconds, and in some cases seven and a half seconds.

245. The distance between the transit stations is only a part of the length required for the flume; a certain length above the upper transit station is necessary to give room for putting the tubes into the water, and to permit them to attain, sensibly, the same velocity as the water before they arrive at the transit station. By reference to art. 195 it will be seen that a tube two inches in diameter, floating twenty feet, attains $\frac{43}{44}$ of the velocity of the current. Twenty feet was about the distance the tubes floated before they reached the upper transit station, in the experiments given in table XXII., from which the formula for the correction of flume measurements was determined, and the correction for the very small error, resulting from this distance being insufficient, is implicitly included in the formula. Twenty feet may therefore be taken as the proper distance, and if circumstances are such as to require a much less distance, the resulting error can be corrected by means of formula (5.), article 194.

246. The same method may be extended to gauging natural watercourses. A favorable place for the purpose should be selected; that is, one free from reverse currents, the bottom smooth, the section uniform for a sufficient distance,

* Methods for making and recording observations of time are practised in some astronomical observatories, by means of which the one-hundredth part of a second is estimated; these methods could undoubtedly be adapted to our purpose if required.

and with as long a reach above, free from bends, great irregularities of section and obstructions, as can be found. Two parallel sections, in planes at right angles to the direction of the current, or nearly so, should be carefully measured, so that the depth at every point may be known. The proper distance between the sections will depend much on the regularity of the channel; it will usually be desirable that they should be far enough apart to permit the observations for the velocity to be made, without resorting to the use of the electric telegraph; excepting in very large rivers, a distance of from fifty to one hundred feet, depending on the width, would usually permit this to be done with sufficient accuracy for most purposes, although a greater distance would usually be desirable.

The loaded poles or tubes must not touch the bottom while passing from one transit station to the other. It will probably rarely occur that one hundred feet in length of the channel of a river will be found of such regularity that the poles could be immersed to an average depth of six inches from the bottom. By resorting to the more exact mode of observing the transits, the sections might be within twenty feet of each other, or even half that distance if necessary. There would seldom be any difficulty in finding a suitable place for a gauge made in this manner, in any river confined within regular banks. Something could be done, in so short a length, towards removing obstructions and filling up depressions. In making the observations, loaded tubes or poles, of lengths adapted to the different parts of the section, should be provided; they may be put into and taken out of the water from boats or rafts. Theodolites should be placed in the planes of the sections, on the same bank; the observer at each should have the key of a break-circuit within his reach, while observing the transit of the floating pole. The observations of the times of the transits may be made in the same manner as at the Tremont measuring flume (art. 205). If the sections are very near together, a separate observer may be necessary for the transit at each station, both, however, using the same chronometer. The distance from fixed points on the bank, at which the floats pass the transits, corresponding to the distances from the left-hand side of the flume, in the flume measurements, can be observed by means of marked cords, stretched across the river, just over the water, and at short distances above and below the sections, and supported from the bottom at intervals, if necessary; or it may be done by means of a system of signals and triangulations.

The section of the river not being rectangular, it will usually be most convenient to divide it into several parts, finding the area of the section, the mean velocity of the poles, computing the quantity and making the correction by formula (2.), article 233, for each part separately. The sum will of course be the gauge of the whole river.

The degree of accuracy attainable in gauging a natural watercourse, by this method, will depend entirely upon the regularity and smoothness of the part of the channel selected for the operation, and of the immediate approach to the same. If the bottom is covered with large stones or sunken timber, it will prevent the attainment of much precision. In such cases, if the greatest attainable precision is desired, either the obstructions must be removed or the bed of the channel filled up with some sort of material suitable for the purpose, to the level of the top of the obstructions. In any case, the degree of precision attainable will depend on the degree of approximation in the channel to the regularity and smoothness of the measuring flumes.

EXPERIMENTS ON THE FLOW OF WATER THROUGH SUBMERGED ORIFICES AND DIVERGING TUBES.

247. Daniel Bernoulli proved,—on the hypothesis *that no force is lost,—that the fluid in all parts of the same section has the same velocity, and remains in one mass;* that the velocity of the discharge from a vessel, by an orifice of small area relatively to that of the vessel, is that due to the head above the orifice from which the fluid is finally discharged, whether such orifice is in the side or bottom of the vessel itself, or at the end of a tube projecting from the side or bottom of the vessel, the sides of the tube being either parallel, converging, or diverging.* This being established, it follows, if the conditions of the hypothesis can be complied with, that the velocity of discharge from a simple orifice in a vessel may be increased to any extent by the application of a tube with diverging sides; for the area of the orifice at the end of the tube from which the fluid is finally discharged may be as many times larger than the orifice in the side or bottom of the vessel as we please, and as the same quantity must pass through both orifices in the same time, the velocity through the orifice in the vessel will be as much greater than the velocity through the orifice at the end of the tube as its area is less.

248. The fact that the flow through an orifice could be increased by the application of a diverging tube appears to have been known to the ancient Romans. Experiments have been made upon them in modern times by Gravesande, Bernoulli, Venturi, and Eytelwein, and perhaps others. And experiments on the discharge of air between two discs, which afford an aperture similar in effect to a diverging tube, have been made by Thomas Hopkins.† Most of our experimental knowledge on the flow of water through diverging tubes is due to Venturi, whose experiments were made at Modena about the year 1791, and published in

* Hydrodynamica. Strasburg, 1738.

† Memoirs of the Literary and Philosophical Society of Manchester. Vol. V., Second Series. London, 1831.

Paris in 1797, under the title, *Recherches experimentales sur le Principe de la Communication latérale du Mouvement dans les Fluids.**

Venturi experimented on many forms of diverging tubes; in pipes of regular form the maximum increase of velocity was obtained with a conical tube in which the sides diverged from each other at an angle of 4° 27′; this tube was applied to a mouth-piece having nearly the form of the contracted vein; a certain volume of water under a constant head was discharged through the mouth-piece alone in forty-two seconds; when the diverging tube was applied to the mouth-piece, the same quantity of water was discharged, under the same head, in twenty-one seconds; increasing the velocity through the mouth-piece in the ratio of 1 to 2. In a similar tube of greater length the water did not fill the tube throughout its whole length unless a prominence was made on the inside of the tube, at the bottom, which caused the water to fly upward and fill the down-stream end of the tube; with this tube, the same volume of water was discharged in nineteen seconds, increasing the discharge through the mouth-piece in the ratio of 1 to 2.21.

Eytelwein made some similar experiments with a mouth-piece and a tube whose sides diverged at an angle of 5° 9′. He found that the application of the tube to the mouth-piece increased the velocity through the latter in the ratio of 1 to 1.69.

249. According to Bernoulli's theory, in Venturi's experiment, last above quoted, the velocity through the smallest section of the mouth-piece should be increased by the diverging tube, in the ratio of 1 to 3.03. In Eytelwein's experiment the increase should be in the ratio of 1 to 3.21. In both these experiments the water in the tube undoubtedly remained in unbroken masses. There must, consequently, have been considerable losses of force. The increased flow appears to be due to what is termed by Venturi *the lateral communication of motion in fluids*, and to the pressure of the atmosphere. According to the principle of Venturi, a column of water flowing through a mass of water at rest tends to communicate a portion of its velocity to the mass, and to cause it to move along with it; and if the column of water is moving in a pipe a little larger than itself, it will communicate motion to the entire shell of water surrounding it. If the water is flowing through a conical tube whose sides diverge at a small angle, the section of the pipe is continually enlarging by insensible degrees; but by the principle of Venturi the stream must fill each successive section, and the mean velocity

* See a translation of Venturi's work, in Nicholson's Journal of Natural Philosophy, Vol. III, London, 1802. Also, in Tracts on Hydraulics, by Thomas Tredgold, 2d Edition. London, 1836.

must diminish in the ratios that the areas of the sections increase. The pressure of the atmosphere on the surface of the water in the vessel and on the orifice from which the water escapes may for this purpose be called the same, and equal to a column of water thirty-four feet high. Supposing the mass of water flowing through the pipe to be divided into very thin slices, by planes at right angles to the direction of the current; from its inertia, each slice will tend to retain its velocity, but on account of the enlarging sections it cannot do this, but tends to separate itself from the slice immediately following it; this is prevented by the pressure of the atmosphere, and the effect is to balance a portion of the pressure of the atmosphere on its down-stream side; the entire pressure of the atmosphere remains on the up-stream side of the slice, and the difference between the effective pressures on the up-stream and down-stream sides accelerates the motion of the slice. All the slices are acted on in a similar manner, and the increased discharge is due to the sum of the actions upon them.

In the experiment above quoted of Venturi, with a pipe of regular form, the discharge through the orifice took place under a head of 2.887 feet; the head being as the square of the velocity, the equivalent head, under which the discharge took place with the diverging tube, was $2.887 \times 2^2 = 11.548$ feet, which exceeds the actual head of water in the experiment by 8.661 feet, which is the portion of the total pressure of the atmosphere on the surface of the water in the reservoir rendered active in that experiment.

250. Venturi found no increased discharge by increasing the length of his diverging tube beyond 1.096 feet, on account of the water not filling the whole section of the part of the tube added beyond that length. This difficulty, however, can be obviated by submerging the diverging tube; for in that case it must remain full of water, whatever may be its length or the angle of divergency of its sides.

In these experiments the tubes were submerged, which distinguishes them from any previously recorded, and greater effects were produced. The diffuser applied by Mr. Boyden to turbine water-wheels, to increase their efficiency (art. 12), acts on the same principle as the diverging tube; this apparatus has been extensively applied in Lowell, and it has thus become a matter of great interest to ascertain to what extent a conical diverging tube, discharging under water, could be made to increase the discharge through a simple orifice. For this purpose the following experiments were made.

DESCRIPTION OF THE APPARATUS.

251. The tube used in these experiments is represented by figures 1, 2, 3, and 4, plate XXI. It is composed of cast iron and is made in five pieces, *A*, *B*, *C*, *D*, and *E*, which when screwed together, as represented in figures 1 and 2, form a compound tube, consisting of a mouth-piece of a form to avoid contraction, and a diverging tube, in which the diameter increases from about 0.1 foot, at its junction *b* with the mouth-piece, to about 0.4 foot at *f*. The part of the mouth-piece between *a* and *g* is formed by the revolution of a common cycloid about the axis of the tube; from *a* to *b* it is cylindrical. The interior of the parts *C*, *D*, *E* are frustums of a cone; a portion of the part *B* is also a frustum of the same cone; but, to avoid any angle in passing from the cylinder *a b* to the frustum of the cone, a portion of the part *B* is formed by the revolution of an arc of a circle of about 22.69 feet radius, the sides of the cylinder *a b* and of the cone both being tangent to this arc. The parts of the compound tube being screwed together could be readily taken apart and the mouth-piece used by itself, or with one or more of the conical parts attached. The interior of the mouth-piece and diverging tubes were first turned separately, they were then screwed together and ground on a mandril with emery until they became quite smooth, without, however, having a bright polish. This mode of finishing insured the smallest possible degree of irregularity at the junctions of the several parts.

252. For the purpose of making the experiments, the compound tube was mounted in a cistern (figures 1, 2, and 3, plate XX.) constructed for the purpose. The cistern was made of white-pine wood, very strongly framed, and supported on brick piers, which were built up several feet in height from a solid foundation The cistern consists of three compartments; the upper compartment, *E*, is the reservoir supplying the mouth-piece *M*, and the diverging tube attached to it. *F*, the middle compartment, receives the water discharged through the tube. *G* is the lower compartment, in the end of which is placed the weir, *W*, at which the quantity of water discharged was gauged.

The supply of water for the experiments was obtained from the main pipes laid down by the manufacturing companies at Lowell for conveying water from an elevated reservoir to their several establishments mainly for the purpose of extinguishing fire. For these experiments, it was important that the supply of water flowing into the reservoir *E* should be as nearly uniform as possible, but the effective pressure in the main pipes was subject to some irregularity, which of course caused a corresponding irregularity in the discharge from the orifice through which the supply of water was drawn. To eliminate this source of irregularity, the water was first drawn into the cistern *I*, figures 2 and 3, plate XX., in considerably greater volume than was required to be admitted into the reservoir *E*; the excess passed over a weir in the side of the cistern *I*, and from thence was discharged through the waste-pipe *K*. The supply for the reservoir *E* was drawn from the cistern *I* through the pipe *H*, the quantity being regulated by the cock *L*. By this arrangement, it will be seen that, so long as the water was admitted into the cistern *I* in excess of that admitted into the reservoir *E*, the head acting on the cock *L* must have been subject to only very small variations, and consequently the discharge through a constant orifice in the cock *L* must have been very nearly uniform. It was important that the water in the part of the reservoir *E*, near the side containing the mouth-piece, should be as nearly quiescent as possible. The water was admitted under a head of about 18 feet, which necessarily produced a great commotion in the part of the reservoir where it entered, and to prevent this from extending to the side containing the mouth-piece, it was made to pass through six diaphragms, *R*, *R*, *R*, &c., figures 1 and 2, plate XX. The first two diaphragms were made of boards, about one inch thick, containing numerous holes about half an inch in diameter, as shown in figure 4; the other four diaphragms were of strainer-cloth, placed about two inches apart and stretched tightly in a frame. The strainer-cloth used was the well-known fabric sold under that name, made of flax or hemp, with about twenty threads to an inch in both warp and filling, the width of the spaces between the threads being from two to three times the thickness of the thread. The effect of these diaphragms was to prevent any sensible commotion in the part of the reservoir between the lower diaphragm and the side containing the mouth-piece. The part of the reservoir *E*, between the down-stream diaphragm and the mouth-piece, was about 2.34 feet long in the direction of the current, 3 feet wide, and 4.5 feet deep. The division *F* was about 6.75 feet long, 3 feet wide, and 3.35 feet deep; the water passed from this division to the division *G* through the diaphragm *N*, similar to the wooden diaphragms in division *E*, above described; and also through the diaphragm *P*, consisting of a single thickness of strainer-cloth. The dimensions of the part of the reservoir *G*, between the diaphragm *P* and the end containing the weir *W*, is about 3.6 feet long in the direction of the current, 3 feet wide, and 3.20 feet deep. The disturbance in the division *F* was slight, and as the apparatus was first designed, the weir was placed in the partition *N*, but on trial the agitation was found to be too great to admit of an unexceptionable gauge at the weir; the division *G* was then added, which, with the diaphragms, removed all difficulty from this cause.

253. A weir was adopted to gauge the quantity of water passing through the tube, in preference to any other kind of orifice, because it admitted of greater variations in the quantity of water discharged, with any admissible variation in the height of the water in the reservoir in the side of which it is placed; and by adopting a weir of the same dimensions and form as that used by Poncelet and Lesbros (art. 161), the quantity could be computed with great precision.

The weir *W*, figures 1, 2, and 6, plate XX., is represented on a larger scale by figures 5, 6, and 7, plate XXI., and a section of the crest of the weir is given, full size, in figure 8, plate XXI. The crest and sides of the weir were made of plates of cast iron, planed and finished with great care, the up-stream edges presented to the current having sharp corners, or as nearly so as could be made with that metal. The only material variation from the weir used by Poncelet and Lesbros is in the thickness of the crest, which in their weir was an edge, whereas in our weir it had a thickness of about 0.02 inch; this variation was made to enable the zero points of the several gauges, used for measuring the heights of the water in the different compartments of the apparatus, to be made in a particular manner, which will be described hereafter. This difference in the thickness of the crest of the weir could have affected the accuracy of the gauge in only a few of the experiments, namely, those in which the depth on the weir was less than 0.05 foot, as at this depth and all greater depths it was observed that the contraction was complete; that is to say, at this depth the stream passing over the weir touched the orifice only at the up-stream edge, as represented in figure 3, plate XVIII., and the flow was the same as if the crest of the weir had no sensible thickness. With depths on the weir less than 0.05 foot, the stream of water was in contact with the whole width of the crest of the weir; which, if it had any sensible effect, would tend to increase the discharge, with the same depth on the weir, in consequence of an action similar to that produced by a short additional tube attached to the down-stream side of an orifice in a thin plate.

The length of the curved part of the mouth-piece *A*, figure 2, plate XXI., measured on the axis *a g*, is . . . . 1.00 foot.
The length of the cylindrical part of the mouth-piece, measured on the axis *a b*, is . . . . 0.10 "
The effective lengths of the parts *B*, *C*, *D*, and *E*, of the diverging tube, are each . . . . 1.00 "
The diameter of the circle generating the semi-cycloid of the mouth-piece is . . . . 0.635 "

The diameters of the several parts of the mouth-piece and diverging tube are given in column 15, table XXVII.

The angle of divergence of the sides of the conical part of the compound tube is . . . . 5° 1′.

254. The elevations of the surface of the water in the several compartments of the cistern were measured by means of the hook gauges represented by figures 9, 10, and 11, plate XXI., and described in articles 45 and 143. They were placed in the hook gauge boxes *A*, *B*, *C*, *D*, figures 1 and 2, plate XX. Communication was established between the several hook gauge boxes and the corresponding compartments of the cistern by means of the orifices 0, figures 1, 2, 5, and 6. The orifices affording communication with the compartments *F* and *G* were 0.10 foot in diameter; the orifice affording communication with the compartment *E* was about five times as large; oscillations in the elevation of the surface being anticipated in this compartment, the amplitude of which it was desirable to measure. There is reason to think that the flow through a diverging tube is to a certain extent in a condition of unstable equilibrium. In Venturi's experiments, the water discharging into the air from diverging tubes was observed to have great irregularity of motion, "and even eddies within the tube; whence the jet comes forth by leaps, and with irregular scattering."* These irregularities are undoubtedly due, in part at least, to an unstable equilibrium, and there must be a corresponding irregularity in the exhausting power of the diverging tube, which would be indicated, in our experiments, by oscillations in the elevation of the surface of the water in compartment *E*, which would rise and fall as the exhausting power of the tube was less or greater.

The elevations of the surface of the water in all the compartments is reckoned from the top of the weir. When no water was admitted to the reservoir *E*, the water in all the divisions of the cistern would fall to the level of the crest of the weir. The comparison between the zero points of the several hook gauges and the crest of the weir was made in the manner described in article 143. Two ten-pointed instruments (figure 14, plate XXI.), of slightly different dimensions, were used, which furnished independent results, a mean of which was taken. They were made of steel and magnetized, which enabled them to maintain their positions when placed on the crest of the weir. Small variations in the apparatus were expected to occur, resulting from changes of temperature and in the hygrometric condition of the wood of which the cistern was con-

* Tracts on Hydraulics.

structed; comparisons were accordingly made on each day that experiments were made; the results are given in the following table:—

Date.	Corrections to be applied to the reading of the hook gauges, to give the elevations of the points of the hooks above the top of the weir.			
1854.	Gauge *A*.	Gauge *B*.	Gauge *C*.	Gauge *D*.
September 20.	—1.5535	—1.5490	—1.5451	—0.3921
" 21 A.M.	—1.5519	—1.5476	—1.5439	—0.3916
" 21 P.M.	—1.5525	—1.5484	—1.5449	—0.3920
" 22.	—1.5528	—1.5487	—1.5447	—0.3918
" 25.	—1.5531	—1.5487	—1.5454	—0.3926
" 26.	—1.5535	—1.5490	—1.5458	—0.3930
October 7.	—1.5541	—1.5502	—1.5474	—0.3940
" 10.	—1.5541	—1.5502	—1.5476	—0.3938
" 12.	—1.5541	—1.5502	—1.5476	—0.3942
" 16.	—1.5536	—1.5500	—1.5472	—0.3935

MODE OF CONDUCTING THE EXPERIMENTS.

255. Water was admitted through the leathern hose *Q* into the cistern *I*, figures 2 and 3, plate XX., in excess of the supply required for the experiment. The index of the cock *L*, figures 2 and 3, was set in the desired position. When it was supposed that the flow had become permanent throughout all the divisions of the cistern, observations of the elevations of the surface of the water in the several compartments were commenced; they were taken by a separate observer at each hook gauge, every thirty seconds, and were continued until some minutes after the elevation of the surface in the compartments *F* and *G* had become stationary, which indicated that a permanent flow had been obtained. The watches used by the several observers were set to indicate the same time, and the time when each observation was made being recorded, a subsequent comparison of the records of the observations made at the several hook gauges enabled those to be selected in which the permanence of the flow was the most perfect. Not less than five, and usually more than ten, successive observations, made at the same times at each hook gauge, were used, from which the mean elevations in the several compartments during the experiment were deduced.

256. Experiments 1 to 18, table XXVII., were made with the mouth-piece *A* alone. Experiments 19 to 38 were made with the mouth-piece *A* and the first joint *B* of the diverging tube. Experiments 39 to 50 were made with the mouth-piece and the two joints *B* and *C* of the diverging tube. Experiments 51 to 64

were made with the mouth-piece and the three joints *B*, *C*, and *D* of the diverging tube. Experiments. 65 to 90 were made with the complete compound tube, represented by figures 1 and 2, plate XXI., and in figures 1 and 2, plate XX. Experiment 91 was made with the mouth-piece alone. Experiment 92, with the complete compound tube. Experiments 93 to 101 were made with an orifice in a thin plate represented by figures 12 and 13, plate XXI. This plate is of cast iron 0.042 foot in thickness, but the orifice is chamfered off on the down-stream side, so that the effective thickness of the plate at the orifice is 0.0014 foot, or about one sixtieth of an inch.

257. The mouth-piece, diverging tubes, and plate were all of cast iron; this metal was adopted instead of brass as being the cheapest, and experience having shown that oxidation of cast iron immersed in the water of Merrimack River proceeds very slowly, and expecting to be able to find, readily, some substance, a coating of which, of imperceptible thickness on the surface of the metal, would entirely prevent it; no such substance was found, however. Drying oils of several kinds were tried, also a mixture of grease and mercury, also collodion, but without satisfactory effect. The plan finally adopted was to keep the interior of the orifices and tubes and the accessible parts of the weir, when not in use, covered with a thick coating of grease. Previous to each session of the experiments this was removed as completely as possible by rubbing with cotton-waste and woollen cloth, until on rubbing with a clean white cloth no sensible mark was made on it. Of course the whole of the grease was not removed by this operation; the quantity remaining, however, must have been extremely small, but it was sufficient to protect the iron from oxidation for some time, or until it was partially washed off. With this process, however, there must have been constant changes going on in the state of the interior surface of the tube, which might affect the flow of the water in some degree. I accordingly noted carefully the circumstances and indications relating to the application and removal of the grease; and under the head of Remarks in the table of experiments I have stated the essential parts of my observations on this matter.

EXPERIMENTS ON THE FLOW OF WATER

No. of the Exp.	1 Date. 1854.	2 Time of making the observations from which the mean heights given in this table are deduced. Beginning.			Ending.			3 Temperature, in degrees of Fahrenheit's thermometer;		4 Reference to figure 2, plate XXI., indicating the parts of the compound tube used.	5 Position of the index of the index-card.	6 Mean depth of water on the weir, by gauge A. h	7 Value of C in the formula in the next column.	8 Quantity of water discharged, calculated by the formula $D = C\,l\,h\sqrt{2\,g\,h}$	9 Height of the surface of the water in compartment F, figures 1 and 2, plate XX. by gauge B.	by gauge C.	Mean.
		H.	Min.	Sec.	H.	Min.	Sec.	of the air.	of the water.		Degrees.	Feet.		Cubic feet per second.	Feet.	Feet.	Feet.
1	Sept. 20, P.M.	3	37	15	3	40	45		64.6	A	32.50	0.0269	0.4219	0.00980	0.0268	0.0269	0.0269
2	" " "	3	57	0	3	59	0		64.6	"	32.50	0.0270	0.4219	0.00985	0.0269	0.0270	0.0269
3	" " "	4	22	30	4	25	45		64.6	"	34.25	0.0388	0.4202	0.01690	0.0391	0.0392	0.0391
4	" " "	4	31	30	4	35	30		64.6	"	34.25	0.0383	0.4203	0.01658	0.0380	0.0384	0.0382
5	" " "	4	53	15	4	58	15		64.6	"	35.50	0.0467	0.4191	0.02226	0.0469	0.0471	0.0470
6	" " "	5	20	50	5	26	0	62.6	64.6	"	36.50	0.0532	0.4182	0.02701	0.0536	0.0535	0.0535
7	" " "	5	38	40	5	41	50	62.6	64.6	"	37.50	0.0607	0.4172	0.03283	0.0612	0.0612	0.0612
8	" 21, A.M.	9	11	40	9	17	40	56.4	62.8	"	37.50	0.0616	0.4170	0.03355	0.0614	0.0620	0.0617
9	" " "	9	41	40	9	45	40	58.5	62.9	"	38.50	0.0680	0.4162	0.03884	0.0683	0.0686	0.0684
10	" " "	10	11	30	10	20	0	60.5	63.2	"	39.50	0.0739	0.4153	0.04391	0.0738	0.0746	0.0742
11	" " "	10	46	50	10	54	20	60.9	63.3	"	40.50	0.0803	0.4145	0.04964	0.0799	0.0802	0.0800
12	" " "	11	15	40	11	19	0	61.0	63.4	"	41.50	0.0848	0.4138	0.05378	0.0846	0.0852	0.0849
13	" 21, P.M.	2	16	40	2	23	50	61.0	63.4	"	42.50	0.0906	0.4130	0.05927	0.0905	0.0910	0.0907
14	" " "	2	39	30	2	45	20	62.0	63.6	"	43.25	0.0945	0.4125	0.06306	0.0948	0.0951	0.0949
15	" " "	3	5	0	3	9	0	62.9	63.7	"	44.00	0.0991	0.4118	0.06761	0.0992	0.0999	0.0995
16	" " "	3	30	30	3	33	40	67.7	63.7	"	44.75	0.1037	0.4112	0.07227	0.1036	0.1045	0.1040
17	" " "	3	46	40	3	52	0	67.1	63.8	"	45.50	0.1069	0.4108	0.07556	0.1071	0.1077	0.1074
18	" 22, A.M.	9	13	30	9	17	20	58.6	62.7	"	45.67	0.1072	0.4107	0.07586	0.1063	0.1085	0.1074
19	" " "	10	29	15	10	34	0	60.9	62.9	A B	54.50	0.1505	0.4049	0.12441	0.1531	0.1559	0.1545
20	" 25, A.M.	8	59	20	9	3	30	60.8	62.4	"	32.50	0.0285	0.4216	0.01068	0.0284	0.0282	0.0283
21	" " "	9	16	0	9	19	0	61.0	62.3	"	34.25	0.0393	0.4201	0.01722	0.0393	0.0392	0.0392
22	" " "	9	28	0	9	31	0	61.3	62.3	"	35.50	0.0472	0.4190	0.02261	0.0478	0.0476	0.0477
23	" " "	9	45	30	9	45	50	61.7	62.3	"	36.50	0.0555	0.4179	0.02876	0.0556	0.0557	0.0556
24	" " "	9	58	0	10	1	25	62.2	62.3	"	37.50	0.0618	0.4170	0.03372	0.0621	0.0622	0.0621
25	" " "	10	14	10	10	17	0	63.8	62.4	"	38.50	0.0678	0.4162	0.03867	0.0682	0.0686	0.0684
26	" " "	10	30	0	10	33	20	63.7	62.4	"	39.50	0.0732	0.4154	0.04330	0.0740	0.0740	0.0740
27	" " "	10	45	30	10	50	0	64.2	62.4	"	40.50	0.0796	0.4146	0.04900	0.0803	0.0805	0.0804
28	" " "	11	5	50	11	8	20	64.8	62.5	"	41.50	0.0849	0.4138	0.05387	0.0860	0.0865	0.0862
29	" " "	11	19	10	11	25	10	65.0	62.6	"	42.50	0.0901	0.4131	0.05880	0.0911	0.0914	0.0912
30	" " "	11	40	40	11	43	40	65.4	62.6	"	43.25	0.0946	0.4125	0.06316	0.0957	0.0963	0.0960
31	" 25, P.M.	0	15	0	0	20	0	66.3	62.8	"	54.67	0.1517	0.4048	0.12587	0.1547	0.1578	0.1562
32	" " "	3	27	0	3	31	30	70.5	63.6	"	54.67	0.1512	0.4048	0.12525	0.1549	0.1575	0.1562
33	" " "	3	49	10	3	53	0	70.7	63.6	"	44.00	0.0998	0.4118	0.06833	0.1013	0.1017	0.1015
34	" " "	4	10	0	4	13	30	70.5	63.7	"	44.75	0.1031	0.4113	0.07166	0.1050	0.1057	0.1053
35	" " "	4	31	0	4	34	10	70.5	63.7	"	45.50	0.1080	0.4106	0.07669	0.1100	0.1105	0.1102
36	" " "	4	57	30	5	3	30	70.7		"	47.00	0.1155	0.4096	0.08461	0.1176	0.1184	0.1180
37	" " "	5	20	0	5	23	0	70.0	63.9	"	50.00	0.1260	0.4081	0.09606	0.1284	0.1300	0.1292
38	" " "	5	41	30	5	47	0	70.2	64.0	"	54.67	0.1507	0.4049	0.12466	0.1534	0.1572	0.1553
39	" 26, P.M.	2	32	10	2	35	30	74.7	64.6	A B C	60.00	0.1775	0.4023	0.15833	0.1889	0.1897	0.1893
40	" " "	3	29	0	3	32	0	75.0	64.6	"	32.50	0.0292	0.4215	0.01107	0.0297	0.0296	0.0296
41	" " "	3	38	30	3	41	40	75.1	64.6	"	34.25	0.0396	0.4201	0.01742	0.0402	0.0401	0.0401
42	" " "	3	52	30	3	56	25	75.1	64.6	"	36.50	0.0557	0.4179	0.02891	0.0566	0.0567	0.0566
43	" " "	4	5	0	4	10	40	75.4	64.6	"	38.50	0.0677	0.4162	0.03858	0.0694	0.0690	0.0692
44	" " "	4	28	40	4	32	40	75.5	64.7	"	40.50	0.0795	0.4146	0.04891	0.0814	0.0812	0.0813
45	" " "	4	44	10	4	46	50	76.0	64.8	"	42.50	0.0914	0.4129	0.06004	0.0939	0.0940	0.0939
46	" " "	4	59	30	5	3	20	76.6	64.9	"	45.50	0.1067	0.4108	0.07535	0.1103	0.1103	0.1103
47	" " "	5	17	40	5	22	40	75.6	65.0	"	50.00	0.1271	0.4080	0.09729	0.1321	0.1319	0.1320
48	" " "	5	41	10	5	46	0			"	54.67	0.1458	0.4054	0.11878	0.1529	0.1529	0.1529
49	" " "	6	10	30	6	18	50			"		0.1696	0.4031	0.14817	0.1804	0.1808	0.1806
50	Oct. 7, A.M.	9	16	30	9	19	0	58.2	59.5	"	60.00	0.1778	0.4023	0.15873	0.1897	0.1908	0.1902

No. of the Exp.	10 Mean height of the surface of the water in compartment K, figures 1 and 2, plate XX, by gauge D. Feet.	11 Effective head producing the discharge. H Feet.	12 Velocity due the head in the preceding column. V Feet per second.	13 Mean velocity by experiment through the smallest section of the tube or orifice. v Feet per second.	14 Ratio of the velocity at the smallest section to the velocity due the head. $\frac{v}{V}$	15 Diameter of the tube or orifice at the place of final discharge. Feet.	16 Mean velocity by experiment at the final discharge from the tube. v' Feet per second.	17 Ratio of the velocity at the final discharge to the velocity due the head. $\frac{v'}{V}$	18 Remarks.
1	0.0608	0.0339	1.4767	1.2035	0.8150	0.1018	1.2035	0.8150	On the completion of experiment 7, the water was drawn out of the cistern, and the interior of the mouth-piece examined. Only slight traces of oxidation were observed. In order to prevent oxidation before the experiments were resumed, the interior was wiped dry, and smeared with a grease consisting of about 20 parts of beef tallow, 19 parts of fine sperm oil, and 1 part of beeswax. The cistern remained empty until the experiments were resumed, September 21st, when, previous to experiment 8, the grease was removed by thoroughly rubbing the surface with cloth and cotton-waste. Experiment 8 was a repetition of experiment 7; the increased discharge observed in experiment 8 must be attributed to the change in the state of the surface, due to the greasing and wiping previously described. At 5h 30m P.M., September 21st, the water was drawn out of the cistern and the interior of the mouth-piece examined. A large part of the surface at and near the smallest section, where the velocity of the water was greatest, was covered with oxidation; this was rubbed off with a cloth, when the previous lustre of the surface was observed to be tarnished. It was then greased anew. The water was left out of the cistern until the experiments were resumed September 23d, A.M., previous to which the grease was wiped off. Experiment 18 was a repetition of experiment 17, for the purpose of ascertaining the effect of the change in the state of the surface. There was no change in the discharge, however, that could be attributed to the change in the state of the surface.
2	0.0609	0.0310	1.4789	1.2103	0.8183	"	1.2103	0.8183	
3	0.1389	0.0998	2.5337	2.0765	0.8195	"	2.0765	0.8195	
4	0.1384	0.1002	2.5387	2.0369	0.8024	"	2.0369	0.8024	
5	0.2117	0.1647	3.2549	2.7347	0.8402	"	2.7347	0.8402	
6	0.2835	0.2300	3.8464	3.3180	0.8626	"	3.3180	0.8626	
7	0.3790	0.3178	4.5213	4.0341	0.8923	"	4.0341	0.8923	
8	0.3785	0.3118	4.4784	4.1222	0.9205	"	4.1222	0.9205	
9	0.4838	0.4154	5.1691	4.7719	0.9232	"	4.7719	0.9232	
10	0.6010	0.5269	5.8217	5.3945	0.9266	"	5.3945	0.9266	
11	0.7390	0.6590	6.5107	6.0985	0.9367	"	6.0985	0.9367	
12	0.8816	0.7767	7.0682	6.6070	0.9348	"	6.6070	0.9348	
13	1.0486	0.9579	7.8495	7.2822	0.9277	"	7.2822	0.9277	
14	1.1782	1.0833	8.3475	7.7481	0.9282	"	7.7481	0.9282	
15	1.3324	1.2327	8.9046	8.3065	0.9328	"	8.3065	0.9328	
16	1.5098	1.3968	9.4788	8.8786	0.9367	"	8.8786	0.9367	
17	1.6214	1.5140	9.8681	9.2837	0.9407	"	9.2837	0.9407	
18	1.6232	1.5158	9.8742	9.3205	0.9439	"	9.3205	0.9439	
19	1.6285	1.4690	9.7207	15.2853	1.5725	0.1454	7.4928	0.7708	After the conclusion of the experiments September 23d, the water was drawn out of the cistern and the mouth-piece and the first joint of the diverging tube were greased. The cistern remained empty until 9 A.M., September 24th, when it was filled. September 25th, A.M., previous to the commencement of the experiments, the cistern was emptied and the grease wiped off the interior of the mouth-piece and first joint of the diverging tube.
20	0.0485	0.0207	1.1399	1.3116	1.1506	"	0.6429	0.5640	
21	0.0873	0.0181	1.7590	2.1162	1.2031	"	1.0374	0.5897	
22	0.1204	0.0727	2.1625	2.7781	1.2847	"	1.3618	0.6297	
23	0.1552	0.0996	2.5311	3.5329	1.3958	"	1.7318	0.6842	
24	0.1923	0.1302	2.8939	4.1423	1.4314	"	2.0305	0.7017	
25	0.2327	0.1643	3.2509	4.7508	1.4614	"	2.3288	0.7164	
26	0.2745	0.2005	3.5912	5.3193	1.4812	"	2.6075	0.7261	At 2h 25m P.M., September 25th, the cistern was emptied and the interior of the pipes examined. The mouth-piece was free from oxidation, the first joint of the diverging tube was oxidated sufficiently to redden the fingers when rubbed upon it; both the pipes were wiped clean and dry, then coated with grease which was afterwards wiped off as much as practicable by rubbing with a cloth. Experiment 32 was a repetition of 31, to ascertain the effect due to the state of the surface caused by cleaning and greasing. The change in the discharge, however, due to this cause, was, if any, extremely small. After the conclusion of the experiments September 25th, P.M. the cistern was emptied; the mouth-piece was found free from oxidation, and the first joint of the diverging pipe was only slightly oxidated; both pipes were greased and the cistern filled with water.
27	0.3286	0.2482	3.9958	6.0203	1.5067	"	2.9511	0.7386	
28	0.3836	0.2974	4.3738	6.6187	1.5133	"	3.2445	0.7418	
29	0.4370	0.3458	4.7163	7.2238	1.5317	"	3.5410	0.7508	
30	0.4920	0.3980	5.0470	7.7604	1.5376	"	3.8041	0.7537	
31	1.6179	1.4617	9.6965	15.4647	1.5919	"	7.5807	0.7818	
32	1.6025	1.4461	9.6446	15.3885	1.5955	"	7.5432	0.7821	
33	0.5470	0.4155	5.3331	8.3347	1.5682	"	4.1150	0.7687	
34	0.5971	0.4918	5.6244	8.8038	1.5653	"	4.3156	0.7673	
35	0.6660	0.5538	5.9792	9.4227	1.5759	"	4.6190	0.7725	
36	0.7850	0.6670	6.5501	10.3957	1.5971	"	5.0955	0.7780	
37	0.9836	0.8544	7.4134	11.8017	1.5919	"	5.7851	0.7804	
38	1.6257	1.4701	9.7253	15.3158	1.5748	"	7.5077	0.7720	
39	1.6040	1.4147	9.5393	19.4523	2.0392	0.2339	3.6847	0.3863	September 26. 1h 25m P.M. The cistern has stood full of water since last evening: the water was now drawn off, and the grease wiped off the mouth-piece and first joint of diverging pipe. The second joint was then put on for the experiments of to-day.
40	0.0439	0.0143	0.9591	1.3529	1.4179	"	0.2576	0.2686	
41	0.0710	0.0309	1.4098	2.1405	1.5183	"	0.4055	0.2876	
42	0.1182	0.0616	1.9906	3.5520	1.7844	"	0.6728	0.3380	
43	0.1667	0.0975	2.5013	4.7463	1.8929	"	0.8979	0.3586	
44	0.2241	0.1428	3.0307	6.0090	1.9827	"	1.1383	0.3756	October 7, A.M. The cistern has been kept full of water since September 26th, excepting on two or three occasions, when it was emptied, to permit the tubes to be cleaned and greased anew. This morning, on emptying the cistern and wiping off the grease, no oxidation was observed.
45	0.2998	0.2054	3.6348	7.3771	2.0296	"	1.3974	0.3845	
46	0.4220	0.3117	4.4777	9.2576	2.0675	"	1.7536	0.3916	
47	0.6271	0.4951	5.6427	11.9537	2.1184	"	2.2613	0.4013	
48	0.8673	0.7144	6.7788	14.5029	2.1627	"	2.7643	0.4078	
49	1.2803	1.0999	8.4113	18.2943	2.1643	"	3.4483	0.4100	
50	1.5018	1.3116	9.1851	19.5016	2.1232	"	3.6941	0.4022	

EXPERIMENTS ON THE FLOW OF WATER

No. of the Exp.	1. Date. 1854.	2. Time of making the observations from which the mean heights given in this table are deduced. Beginning. H.	Min.	Sec.	Ending. H.	Min.	Sec.	3. Temperature, in degrees of Fahrenheit's thermometer; of the air.	of the water.	4. Reference to figure 2, plate XXI., indicating the parts of the compound tube used.	5. Position of the index of the inlet cock. Degrees.	6. Mean depth of water on the weir, by gauge A. h Feet.	7. Value of C in the formula in the next column.	8. Quantity of water discharged, calculated by the formula $D = Clh\sqrt{2gh}$ Cubic feet per second.	9. Height of the surface of the water in compartment F, figures 1 and 2, plate XX. by gauge B. Feet.	by gauge C. Feet.	Mean. Feet.
51	Oct. 7, A.M.	10	59	0	11	1	30	66.0	60.5	A B C D	62.00	0.1874	0.4014	0.17137	0.2055	0.2058	0.2056
52	" " "	11	18	0	11	20	30	66.1	60.6	"	32.50	0.0284	0.4217	0.01062	0.0290	0.0288	0.0289
53	" " "	11	40	0	11	42	10	66.1	60.6	"	34.25	0.0394	0.4201	0.01729	0.0405	0.0404	0.0404
54	" " "	11	51	0	11	53	40	66.1	60.6	"	36.50	0.0555	0.4173	0.02876	0.0575	0.0574	0.0574
55	" 7, P.M.	2	16	0	2	18	30	68.6	59.8	"	38.50	0.0668	0.4163	0.03783	0.0701	0.0700	0.0700
56	" " "	2	27	0	2	29	0	69.0	59.6	"	40.50	0.0801	0.4145	0.04945	0.0846	0.0848	0.0847
57	" " "	2	35	40	2	29	0	69.1	59.5	"	42.50	0.0908	0.4130	0.05947	0.0963	0.0962	0.0962
58	" " "	2	47	10	2	51	30	69.1	59.4	"	45.50	0.1083	0.4106	0.07701	0.1157	0.1157	0.1157
59	" " "	3	6	0	3	11	0	69.5	59.3	"	50.00	0.1273	0.4079	0.09750	0.1372	0.1372	0.1372
60	" " "	3	22	0	3	26	40	69.9	59.3	"	54.67	0.1462	0.4053	0.11924	0.1593	0.1595	0.1594
61	" " "	3	42	20	3	47	30	70.1	59.4	"	60.00	0.1700	0.4030	0.14866	0.1875	0.1880	0.1877
62	" " "	4	17	0	4	22	30	70.9	59.5	"	62.00	0.1880	0.4013	0.17215	0.2098	0.2102	0.2100
63	" " "	4	40	10	4	45	0	71.3	59.7	"	63.50	0.1974	0.4004	0.18481	0.2215	0.2216	0.2215
64	Oct. 10, A.M.	8	43	0	8	47	0	61.2	59.0	"	62.00	0.1895	0.4012	0.17417	0.2063	0.2066	0.2064
65	" " "	9	51	30	9	55	30	65.0	59.2	A B C D E	62.50	0.1907	0.4010	0.17574	0.2100	0.2101	0.2100
66	" " "	10	55	30	11	5	30	63.8	59.0	"	62.50	0.1893	0.4012	0.17390	0.2091	0.2094	0.2092
67	" " "	11	17	30	11	22	30	64.0	59.0	"	32.50	0.0292	0.4215	0.01107	0.0300	0.0298	0.0299
68	" " "	11	44	0	11	47	0			"	34.25	0.0390	0.4202	0.01703	0.0404	0.0401	0.0402
69	" " P.M.	2	4	30	2	8	30	64.3	59.1	"	35.50	0.0460	0.4192	0.02177	0.0481	0.0478	0.0479
70	" " "	2	23	30	2	28	30	64.8	59.3	"	36.50	0.0563	0.4178	0.02937	0.0589	0.0587	0.0588
71	" " "	2	43	30	2	46	30			"	37.50	0.0621	0.4170	0.03396	0.0652	0.0649	0.0650
72	" " "	2	58	0	3	2	30	65.0	59.5	"	38.50	0.0680	0.4162	0.03884	0.0716	0.0712	0.0714
73	" " "	3	33	30	3	38	0	65.3	59.6	"	39.50	0.0745	0.4153	0.04444	0.0788	0.0785	0.0786
74	" " "	3	51	30	3	57	30	65.6	59.7	"	40.50	0.0801	0.4145	0.04945	0.0849	0.0847	0.0848
75	" " "	4	14	0	4	21	0	66.1	59.7	"	41.50	0.0848	0.4138	0.05378	0.0901	0.0897	0.0899
76	" " "	4	34	0	4	40	0	66.5	59.8	"	42.50	0.0916	0.4129	0.06024	0.0978	0.0975	0.0976
77	" " "	4	57	0	5	1	0	66.2	59.8	"	43.25	0.0960	0.4123	0.06454	0.1025	0.1023	0.1024
78	" " "	5	40	0	5	42	0			"	62.50	0.1931	0.4008	0.17898	0.2191	0.2184	0.2187
79	" 12, A.M.	8	33	0	8	37	30	62.8	59.5	"	62.50	0.1906	0.4011	0.17565	0.2092	0.2090	0.2091
80	" " "	8	51	30	8	56	30	62.7	59.5	"	44.00	0.1003	0.4117	0.06882	0.1041	0.1042	0.1041
81	" " "	9	9	30	9	17	30	62.8	59.6	"	44.75	0.1042	0.4111	0.07277	0.1090	0.1087	0.1088
82	" " "	9	29	0	9	35	0	63.1	59.6	"	45.50	0.1128	0.4099	0.08172	0.1189	0.1184	0.1186
83	" " "	9	51	30	9	57	30	63.2	59.6	"	47.00	0.1150	0.4096	0.08406	0.1210	0.1206	0.1208
84	" " "	10	12	0	10	17	0	64.0	59.6	"	50.00	0.1275	0.4079	0.09773	0.1348	0.1346	0.1347
85	" " "	10	38	30	10	44	0	65.0	59.7	"	54.67	0.1471	0.4052	0.12031	0.1575	0.1575	0.1575
86	" " "	11	6	30	11	10	0	65.6	59.8	"	60.00	0.1697	0.4031	0.14830	0.1846	0.1850	0.1848
87	" " "	11	34	30	11	37	30	67.6	59.9	"	62.00	0.1896	0.4012	0.17431	0.2095	0.2088	0.2091
88	" " "	11	53	30	11	58	30			"	62.50	0.1911	0.4010	0.17630	0.2114	0.2117	0.2115
89	" " P.M.	2	40	0	2	50	0	69.8	60.3	"	62.50	0.1917	0.4010	0.17713	0.2131	0.2136	0.2133
90	" " "	3	3	0	3	11	30	69.8	60.3	"	62.50	0.1919	0.4009	0.17736	0.2136	0.2143	0.2139
91	" " "	4	20	0	4	25	0	69.3	60.6	A	45.50	0.1077	0.4107	0.07639	0.1124	0.1144	0.1134
92	" " "	5	27	30	5	39	30	71.6	60.7	A B C D E	62.50	0.1917	0.4010	0.17713	0.2204	0.2209	0.2206
93	" 16, A.M.	9	18	30	9	21	30	55.0	57.1		40.00	0.0778	0.4148	0.04737	0.0786	0.0793	0.0789
94	" " "	9	58	0	10	1	30	56.1	57.1		32.50	0.0293	0.4215	0.01113	0.0299	0.0296	0.0297
95	" " "	11	4	0	11	7	30	56.6	57.6		35.50	0.0494	0.4187	0.02412	0.0504	0.0503	0.0503
96	" " "	11	39	0	11	45	30	59.2	58.0		36.50	0.0522	0.4184	0.02626	0.0541	0.0533	0.0537
97	" " P.M.	2	10	0	2	14	0				40.00	0.0777	0.4148	0.04728	0.0796	0.0798	0.0797
98	" " "	2	40	0	2	44	0	65.2			37.50	0.0618	0.4170	0.03372	0.0634	0.0639	0.0636
99	" " "	2	59	0	3	3	0	65.6	57.5		38.50	0.0682	0.4161	0.03900	0.0703	0.0702	0.0702
100	" " "	3	19	0	3	23	0	65.7	57.6		39.50	0.0744	0.4153	0.04433	0.0769	0.0769	0.0769
101	" " "	4	11	0	4	14	0	61.9	57.8		40.00	0.0775	0.4148	0.04710	0.0799	0.0801	0.0801

No. of the Exp.	10 Mean height of the surface of the water in compartment E, figures 1 and 2, plate XX., by gauge D. Feet.	11 Effective head producing the discharge. H Feet.	12 Velocity due the head in the preceding column. V Feet per second.	13 Mean velocity by experiment through the smallest section of the tube or orifice. v Feet per second.	14 Ratio of the velocity at the smallest section to the velocity due the head. $\frac{v}{V}$	15 Diameter of the tube or orifice at the place of final discharge. Feet.	16 Mean velocity by experiment at the final discharge from the tube. v' Feet per second.	17 Ratio of the velocity at the final discharge to the velocity due the head. $\frac{v'}{V}$	18 Remarks.
51	1.6327	1.4271	9.5810	21.0550	2.1976	0.3209	2.1189	0.2212	At 8h 30m A.M., October 7, the diaphragm of strainer cloth in the gauging basin was cleaned; it had become obstructed by an accumulation of gummy matter, apparently an exudation from the new pine planks of which the cistern was constructed. October 7, P.M. After the conclusion of experiment 63, the cistern was emptied and the three joints B, C, and D of the diverging tube taken off and examined; all of them, together with the mouth-piece, were a little oxidated, the mouth-piece the least so, and the joints C and D the most; they were then all wiped clean and coated anew with grease; the diverging tube was not put on again to-day. October 10, A.M. The cistern has been kept full of water since October 7. This morning it was emptied, and the grease wiped off the mouth-piece; the joints B, C, and D were put on, the grease having been first wiped off.
52	0.0427	0.0138	0.9422	1.3050	1.3850	"	0.1313	0.1394	
53	0.0809	0.0405	1.6140	2.1243	1.3162	"	0.2138	0.1325	
54	0.1162	0.0588	1.9448	3.5329	1.8166	"	0.3555	0.1828	
55	0.1581	0.0881	2.3805	4.6472	1.9522	"	0.4677	0.1965	
56	0.2173	0.1326	2.9205	6.0757	2.0804	"	0.6114	0.2094	
57	0.2773	0.1811	3.4131	7.3064	2.1407	"	0.7353	0.2154	
58	0.3901	0.2744	4.2012	9.4620	2.2522	"	0.9522	0.2267	
59	0.5740	0.4368	5.3006	11.9790	2.2599	"	1.2055	0.2274	
60	0.7887	0.6293	6.3622	14.6494	2.3025	"	1.4743	0.2317	
61	1.1048	0.9171	7.6806	18.2642	2.3780	"	1.8380	0.2393	
62	1.3872	1.1772	8.7018	21.1509	2.4306	"	2.1286	0.2446	
63	1.5827	1.3612	9.3572	22.7058	2.4266	"	2.2850	0.2442	
64	1.5952	1.3888	9.4516	21.3992	2.2641	"	2.1535	0.2278	
65	1.6283	1.4183	9.5514	21.5920	2.2606	0.4085	1.3409	0.1404	At 9h 0m October 10, the cistern was emptied and the joint E put on. No change was made in the apparatus between experiments 65 and 66; the water flowed continuously from 9h 15m until after the conclusion of experiment 66. October 10, P.M. After the conclusion of experiment 78 the cistern was emptied, and the four joints of the diverging tube taken off. There were only a few slight streaks of oxidation on the mouth-piece; the joints B and C of the diverging tube were oxidated in longitudinal streaks; joints D and E were nearly covered with oxidation, which was however rubbed off with ease, leaving the surface, apparently, as smooth as before. The interior of the mouth-piece and of the four joints of the diverging tube were wiped clean and coated with grease; the diverging tube was not put on again to-day. October 12, A.M. The apparatus was prepared for the experiments of to-day by removing the grease from the interior of the mouth-piece and four joints of the diverging tube, and putting the latter in their places. At 3h 15m P.M., October 12, the cistern was emptied and the tube examined; the interior of the mouth-piece and all the joints were oxidated, and in a little greater degree than after experiment 78 as noted above. The four joints of the diverging tube were taken off, and together with the mouth-piece were well rubbed with a cloth, which removed all the red oxide.
66	1.6165	1.4073	9.5143	21.3653	2.2456	"	1.3268	0.1395	
67	0.0438	0.0139	0.9456	1.3599	1.4381	"	0.0845	0.0893	
68	0.0687	0.0285	1.3540	2.0925	1.5455	"	0.1300	0.0960	
69	0.0858	0.0379	1.5614	2.6741	1.7126	"	0.1661	0.1064	
70	0.1163	0.0575	1.9232	3.6087	1.8764	"	0.2241	0.1165	
71	0.1374	0.0724	2.1580	4.1725	1.9335	"	0.2591	0.1201	
72	0.1596	0.0882	2.3819	4.7719	2.0034	"	0.2963	0.1244	
73	0.1884	0.1098	2.6576	5.4608	2.0546	"	0.3391	0.1276	
74	0.2163	0.1315	2.9084	6.0757	2.0890	"	0.3773	0.1297	
75	0.2423	0.1524	3.1310	6.6070	2.1102	"	0.4103	0.1310	
76	0.2848	0.1872	3.4701	7.4013	2.1329	"	0.4596	0.1325	
77	0.3101	0.2080	3.6578	7.9294	2.1678	"	0.4924	0.1346	
78	1.5010	1.2823	9.0820	21.9899	2.4213	"	1.3656	0.1504	
79	1.6176	1.4085	9.5184	21.5804	2.2672	"	1.3402	0.1408	
80	0.3261	0.2220	3.7789	8.4558	2.2376	"	0.5251	0.1390	
81	0.3539	0.2451	3.9706	8.9407	2.2517	"	0.5552	0.1398	
82	0.4248	0.3062	4.4380	10.0407	2.2624	"	0.6236	0.1405	
83	0.4397	0.3189	4.5291	10.3283	2.2804	"	0.6414	0.1416	
84	0.5557	0.4210	5.2039	12.0072	2.3073	"	0.7457	0.1433	
85	0.7987	0.6412	6.4222	14.7812	2.3016	"	0.9180	0.1429	
86	1.1483	0.9635	7.8725	18.2201	2.3144	"	1.1315	0.1437	
87	1.5575	1.3484	9.3131	21.4161	2.2996	"	1.3300	0.1428	
88	1.5884	1.3769	9.4110	21.6600	2.3016	"	1.3451	0.1429	
89	1.5745	1.3612	9.3572	21.7621	2.3257	"	1.3515	0.1444	
90	1.5588	1.3449	9.3010	21.7907	2.3428	"	1.3533	0.1455	
91	1.6285	1.5151	9.8720	9.3858	0.9507	0.1018	9.3858	0.9507	
92	1.5069	1.2863	9.0961	21.7621	2.3925	0.4085	1.3515	0.1486	At 4h 30m P.M., October 12, the cistern was emptied again, and the four joints of the diverging tube re-attached. At 6 P.M. the cistern was emptied; the wide part of the mouth-piece was much oxidated, but only slightly so at the smallest section. The diverging tube was oxidated in only a few spots.
93	1.5925	1.5136	9.8671	5.8316	0.5910	0.1017	5.8316	0.5910	Orifice in a thin plate. The plate, Figs. 12 and 13, plate XXI., containing the orifice, was put on October 14; the accessible parts of it were greased, and the cistern filled with water, and so remained until October 16, A.M., when it was emptied, and the grease wiped off. No oxidation was observed. At 0h 15m P.M., October 16, the cistern was emptied and the plate examined; there was a thin coating of oxide over most of the surface; all the accessible parts of the plate were wiped clean and greased anew. At 1h 15m P.M., the grease was wiped off again.
94	0.1213	0.0916	2.4274	1.3695	0.5642	"	1.3695	0.5642	
95	0.4855	0.4352	5.2909	2.9783	0.5629	"	2.9783	0.5629	
96	0.5372	0.4835	5.5768	3.2328	0.5797	"	3.2328	0.5797	
97	1.5784	1.4987	9.8181	5.8203	0.5928	"	5.8203	0.5928	
98	0.8100	0.7764	7.0669	4.1504	0.5873	"	4.1504	0.5873	
99	1.0344	1.0212	8.1167	4.8012	0.5915	"	4.8012	0.5915	
100	1.4004	1.3235	9.2267	5.4601	0.5918	"	5.4601	0.5918	
101	1.5701	1.1903	8.7909	5.7979	0.5922	"	5.7979	0.5922	

DESCRIPTION OF TABLE XXVII., CONTAINING THE EXPERIMENTS ON THE FLOW OF WATER THROUGH SUBMERGED TUBES AND ORIFICES.

258. The greater portion of this table will be intelligible from the headings of the several columns, without further explanation.

As previously stated, the quantity of water flowing was gauged by means of a weir of substantially the same form and dimensions as that used by Poncelet and Lesbros, in their experiments made at Metz in 1827 and 1828. Table X., *Experiences hydrauliques*, &c., previously cited, contains the results of the experiments made in 1828. The quantities E discharged by experiment with certain depths on the weir are given; also the quantities with the same depths, computed by the formula $d = l\,h\,\sqrt{2\,g\,h}$; also the values of $\frac{E}{d}$. These last quantities are the values of the coefficient C, by means of which the real discharge can be deduced from the value of d. We can then compute the real discharge by the formula

$$D = C\,l\,h\,\sqrt{2\,g\,h}.$$

The value of C is not the same for all depths, as may be seen by the following table, which contains the principal results of table X. of Poncelet and Lesbros above cited, changing the unit from metres to English feet. The length of the weir l was 0.10 metres or 0.6562 foot.

Depth of water on the weir, taken 11.48 feet up stream from the weir. h Feet.	Discharge by experiment. E Cubic feet per second.	Discharge computed by the formula $d = lh\sqrt{2gh}$. Cubic feet per second.	Value of C in the formula. $D = Clh\sqrt{2gh}$.
0.6821	1.1528	2.9656	0.3888
0.5351	0.8098	2.0608	0.3930
0.3376	0.4071	1.0327	0.3943
0.1985	0.1864	0.4655	0.4003
0.1463	0.1194	0.2947	0.4053
0.0771	0.0468	0.1127	0.4149

The values of C, given in column 7, are deduced from the values of C in the preceding table, by interpolation. The quantities of water discharged by the tube or orifice given in column 8 are computed by the formula $D = C\,l\,h\,\sqrt{2\,g\,h}$, in which C has the value given in column 7; the length of the weir l, by measurement, = 0.6579 foot; h = the value given in column 6, and $g = 32.1618$, which is its value for the place where the experiments were made (art. 68).

259. As previously stated, according to the first design of the apparatus, the weir was intended to be placed in the partition N, figures 1 and 2, plate XX., and the depth on the weir was intended to be measured by the hook gauge B; on trial, however, it was found that the agitation in the compartment F was too great to admit of a satisfactory gauge being made with the weir in this position, and it was accordingly removed to the position represented in the figures. The hook gauge B was allowed to remain, and the height of the surface of the water in the compartment F was observed by means of both the gauges B and C, and the mean of the two is taken as the elevation of the surface of the water in this compartment. By comparing the heights taken at the two gauges, given in column 9, it will be seen that, when the quantity of water discharged was small, there was little or no difference in the indications of the two gauges; with the larger volumes, the height at gauge B was sensibly the greatest.

The effective head producing the discharge given in column 11 is the difference of the heights of the surface of the water in compartments E and F.

The velocity given in column 12 is computed by the formula $V = \sqrt{2gh}$.

260. The smallest section of the compound tube is in the mouth-piece between a and b, figure 2, plate XXI., and was found, by careful and repeated measurements made by different persons, to be 0.1018 foot. The diameter of the orifice in the thin plate was found in a similar manner to be 0.1017 foot. The area of the orifice in the mouth-piece was consequently 0.0081393 square foot, and the area of the orifice in the thin plate was 0.0081233 square foot. The velocities given in column 13 are obtained by dividing the quantities given in column 8 by the area of the smallest section through which the water was discharged.

DEDUCTIONS FROM THE EXPERIMENTS GIVEN IN TABLE XXVII.

261. Confining ourselves, for the present, to the velocities at the smallest section, we find by these experiments that in all the tubes and orifices used the ratio of the velocity at the smallest section to the velocity due the head is least when the heads are very small. Thus with the mouth-piece A alone,

When the effective head is 0.0339 foot (experiment 1), the ratio is 0.8150
" " " 0.2300 " (" 6), " " 0.8626
" " " 0.9579 " (" 13), " " 0.9277
" " " 1.5140 feet (" 17), " " 0.9407

With the mouth-piece *A* and the first joint *B* of the diverging tube,

When the effective head is 0.0202 foot (experiment 20), the ratio is 1.1506
" " " 0.0996 " (" 23), " " 1.3958
" " " 0.8544 " (" 37), " " 1.5919
" " " 1.4704 feet (" 38), " " 1.5748

With the mouth-piece *A* and the two first joints *B* and *C* of the diverging tube,

When the effective head is 0.0143 foot (experiment 40), the ratio is 1.4179
" " " 0.0616 " (" 42), " " 1.7844
" " " 1.0999 feet (" 49), " " 2.1643
" " " 1.3116 " (" 50), " " 2.1232

With the mouth-piece *A* and the three first joints *B*, *C*, and *D* of the diverging tube,

When the effective head is 0.0138 foot (experiment 52), the ratio is 1.3850
" " " 0.0588 " (" 54), " " 1.8166
" " " 1.1772 feet (" 62), " " 2.4306
" " " 1.3612 " (" 63), " " 2.4266

With the complete compound tube,

When the effective head is 0.0139 foot (experiment 67), the ratio is 1.4381
" " " 0.0575 " (" 70), " " 1.8764
" " " 1.2823 feet (" 78), " " 2.4213
" " " 1.4085 " (" 79), " " 2.2672

With the thin plate,

When the effective head is 0.0916 foot (experiment 94), the ratio is 0.5642
" " " 0.4835 " (" 96), " " 0.5797
" " " 1.0242 feet (" 99), " " 0.5915
" " " 1.4903 " (" 101), " " 0.5922

262. By the preceding extracts from table XXVII. it will be seen that *the ratio of the velocity at the smallest section of the tube or orifice to the velocity due the head is the least when the effective head is the least, and in the cases of the mouth-piece and orifice in the thin plate, the ratio is the greatest when the effective head is the greatest.*

In the case of the diverging tube, the value of the ratio is a maximum when the effective head is somewhat less than the greatest.

It is the general result of the great number of experiments on record, on the flow of water through orifices in a thin plate, discharging freely into the air, that the coefficient of discharge (which in simple orifices is the same thing as the ratio of the velocity at the smallest section of the orifice to the velocity due the head) is greatest for very small heads. In these experiments where the discharge takes place under water, the coefficient of discharge is least with the very small heads. This result is so marked and uniform that there can be no doubt of the fact.

263. As to the value of the coefficient of discharge for the mouth-piece *A*, a mean of all the experiments in which the effective head is not less than 1.5 feet gives 0.9451, the mean effective head being 1.5150 feet. This is nearly the same as the greatest value of the coefficient of discharge found by Castel for the smallest section of an orifice in a converging conical tube, namely, 0.956, which is for a tube in which the sides converge at an angle of 13° 40′, and discharging freely into the air.* Michelotti, in one of his experiments, by employing a cycloidal tube, found it 0.983.† Eytelwein found 0.9798.‡ Other experimenters have found from 0.96 to 0.98. We must, therefore, conclude that *the coefficient of discharge for the mouth-piece A, when discharging under water, is about 3 per cent less than has been found for similar orifices when discharging freely into the air.*

264. The value of the coefficient of discharge for the orifice in a thin plate, taking the mean of the three experiments in which the effective head is near 1.5 feet, is 0.5920, the mean effective head being 1.5009 feet. This is less than has been found for circular orifices in a thin plate discharging freely into the air. There are great numbers of these experiments on record, made with orifices of various diameters and under various heads. The general result for the coefficient of discharge is very nearly 0.62. We must, therefore, conclude that *the flow through a submerged orifice in a thin plate is less than when the discharge takes place freely into the air, in the ratio of 0.59 to 0.62, or about 5 per cent less.*

265. The values of the ratio of the velocity at the smallest section to the velocity due the head, for the several combinations of the mouth-piece and the diverging tube, taking the largest values found in these experiments, are as follows: —

* D'Aubuisson's Hydraulics, Bennett's translation, page 56.

† Mémoires de l'Académie Royale des Sciences de Turin, 1784–85.

‡ Handbuch der Mechanik und der Hydraulik.

For the mouth-piece A alone . . . . (exp. 91) 0.9507

For the mouth-piece A and the first joint B of the diverging tube . . . . (" 32) 1.5955

For the mouth-piece A and the first two joints B and C of the diverging tube . . . . (" 49) 2.1643

For the mouth-piece A and the first three joints B, C, and D of the diverging tube . . . . (" 62) 2.4306

For the complete compound tube as represented by figure 2, plate XXI. . . . . (" 78) 2.4213

The maximum effect was produced with the mouth-piece and first three joints of the diverging tube, the addition of the fourth joint caused a slight diminution. In experiment 62, giving the greatest effect, the increase in the velocity of the water in the smallest section due to the diverging tubes is in the ratio of 0.9507 to 2.4306, or as 1 to 2.5566. To produce this increased velocity in the smallest section without using the diverging tube the head must be increased in the ratio of 1 to $(2.5566)^2$ or as 1 to 6.5364. The effective head in experiment 62 was 1.1772 feet. To give the velocity in the same experiment, if the diverging tube had not been attached, would have required an effective head of $1.1772 \times 6.5364 = 7.6947$ feet. The difference in these heads is $7.6947 - 1.1772 = 6.5175$ feet. A portion of the pressure of the atmosphere on the surface of the water in the upper division E of the cistern, figures 1 and 2, plate XX., equivalent to this head of water, is rendered active by the addition of the diverging tube to the mouth-piece.

266. According to Bernoulli's theory, the velocity of the water at its final discharge from the tube should be that due to the head;* in experiment 62 this

* Call A the area of the section and V the velocity of the water at $a\,b$, figure 2, plate XX. B the area of the section and v the velocity at $c\,d$; h = the head or difference of height of the surface of the water in compartments E and F. The motion having become permanent, we have

$$A\,V = B\,v.$$

The volume of water included between the sections $a\,b$ and $c\,d$ in the small time t will move to $a'\,b'\,c'\,d'$; the volume included between the sections $a'\,b'$ and $c\,d$ is common to both positions, every particle in one having its counterpart in the other, both in position and velocity. In finding the change in the living force in the two positions, we need only consider the volumes $a\,a'\,b\,b'$ and $c\,c'\,d\,d'$. These volumes are equal, and assuming the water to be pure and at its maximum density, the weight of each is $62.382\,A\,V\,t$.

velocity is 8.7018 feet per second; the velocity at other parts of the compound tube would be inversely as the squares of the diameters; at the smallest section the velocity must be greater than at the final discharge in the ratio of 1 to $\left(\frac{0.3209}{0.1018}\right)^2 = 9.9367$. To give this velocity at the smallest section without the diverging tube would require the effective head of water to be increased from 1.1772 feet to $1.1772 \times (9.9367)^2 = 116.24$ feet; the increase being 115.06 feet; if the pressure of the atmosphere was great enough, its pressure, to this extent, would be rendered active. The total pressure of the atmosphere is usually about 34 feet, and this of course is the limit to which it can be rendered active. Abstracting from the effects of vaporization, whenever the exhausting effect of the diverging tube exceeds the pressure of the atmosphere, (added to the pressure due to the actual head of water at the smallest section,) breaks must occur in the mass of water in the compound tube, at or near the smallest section, and the flow through the smallest section will be the same as if the discharge took place in a vacuum. In experiment 62, the exhausting effect of the diverging tube,

---

The living force of the volume $a a' b b'$ is $\frac{62.382\, A\, V t}{g} V^2$

" " " " " " $c c' d d'$ is $\frac{62.382\, A\, V t}{g} v^2$

The increase of living force in passing from one position to the other being

$$\frac{62.382\, A\, V t}{g} (v^2 - V^2) \tag{1.}$$

This increase of living force is produced by the action of gravity on the volume of water $A\,Vt$ descending through the height h, which is equivalent to an amount of work represented by

$$62.382\, A\, V t\, h. \tag{2.}$$

By the doctrine of living forces, the living force (1.) is equivalent to the amount of work represented by

$$\frac{62.382\, A\, V t}{2 g} (v^2 - V^2) \tag{3.}$$

The amount of work in (2.) and (3.) must be equal; we have, therefore,

$$62.382\, A\, V t\, h = \frac{62.382\, A\, V t}{2 g} (v^2 - V^2);$$

from which we deduce

$$h = \frac{v^2 - V^2}{2 g}$$

If V is very small relatively to v, it may be neglected, and we have

$$h = \frac{v^2}{2 g}, \text{ and } v = \sqrt{2 g h}.$$

according to Bernoulli's theory, exceeds three times the actual pressure at the smallest section, and if it had produced its full effect according to theory or even one third of that effect, breaks must have occurred in the mass of water near the smallest section.

The ratio of the actual velocity of the water at its final discharge to the velocity according to Bernoulli's theory is given in column 17. In experiment 62 it is 0.2446, or about one quarter of the velocity due the head, indicating a loss of about fifteen sixteenths of the living force. It is difficult to see how so much can be lost. There are no abrupt changes in velocity, and the interior surfaces of the mouth-piece and diverging tube are smooth and free from sensible irregularity. The slight oxidation observable after some of the experiments appears to have produced no sensible loss, as in experiment 62, which gave the greatest result, there was considerable oxidation, while in other experiments giving a less effect there was no oxidation.

The chief discrepancy between the hypothesis on which Bernoulli's theory is founded and the real conditions of the motion appears to be due to the retarding effects of the walls of the tube. According to the hypothesis, the velocity in all parts of the same section is the same; Prony's well-known formula for the motion of water in pipes is founded upon the idea that the principal retardation is due to the sides; whence it follows, that the velocity must be least at the sides and greatest at the centre. Darcy * made many experiments on the subject by means of Pitot's tube, and found that in long straight pipes there was a material variation in the velocities at different distances from the centre, and determined a formula expressing the law of the variation. It would not be safe to apply this formula to these experiments on account of the short length and varying diameter of the compound tube, but it is clear that variations in the velocity must exist to an extent which must greatly modify the results deduced from Bernoulli's theory.

267. As previously stated, Venturi, by adding a diverging tube increased the discharge of an orifice having nearly the form of the contracted vein, and discharging freely into the air, in the ratio of 1 to 2.21. In these experiments, in an orifice without contraction discharging under water the discharge was increased by adding a diverging tube in the ratio of 1 to 2.56. Making the comparison with an orifice in a thin plate, the maximum coefficient of discharge with the thin plate is 0.5928, and with the mouth-piece of cycloidal form and diverging tube, the maximum coefficient

* *Recherches expérimentales relatives au Mouvement de l'Eau dans les Tuyaux*, par Henry Darcy. Paris, 1857.

is 2.4306; the discharge with the same area of orifice and the same head being increased in the ratio of 1 to 4.12.

268. Considerable irregularities will be observed in the value of the ratio of the velocity in the smallest section to the velocity due the head, given in column 14. Thus, in the experiments with the complete compound tube, we have the following, which were intended to be identical, the repetitions being made for the purpose of detecting such variations, if any should occur. In all these experiments the index of the inlet cock, *L*, figures 2 and 3, plate XX., was set at the same point, viz. 62.5°, or as nearly so as practicable, in order to admit the same quantity of water.

Number of the experiment in Table XXVII.	Quantity of water discharged; in Cubic feet per second.	Effective head producing the discharge; in feet.	Ratio of the velocity at the smallest section to the velocity due the head.
65	0.17574	1.4183	2.2606
66	0.17390	1.4073	2.2456
78	0.17898	1.2823	2.4213
79	0.17565	1.4085	2.2672
88	0.17630	1.3769	2.3016
89	0.17713	1.3612	2.3257
90	0.17736	1.3449	2 3428
92	0.17713	1.2863	2.3925

In the preceding table, the small irregularities in the quantities of water discharged are due to corresponding small variations in setting the index of the inlet cock. The irregularities in the effective head are mainly due to changes in the efficiency of the diverging tube. The only known variation on which these changes could depend is in the state of the interior surface of the tube. Thus No. 65 was the second experiment made after the grease was wiped off. Twelve experiments were made between Nos. 65 and 78, no change being made in the state of the surface, except that caused by the action of the water, which undoubtedly had washed off, before No. 78 was made, a part or the whole of the grease not removed by wiping. In the experiments made soon after wiping the surface, it is probable that the water was repelled from it by the grease, but after the water had run through the tube for some hours the grease was washed off sufficiently to permit the water to come in contact with the iron, which appears to have increased, materially, the exhausting effect of the diverging tube.

269. Previous to making the experiments, it was anticipated that when the diverging tube was used there would be sensible oscillations in the elevation of the surface of the water in compartment *E*, figures 1 and 2, plate XX., due to the unstable equilibrium of the stream. Although the amplitudes of the oscillations

of the surface were much less than was expected, they were quite sensible. Thus we find, by referring to the original notes, that with the mouth-piece alone, the amplitude of the oscillations,

when	the	effective	head	was	0.10 foot,	was	about	0.0003	foot.
"	"	"	"	"	1.00 "	"	"	0.0006	"
"	"	"	"	"	1.40 feet	"	"	0.0007	"

With the complete compound tube the amplitude of the oscillations,

when	the	effective	head	was	0.10 foot,	was	about	0.0021	foot.
"	"	"	"	"	1.00 "	"	"	0.0103	"
"	"	"	"	"	1.40 feet	"	"	0.0117	"

The variation with heads from 1.00 foot to 1.40 feet being about 17 times as great with the complete diverging tube as with the mouth-piece alone.

270. As previously stated, the principles involved in the flow of water through a diverging tube find a useful application in Mr. Boyden's Diffuser. This invention, applied to a turbine water-wheel 104.25 inches in diameter and about seven hundred horse power, is represented in plates XXII. and XXIII. This turbine is one of four of the same power constructed from the designs of the author for the cotton-mills of the Merrimack Manufacturing Company in Lowell. Plate XXII. is a sectional elevation through the axis, showing the lower parts of the apparatus. *a, a, a, a* is the wheel, carrying 60 floats of Russian sheet iron, 0.15 inch thick; *b* the main shaft, which is suspended from the top, in a similar manner to the Tremont turbines (plate I.); *c, c* is the disc, carrying 33 guides, *c', c', c', c'*, of Russian sheet iron, 0.125 inch thick, which lean one horizontally to six vertically; *d, d*, the disc pipe, which hangs at its upper end, upon a part of the curved pipe or curb *e, e*, not represented in the plate; *f, f*, the garniture, which supports the upper part of the guides, and is curved at its lower edge, in order to afford a favorable aperture for the flow of the water entering the wheel; *g, g*, the lower curb; *h, h*, the speed gate, which is represented as raised to its greatest height; *i*, a gate rod, which with two others, not represented in the plate, enables the gate to be moved by the governor or by hand; *k, k*, beams extending from the granite walls of the wheel-pit to the lower curb and supporting the latter; *l, l*, pillars resting upon granite blocks in the floor of the wheel-pit, and supporting the beams *k, k*; *m, m*, the diffuser, which is supported by the pillars *l, l*, by means of the curved beams *n, n, n, n*; *w, w*, low water level of the surface of the water in the wheel-pit. The wheel is placed suf-

ficiently low, to permit the diffuser to be submerged at all times when the wheel is in operation, that being essential to the most advantageous operation of the diffuser. Figure 1, plate XXIII., is a horizontal section through the wheel, showing also the disc, guides, and garniture, and also the lower part of the diffuser. Figure 2 is a horizontal section on a larger scale, showing part of the wheel, guides, and diffuser. Figure 3 is a vertical section, showing part of the wheel, diffuser, &c.

When the speed gate is fully raised, and the wheel is moving with the velocity giving its greatest coefficient of useful effect, the water passes through the wheel in a path, which is nearly represented by the dotted line *a*, *b*, figure 2, plate XXIII. On leaving the wheel it necessarily has considerable velocity, which would involve a corresponding loss of power, except for the effect of the diffuser, which utilizes a portion of it. When operating under a fall of 33 feet and the speed gate raised to its full height, this wheel discharges about 219 cubic feet of water per second. The area of the annular space *o*, *o*, *o*, *o*, plate XXII., where the water enters the diffuser, is $0.802 \times 8.792\,\pi = 22.152$ square feet; and if the stream passes through this section radially, its mean velocity must be $\frac{219}{22.152} = 9.886$ feet per second, which is due to a head of 1.519 feet. The area of the annular space *p*, *p*, *p*, *p*, where the water leaves the diffuser, is $1.5 \times 15.333\,\pi = 72.255$ square feet, and the mean velocity $\frac{219}{72.255} = 3.031$ feet per second, which is due to a head of 0.143 feet. According to this, the saving of head, due to the diffuser is $1.519 - 0.143 = 1.376$ feet, being $\frac{1.376}{33. - 1.519}$, or about $4\frac{3}{8}$ per cent of the head available without the diffuser, which is equivalent to a gain in the coefficient of useful effect to the same extent. As previously stated (art. 12), experiments on the same turbine, with and without a diffuser, have shown a gain due to the latter, of about 3 per cent in the coefficient of useful effect. The diffuser adds to the coefficient of useful effect by increasing the velocity of the water passing through the wheel, and it must of course increase the quantity of water discharged in the same proportion. If it increases the available head 3 per cent, the velocity, which varies as the square root of the head, must be increased about 1.5 per cent, and the quantity discharged must be increased in the same proportion. The power of the wheel, which varies as the product of the head into the quantity of water discharged, must be increased about 4.5 per cent.

EXPLANATION OF TABLES XXVIII., XXIX., AND XXX.

These tables have been prepared in the office of the Proprietors of the Locks and Canals on Merrimack River, for the purpose of facilitating the computations connected with gauging the quantities of water drawn from their canals at Lowell.

Table XXVIII. gives the velocities of floats for eight different distances between the transit stations, and for times of passage between them for every tenth of a second, from 20 to 100 seconds.

The use of the table may be extended to such other distances between the transit stations as are multiplies or submultiplies of the distances given in the table, by taking the time the same multiple or submultiple as the distance.

Table XXIX. gives the values of the coefficient $\left(1 - 0.116\ (\sqrt{D} - 0.1)\right)$ for values of D for every 0.001 from 0.000 to 0.100, with the logarithms of the same. (See art. 233.)

Table XXX. gives the velocities, in feet per second, due to every 0.01 foot head, from 0.00 to 49.99 feet, computed for Lowell, by the formulas given in art. 68. These formulas, reduced to the English foot as the unit, become

$$g = 32.1695\ (1 - 0.00284 \cos. 2\ l)\left(1 - \frac{2\,e}{r}\right)$$
$$r = 20887540\ (1 + 0.00164 \cos. 2\ l).$$

The values of g by these formulas for several latitudes and heights above the sea are given in the following table:—

Height above the Sea.	Latitude.						
Feet.	30°	35°	40°	45°	50°	55°	60°
0	32.1239	32.1383	32.1537	32.1695	32.1854	32.2008	32.2152
100	32.1236	32.1380	32.1534	32.1692	32.1851	32.2005	32.2149
200	32.1233	32.1377	32.1531	32.1689	32.1848	32.2002	32.2146
300	32.1229	32.1374	32.1528	32.1686	32.1845	32.1998	32.2143
400	32.1226	32.1371	32.1524	32.1683	32.1842	32.1995	32.2140
500	32.1223	32.1368	32.1521	32.1680	32.1839	32.1992	32.2137
600	32.1220	32.1364	32.1518	32.1677	32.1835	32.1989	32.2134
700	32.1217	32.1361	32.1515	32.1674	32.1832	32.1986	32.2131
800	32.1214	32.1358	32.1512	32.1671	32.1829	32.1983	32.2128
900	32.1211	32.1355	32.1509	32.1668	32.1826	32.1980	32.2125
1000	32.1208	32.1352	32.1506	32.1665	32.1823	32.1977	32.2121
1100	32.1205	32.1349	32.1503	32.1662	32.1820	32.1974	32.2118

TABLE XXVIII.

TABLE OF VELOCITIES OF TUBES IN MEASURING FLUMES, IN FEET PER SECOND. THE TIME OCCUPIED IN PASSING FROM THE UPSTREAM TO THE DOWNSTREAM TRANSIT STATION, AND THE DISTANCE BETWEEN THEM, BEING GIVEN.

TIME.	DISTANCE BETWEEN THE TRANSIT STATIONS, IN FEET.								TIME.	DISTANCE BETWEEN THE TRANSIT STATIONS, IN FEET.							
Sec's.	50.	60.	70.	80.	90.	100.	110.	120.	Sec's.	50.	60.	70.	80.	90.	100.	110.	120.
20.0	2.500	3.000	3.500	4.000	4.500	5.000	5.500	6.000	25.0	2.000	2.400	2.800	3.200	3.600	4.000	4.400	4.800
20.1	2.488	2.985	3.483	3.980	4.478	4.975	5.473	5.970	25.1	1.992	2.390	2.789	3.187	3.586	3.984	4.382	4.781
20.2	2.475	2.970	3.465	3.960	4.455	4.950	5.446	5.941	25.2	1.984	2.381	2.778	3.175	3.571	3.968	4.365	4.762
20.3	2.463	2.956	3.448	3.941	4.433	4.926	5.419	5.911	25.3	1.976	2.372	2.767	3.162	3.557	3.953	4.348	4.743
20.4	2.451	2.941	3.431	3.922	4.412	4.902	5.392	5.882	25.4	1.969	2.362	2.756	3.150	3.543	3.937	4.331	4.724
20.5	2.439	2.927	3.415	3.902	4.390	4.878	5.366	5.854	25.5	1.961	2.353	2.745	3.137	3.529	3.922	4.314	4.706
20.6	2.427	2.913	3.398	3.883	4.369	4.854	5.340	5.825	25.6	1.953	2.344	2.734	3.125	3.516	3.906	4.297	4.687
20.7	2.415	2.899	3.382	3.865	4.348	4.831	5.314	5.797	25.7	1.946	2.335	2.724	3.113	3.502	3.891	4.280	4.669
20.8	2.404	2.885	3.365	3.846	4.327	4.808	5.288	5.769	25.8	1.938	2.326	2.713	3.101	3.488	3.876	4.264	4.651
20.9	2.392	2.871	3.349	3.828	4.306	4.785	5.263	5.742	25.9	1.931	2.317	2.703	3.089	3.475	3.861	4.247	4.633
21.0	2.381	2.857	3.333	3.810	4.286	4.762	5.238	5.714	26.0	1.923	2.308	2.692	3.077	3.462	3.846	4.231	4.615
21.1	2.370	2.844	3.318	3.791	4.265	4.739	5.213	5.687	26.1	1.916	2.299	2.682	3.065	3.448	3.831	4.215	4.598
21.2	2.358	2.830	3.302	3.774	4.245	4.717	5.189	5.660	26.2	1.908	2.290	2.672	3.053	3.435	3.817	4.198	4.580
21.3	2.347	2.817	3.286	3.756	4.225	4.695	5.164	5.634	26.3	1.901	2.281	2.662	3.042	3.422	3.802	4.183	4.563
21.4	2.336	2.804	3.271	3.738	4.206	4.673	5.140	5.607	26.4	1.894	2.273	2.652	3.030	3.409	3.788	4.167	4.545
21.5	2.326	2.791	3.256	3.721	4.186	4.651	5.116	5.581	26.5	1.887	2.264	2.642	3.019	3.396	3.774	4.151	4.528
21.6	2.315	2.778	3.241	3.704	4.167	4.630	5.093	5.556	26.6	1.880	2.256	2.632	3.008	3.383	3.759	4.135	4.511
21.7	2.304	2.765	3.226	3.687	4.147	4.608	5.069	5.530	26.7	1.873	2.247	2.622	2.996	3.371	3.745	4.120	4.494
21.8	2.294	2.752	3.211	3.670	4.128	4.587	5.046	5.505	26.8	1.866	2.239	2.612	2.985	3.358	3.731	4.104	4.478
21.9	2.283	2.740	3.196	3.653	4.110	4.566	5.023	5.479	26.9	1.859	2.230	2.602	2.974	3.346	3.717	4.089	4.461
22.0	2.273	2.727	3.182	3.636	4.091	4.545	5.000	5.455	27.0	1.852	2.222	2.593	2.963	3.333	3.704	4.074	4.444
22.1	2.262	2.715	3.167	3.620	4.072	4.525	4.977	5.430	27.1	1.845	2.214	2.583	2.952	3.321	3.690	4.059	4.428
22.2	2.252	2.703	3.153	3.604	4.054	4.505	4.955	5.405	27.2	1.838	2.206	2.574	2.941	3.309	3.676	4.044	4.412
22.3	2.242	2.691	3.139	3.587	4.036	4.484	4.933	5.381	27.3	1.832	2.198	2.564	2.930	3.297	3.663	4.029	4.396
22.4	2.232	2.679	3.125	3.571	4.018	4.464	4.911	5.357	27.4	1.825	2.190	2.555	2.920	3.285	3.650	4.015	4.380
22.5	2.222	2.667	3.111	3.556	4.000	4.444	4.889	5.333	27.5	1.818	2.182	2.545	2.909	3.273	3.636	4.000	4.364
22.6	2.212	2.655	3.097	3.540	3.982	4.425	4.867	5.310	27.6	1.812	2.174	2.536	2.899	3.261	3.623	3.986	4.348
22.7	2.203	2.643	3.084	3.524	3.965	4.405	4.846	5.286	27.7	1.805	2.166	2.527	2.888	3.249	3.610	3.971	4.332
22.8	2.193	2.632	3.070	3.509	3.947	4.386	4.825	5.263	27.8	1.799	2.158	2.518	2.878	3.237	3.597	3.957	4.317
22.9	2.183	2.620	3.057	3.493	3.930	4.367	4.803	5.240	27.9	1.792	2.151	2.509	2.867	3.226	3.584	3.943	4.301
23.0	2.174	2.609	3.043	3.478	3.913	4.348	4.783	5.217	28.0	1.786	2.143	2.500	2.857	3.214	3.571	3.929	4.286
23.1	2.165	2.597	3.030	3.463	3.896	4.329	4.762	5.195	28.1	1.779	2.135	2.491	2.847	3.203	3.559	3.915	4.270
23.2	2.155	2.586	3.017	3.448	3.879	4.310	4.741	5.172	28.2	1.773	2.128	2.482	2.837	3.191	6.546	3.901	4.255
23.3	2.146	2.575	3.004	3.433	3.863	4.292	4.721	5.150	28.3	1.767	2.120	2.473	2.827	3.180	3.534	3.887	4.240
23.4	2.137	2.564	2.991	3.419	3.846	4.274	4.701	5.128	28.4	1.761	2.113	2.465	2.817	3.169	3.521	3.873	4.225
23.5	2.128	2.553	2.979	3.404	3.830	4.255	4.681	5.106	28.5	1.754	2.105	2.456	2.807	3.158	3.509	3.860	4.211
23.6	2.119	2.542	2.966	3.390	3.814	4.237	4.661	5.085	28.6	1.748	2.098	2.448	2.797	3.147	3.497	3.846	4.196
23.7	2.110	2.532	2.954	3.376	3.797	4.219	4.641	5.063	28.7	1.742	2.091	2.439	2.787	3.136	3.484	3.833	4.181
23.8	2.101	2.521	2.941	3.361	3.782	4.202	4.622	5.042	28.8	1.736	2.083	2.431	2.778	3.125	3.472	3.819	4.167
23.9	2.092	2.510	2.929	3.347	3.766	4.184	4.603	5.021	28.9	1.730	2.076	2.422	2.768	3.114	3.460	3.806	4.152
24.0	2.083	2.500	2.917	3.333	3.750	4.167	4.583	5.000	29.0	1.724	2.069	2.414	2.759	3.103	3.448	3.793	4.138
24.1	2.075	2.490	2.905	3.320	3.734	4.149	4.564	4.979	29.1	1.718	2.062	2.405	2.749	3.093	3.436	3.780	4.124
24.2	2.066	2.479	2.893	3.306	3.719	4.132	4.545	4.959	29.2	1.712	2.055	2.397	2.740	3.082	3.425	3.767	4.110
24.3	2.058	2.469	2.881	3.292	3.704	4.115	4.527	4.938	29.3	1.706	2.048	2.389	2.730	3.072	3.413	3.754	4.096
24.4	2.049	2.459	2.869	3.279	3.689	4.098	4.508	4.918	29.4	1.701	2.041	2.381	2.721	3.061	3.401	3.741	4.082
24.5	2.041	2.449	2.857	3.265	3.673	4.082	4.490	4.898	29.5	1.695	2.034	2.373	2.712	3.051	3.390	3.729	4.068
24.6	2.033	2.439	2.846	3.252	3.659	4.065	4.472	4.878	29.6	1.689	2.027	2.365	2.703	3.041	3.378	3.716	4.054
24.7	2.024	2.429	2.834	3.239	3.644	4.049	4.453	4.858	29.7	1.684	2.020	2.357	2.694	3.030	3.367	3.704	4.040
24.8	2.016	2.419	2.823	3.226	3.629	4.032	4.435	4.839	29.8	1.678	2.013	2.349	2.685	3.020	3.356	3.691	4.027
24.9	2.008	2.410	2.811	3.213	3.614	4.016	4.418	4.819	29.9	1.672	2.007	2.341	2.676	3.010	3.344	3.679	4.013

TABLE XXVIII — CONTINUED.

TABLE OF VELOCITIES OF TUBES IN MEASURING FLUMES, IN FEET PER SECOND. THE TIME OCCUPIED IN PASSING FROM THE UPSTREAM TO THE DOWNSTREAM TRANSIT STATION, AND THE DISTANCE BETWEEN THEM, BEING GIVEN.

TIME.	DISTANCE BETWEEN THE TRANSIT STATIONS, IN FEET.								TIME.	DISTANCE BETWEEN THE TRANSIT STATIONS, IN FEET.							
Sec's.	50.	60.	70.	80.	90.	100.	110.	120.	Sec's.	50.	60.	70.	80.	90.	100.	110.	120.
30.0	1.667	2.000	2.333	2.667	3.000	3.333	3.667	4.000	35.0	1.429	1.714	2.000	2.286	2.571	2.857	3.143	3.429
30.1	1.661	1.993	2.326	2.658	2.990	3.322	3.654	3.987	35.1	1.425	1.709	1.994	2.279	2.564	2.849	3.134	3.419
30.2	1.656	1.987	2.318	2.649	2.980	3.311	3.642	3.974	35.2	1.420	1.705	1.989	2.273	2.557	2.841	3.125	3.409
30.3	1.650	1.980	2.310	2.640	2.970	3.300	3.630	3.960	35.3	1.416	1.700	1.983	2.266	2.550	2.833	3.116	3.399
30.4	1.645	1.974	2.303	2.632	2.961	3.289	3.618	3.947	35.4	1.412	1.695	1.977	2.260	2.542	2.825	3.107	3.390
30.5	1.639	1.967	2.295	2.623	2.951	3.279	3.607	3.934	35.5	1.408	1.690	1.972	2.254	2.535	2.817	3.099	3.380
30.6	1.634	1.961	2.288	2.614	2.941	3.268	3.595	3.922	35.6	1.404	1.685	1.966	2.247	2.528	2.809	3.090	3.371
30.7	1.629	1.954	2.280	2.606	2.932	3.257	3.583	3.909	35.7	1.401	1.681	1.961	2.241	2.521	2.801	3.081	3.361
30.8	1.623	1.948	2.273	2.597	2.922	3.247	3.571	3.896	35.8	1.397	1.676	1.955	2.235	2.514	2.793	3.073	3.352
30.9	1.618	1.942	2.265	2.589	2.913	3.236	3.560	3.883	35.9	1.393	1.671	1.950	2.228	2.507	2.786	3.064	3.343
31.0	1.613	1.935	2.258	2.581	2.903	3.226	3.548	3.871	36.0	1.389	1.667	1.944	2.222	2.500	2.778	3.056	3.333
31.1	1.608	1.929	2.251	2.572	2.894	3.215	3.537	3.859	36.1	1.385	1.662	1.939	2.216	2.493	2.770	3.047	3.324
31.2	1.603	1.923	2.244	2.564	2.885	3.205	3.526	3.846	36.2	1.381	1.657	1.934	2.210	2.486	2.762	3.039	3.315
31.3	1.597	1.917	2.236	2.556	2.875	3.195	3.514	3.834	36.3	1.377	1.653	1.928	2.204	2.479	2.755	3.030	3.306
31.4	1.592	1.911	2.229	2.548	2.866	3.185	3.503	3.822	36.4	1.374	1.648	1.923	2.198	2.473	2.747	3.022	3.297
31.5	1.587	1.905	2.222	2.540	2.857	3.175	3.492	3.810	36.5	1.370	1.644	1.918	2.192	2.466	2.740	3.014	3.288
31.6	1.582	1.899	2.215	2.532	2.848	3.165	3.481	3.797	36.6	1.366	1.639	1.913	2.186	2.459	2.732	3.005	3.279
31.7	1.577	1.893	2.208	2.524	2.839	3.155	3.470	3.785	36.7	1.362	1.635	1.907	2.180	2.452	2.725	2.997	3.270
31.8	1.572	1.887	2.201	2.516	2.830	3.145	3.459	3.774	36.8	1.359	1.630	1.902	2.174	2.446	2.717	2.989	3.261
31.9	1.567	1.881	2.194	2.508	2.821	3.135	3.448	3.762	36.9	1.355	1.626	1.897	2.168	2.439	2.710	2.981	3.252
32.0	1.562	1.875	2.187	2.500	2.812	3.125	3.437	3.750	37.0	1.351	1.622	1.892	2.162	2.432	2.703	2.973	3.243
32.1	1.558	1.869	2.181	2.492	2.804	3.115	3.427	3.738	37.1	1.348	1.617	1.887	2.156	2.426	2.695	2.965	3.235
32.2	1.553	1.863	2.174	2.484	2.795	3.106	3.416	3.727	37.2	1.344	1.613	1.882	2.151	2.419	2.688	2.957	3.226
32.3	1.548	1.858	2.167	2.477	2.786	3.096	3.406	3.715	37.3	1.340	1.609	1.877	2.145	2.413	2.681	2.949	3.217
32.4	1.543	1.852	2.160	2.469	2.778	3.086	3.395	3.704	37.4	1.337	1.604	1.872	2.139	2.406	2.674	2.941	3.209
32.5	1.538	1.846	2.154	2.462	2.769	3.077	3.385	3.692	37.5	1.333	1.600	1.867	2.133	2.400	2.667	2.933	3.200
32.6	1.534	1.840	2.147	2.454	2.761	3.067	3.374	3.681	37.6	1.330	1.596	1.862	2.128	2.394	2.660	2.926	3.191
32.7	1.529	1.835	2.141	2.446	2.752	3.058	3.364	3.670	37.7	1.326	1.592	1.857	2.122	2.387	2.653	2.918	3.183
32.8	1.524	1.829	2.134	2.439	2.744	3.049	3.354	3.659	37.8	1.323	1.587	1.852	2.116	2.381	2.646	2.910	3.175
32.9	1.520	1.824	2.128	2.432	2.736	3.040	3.343	3.647	37.9	1.319	1.583	1.847	2.111	2.375	2.639	2.902	3.166
33.0	1.515	1.818	2.121	2.424	2.727	3.030	3.333	3.636	38.0	1.316	1.579	1.842	2.105	2.368	2.632	2.895	3.158
33.1	1.511	1.813	2.115	2.417	2.719	3.021	3.323	3.625	38.1	1.312	1.575	1.837	2.100	2.362	2.625	2.887	3.150
33.2	1.506	1.807	2.108	2.410	2.711	3.012	3.313	3.614	38.2	1.309	1.571	1.832	2.094	2.356	2.618	2.880	3.141
33.3	1.502	1.802	2.102	2.402	2.703	3.003	3.303	3.604	38.3	1.305	1.567	1.828	2.089	2.350	2.611	2.872	3.133
33.4	1.497	1.796	2.096	2.395	2.695	2.994	3.293	3.593	38.4	1.302	1.563	1.823	2.083	2.344	2.604	2.865	3.125
33.5	1.493	1.791	2.090	2.388	2.687	2.985	3.284	3.582	38.5	1.299	1.558	1.818	2.078	2.338	2.597	2.857	3.117
33.6	1.488	1.786	2.083	2.381	2.679	2.976	3.274	3.571	38.6	1.295	1.554	1.813	2.073	2.332	2.591	2.850	3.109
33.7	1.484	1.780	2.077	2.374	2.671	2.967	3.264	3.561	38.7	1.292	1.550	1.809	2.067	2.326	2.584	2.842	3.101
33.8	1.479	1.775	2.071	2.367	2.663	2.959	3.254	3.550	38.8	1.289	1.546	1.804	2.062	2.320	2.577	2.835	3.093
33.9	1.475	1.770	2.065	2.360	2.655	2.950	3.245	3.540	38.9	1.285	1.542	1.799	2.057	2.314	2.571	2.828	3.085
34.0	1.471	1.765	2.059	2.353	2.647	2.941	3.235	3.529	39.0	1.282	1.538	1.795	2.051	2.308	2.564	2.821	3.077
34.1	1.466	1.760	2.053	2.346	2.639	2.933	3.226	3.519	39.1	1.279	1.535	1.790	2.046	2.302	2.558	2.813	3.069
34.2	1.462	1.754	2.047	2.339	2.632	2.924	3.216	3.509	39.2	1.276	1.531	1.786	2.041	2.296	2.551	2.806	3.061
34.3	1.458	1.749	2.041	2.332	2.624	2.915	3.207	3.499	39.3	1.272	1.527	1.781	2.036	2.290	2.545	2.799	3.053
34.4	1.453	1.744	2.035	2.326	2.616	2.907	3.198	3.488	39.4	1.269	1.523	1.777	2.030	2.284	2.538	2.792	3.046
34.5	1.449	1.739	2.029	2.319	2.609	2.899	3.188	3.478	39.5	1.266	1.519	1.772	2.025	2.278	2.532	2.785	3.038
34.6	1.445	1.734	2.023	2.312	2.601	2.890	3.179	3.468	39.6	1.263	1.515	1.768	2.020	2.273	2.525	2.778	3.030
34.7	1.441	1.729	2.017	2.305	2.594	2.882	3.170	3.458	39.7	1.259	1.511	1.763	2.015	2.267	2.519	2.771	3.023
34.8	1.437	1.724	2.011	2.299	2.586	2.874	3.161	3.448	39.8	1.256	1.508	1.759	2.010	2.261	2.513	2.764	3.015
34.9	1.433	1.719	2.006	2.292	2.579	2.865	3.152	3.438	39.9	1.253	1.504	1.754	2.005	2.256	2.506	2.757	3.008

TABLE XXVIII—Continued.

TABLE OF VELOCITIES OF TUBES IN MEASURING FLUMES, IN FEET PER SECOND. THE TIME OCCUPIED IN PASSING FROM THE UPSTREAM TO THE DOWNSTREAM TRANSIT STATION, AND THE DISTANCE BETWEEN THEM, BEING GIVEN.

TIME.	DISTANCE BETWEEN THE TRANSIT STATIONS, IN FEET.								TIME.	DISTANCE BETWEEN THE TRANSIT STATIONS, IN FEET.							
Sec's.	50.	60.	70	80.	90.	100.	110.	120.	Sec's.	50.	60.	70.	80.	90.	100.	110.	120.
40.0	1.250	1.500	1.750	2.000	2.250	2.500	2.750	3.000	45.0	1.111	1.333	1.556	1.778	2.000	2.222	2.444	2.667
40.1	1.247	1.496	1.746	1.995	2.244	2.494	2.743	2.993	45.1	1.109	1.330	1.552	1.774	1.996	2.217	2.439	2.661
40.2	1.244	1.493	1.741	1.990	2.239	2.488	2.736	2.985	45.2	1.106	1.327	1.549	1.770	1.991	2.212	2.434	2.655
40.3	1.241	1.489	1.737	1.985	2.233	2.481	2.730	2.978	45.3	1.104	1.325	1.545	1.766	1.987	2.208	2.428	2.649
40.4	1.238	1.485	1.733	1.980	2.228	2.475	2.723	2.970	45.4	1.101	1.322	1.542	1.762	1.982	2.203	2.423	2.643
40.5	1.235	1.481	1.728	1.975	2.222	2.469	2.716	2.963	45.5	1.099	1.319	1.538	1.758	1.978	2.198	2.418	2.637
40.6	1.232	1.478	1.724	1.970	2.217	2.463	2.709	2.956	45.6	1.096	1.316	1.535	1.754	1.974	2.193	2.412	2.632
40.7	1.229	1.474	1.720	1.966	2.211	2.457	2.703	2.948	45.7	1.094	1.313	1.532	1.751	1.969	2.188	2.407	2.626
40.8	1.225	1.471	1.716	1.961	2.206	2.451	2.696	2.941	45.8	1.092	1.310	1.528	1.747	1.965	2.183	2.402	2.620
40.9	1.222	1.467	1.711	1.956	2.200	2.445	2.689	2.934	45.9	1.089	1.307	1.525	1.743	1.961	2.179	2.397	2.614
41.0	1.220	1.463	1.707	1.951	2.195	2.439	2.683	2.927	46.0	1.087	1.304	1.522	1.739	1.957	2.174	2.391	2.609
41.1	1.217	1.460	1.703	1.946	2.190	2.433	2.676	2.920	46.1	1.085	1.302	1.518	1.735	1.952	2.169	2.386	2.603
41.2	1.214	1.456	1.699	1.942	2.184	2.427	2.670	2.913	46.2	1.082	1.299	1.515	1.732	1.948	2.165	2.381	2.597
41.3	1.211	1.453	1.695	1.937	2.179	2.421	2.663	2.906	46.3	1.080	1.296	1.512	1.728	1.944	2.160	2.376	2.592
41.4	1.208	1.449	1.691	1.932	2.174	2.415	2.657	2.899	46.4	1.078	1.293	1.509	1.724	1.940	2.155	2.371	2.586
41.5	1.205	1.446	1.687	1.928	2.169	2.410	2.651	2.892	46.5	1.075	1.290	1.505	1.720	1.935	2.151	2.366	2.581
41.6	1.202	1.442	1.683	1.923	2.163	2.404	2.644	2.885	46.6	1.073	1.288	1.502	1.717	1.931	2.146	2.361	2.575
41.7	1.199	1.439	1.679	1.918	2.158	2.398	2.638	2.878	46.7	1.071	1.285	1.499	1.713	1.927	2.141	2.355	2.570
41.8	1.196	1.435	1.675	1.914	2.153	2.392	2.632	2.871	46.8	1.068	1.282	1.496	1.709	1.923	2.137	2.350	2.564
41.9	1.193	1.432	1.671	1.909	2.148	2.387	2.625	2.864	46.9	1.066	1.279	1.493	1.706	1.919	2.132	2.345	2.559
42.0	1.190	1.429	1.667	1.905	2.143	2.381	2.619	2.857	47.0	1.064	1.277	1.489	1.702	1.915	2.128	2.340	2.553
42.1	1.188	1.425	1.663	1.900	2.138	2.375	2.613	2.850	47.1	1.062	1.274	1.486	1.699	1.911	2.123	2.335	2.548
42.2	1.185	1.422	1.659	1.896	2.133	2.370	2.607	2.844	47.2	1.059	1.271	1.483	1.695	1.907	2.119	2.331	2.542
42.3	1.182	1.418	1.655	1.891	2.128	2.364	2.600	2.837	47.3	1.057	1.268	1.480	1.691	1.903	2.114	2.326	2.537
42.4	1.179	1.415	1.651	1.887	2.123	2.358	2.594	2.830	47.4	1.055	1.266	1.477	1.688	1.899	2.110	2.321	2.532
42.5	1.176	1.412	1.647	1.882	2.118	2.353	2.588	2.824	47.5	1.053	1.263	1.474	1.684	1.895	2.105	2.316	2.526
42.6	1.174	1.408	1.643	1.878	2.113	2.347	2.582	2.817	47.6	1.050	1.261	1.471	1.681	1.891	2.101	2.311	2.521
42.7	1.171	1.405	1.639	1.874	2.108	2.342	2.576	2.810	47.7	1.048	1.258	1.468	1.677	1.887	2.096	2.306	2.516
42.8	1.168	1.402	1.636	1.869	2.103	2.336	2.570	2.804	47.8	1.046	1.255	1.464	1.674	1.883	2.092	2.301	2.510
42.9	1.166	1.399	1.632	1.865	2.098	2.331	2.564	2.797	47.9	1.044	1.255	1.461	1.670	1.879	2.088	2.296	2.505
43.0	1.163	1.395	1.628	1.860	2.093	2.326	2.558	2.791	48.0	1.042	1.250	1.458	1.667	1.875	2.083	2.292	2.500
43.1	1.160	1.392	1.624	1.856	2.088	2.320	2.552	2.784	48.1	1.040	1.247	1.455	1.663	1.871	2.079	2.287	2.435
43.2	1.157	1.389	1.620	1.852	2.083	2.315	2.546	2.778	48.2	1.037	1.245	1.452	1.660	1.867	2.075	2.282	2.490
43.3	1.155	1.386	1.617	1.848	2.079	2.309	2.540	2.771	48.3	1.035	1.242	1.449	1.656	1.863	2.070	2.277	2.484
43.4	1.152	1.382	1.613	1.843	2.074	2.304	2.535	2.765	48.4	1.033	1.240	1.446	1.653	1.860	2.066	2.273	2.479
43.5	1.149	1.379	1.609	1.839	2.069	2.299	2.529	2.759	48.5	1.031	1.237	1.443	1.649	1.856	2.062	2.268	2.474
43.6	1.147	1.376	1.606	1.835	2.064	2.294	2.523	2.752	48.6	1.029	1.235	1.440	1.646	1.852	2.058	2.263	2.469
43.7	1.144	1.373	1.602	1.831	2.059	2.288	2.517	2.746	48.7	1.027	1.232	1.437	1.643	1.848	2.053	2.259	2.464
43.8	1.142	1.370	1.598	1.826	2.055	2.283	2.511	2.740	48.8	1.025	1.230	1.434	1.639	1.844	2.049	2.254	2.459
43.9	1.139	1.367	1.595	1.822	2.050	2.278	2.506	2.733	48.9	1.022	1.227	1.431	1.636	1.840	2.045	2.249	2.454
44.0	1.136	1.364	1.591	1.818	2.045	2.273	2.500	2.727	49.0	1.020	1.224	1.429	1.633	1.837	2.041	2.245	2.449
44.1	1.134	1.361	1.587	1.814	2.041	2.268	2.494	2.721	49.1	1.018	1.222	1.426	1.629	1.833	2.037	2.240	2.444
44.2	1.131	1.357	1.584	1.810	2.036	2.262	2.489	2.715	49.2	1.016	1.220	1.423	1.626	1.829	2.033	2.236	2.439
44.3	1.129	1.354	1.580	1.806	2.032	2.257	2.483	2.709	49.3	1.014	1.217	1.420	1.623	1.826	2.028	2.231	2.434
44.4	1.126	1.351	1.577	1.802	2.027	2.252	2.477	2.703	49.4	1.012	1.215	1.417	1.619	1.822	2.024	2.227	2.429
44.5	1.124	1.348	1.573	1.798	2.022	2.247	2.472	2.697	49.5	1.010	1.212	1.414	1.616	1.818	2.020	2.222	2.424
44.6	1.121	1.345	1.570	1.794	2.018	2.242	2.466	2.691	49.6	1.008	1.210	1.411	1.613	1.815	2.016	2.218	2.419
44.7	1.119	1.342	1.566	1.790	2.013	2.237	2.461	2.685	49.7	1.006	1.207	1.408	1.610	1.811	2.012	2.213	2.414
44.8	1.116	1.339	1.562	1.786	2.009	2.232	2.455	2.679	49.8	1.004	1.205	1.406	1.606	1.807	2.008	2.209	2.410
44.9	1.114	1.336	1.559	1.782	2.004	2.227	2.450	2.673	49.9	1.002	1.202	1.403	1.603	1.804	2.004	2.204	2.405

TABLE XXVIII — Continued.

TABLE OF VELOCITIES OF TUBES IN MEASURING FLUMES, IN FEET PER SECOND. THE TIME OCCUPIED IN PASSING FROM THE UPSTREAM TO THE DOWNSTREAM TRANSIT STATION, AND THE DISTANCE BETWEEN THEM, BEING GIVEN.

TIME.	DISTANCE BETWEEN THE TRANSIT STATIONS, IN FEET.							
Sec's.	50.	60.	70.	80.	90.	100.	110.	120.
50.0	1.000	1.200	1.400	1.600	1.800	2.000	2.200	2.400
50.1	0.998	1.198	1.397	1.597	1.796	1.996	2.196	2.395
50.2	0.996	1.195	1.394	1.594	1.793	1.992	2.191	2.390
50.3	0.994	1.193	1.392	1.590	1.789	1.988	2.187	2.386
50.4	0.992	1.190	1.389	1.587	1.786	1.984	2.183	2.381
50.5	0.990	1.188	1.386	1.584	1.782	1.980	2.178	2.376
50.6	0.988	1.186	1.383	1.581	1.779	1.976	2.174	2.372
50.7	0.986	1.183	1.381	1.578	1.775	1.972	2.170	2.367
50.8	0.984	1.181	1.378	1.575	1.772	1.969	2.165	2.362
50.9	0.982	1.179	1.375	1.572	1.768	1.965	2.161	2.358
51.0	0.980	1.176	1.373	1.569	1.765	1.961	2.157	2.353
51.1	0.978	1.174	1.370	1.566	1.761	1.957	2.153	2.348
51.2	0.977	1.172	1.367	1.562	1.758	1.953	2.148	2.344
51.3	0.975	1.170	1.365	1.559	1.754	1.949	2.144	2.339
51.4	0.973	1.167	1.362	1.556	1.751	1.946	2.140	2.335
51.5	0.971	1.165	1.359	1.553	1.748	1.942	2.136	2.330
51.6	0.969	1.163	1.357	1.550	1.744	1.938	2.132	2.326
51.7	0.967	1.161	1.354	1.547	1.741	1.934	2.128	2.321
51.8	0.965	1.158	1.351	1.544	1.737	1.931	2.124	2.317
51.9	0.963	1.156	1.349	1.541	1.734	1.927	2.119	2.312
52.0	0.962	1.154	1.346	1.538	1.731	1.923	2.115	2.308
52.1	0.960	1.152	1.344	1.536	1.727	1.919	2.111	2.303
52.2	0.958	1.149	1.341	1.533	1.724	1.916	2.107	2.299
52.3	0.956	1.147	1.338	1.530	1.721	1.912	2.103	2.294
52.4	0.954	1.145	1.336	1.527	1.718	1.908	2.099	2.290
52.5	0.952	1.143	1.333	1.524	1.714	1.905	2.095	2.286
52.6	0.951	1.141	1.331	1.521	1.711	1.901	2.091	2.281
52.7	0.949	1.139	1.328	1.518	1.708	1.898	2.087	2.277
52.8	0.947	1.136	1.326	1.515	1.705	1.894	2.083	2.273
52.9	0.945	1.134	1.323	1.512	1.701	1.890	2.079	2.268
53.0	0.943	1.132	1.321	1.509	1.698	1.887	2.075	2.264
53.1	0.942	1.130	1.318	1.507	1.695	1.883	2.072	2.260
53.2	0.940	1.128	1.316	1.504	1.692	1.880	2.068	2.256
53.3	0.938	1.126	1.313	1.501	1.689	1.876	2.064	2.251
53.4	0.936	1.124	1.311	1.498	1.685	1.873	2.060	2.247
53.5	0.935	1.121	1.308	1.495	1.682	1.869	2.056	2.243
53.6	0.933	1.119	1.306	1.493	1.679	1.866	2.052	2.239
53.7	0.931	1.117	1.304	1.490	1.676	1.862	2.048	2.235
53.8	0.929	1.115	1.301	1.487	1.673	1.859	2.045	2.230
53.9	0.928	1.113	1.299	1.484	1.670	1.855	2.041	2.226
54.0	0.926	1.111	1.296	1.481	1.667	1.852	2.037	2.222
54.1	0.924	1.109	1.294	1.479	1.664	1.848	2.033	2.218
54.2	0.923	1.107	1.292	1.476	1.661	1.845	2.030	2.214
54.3	0.921	1.105	1.289	1.473	1.657	1.842	2.026	2.210
54.4	0.919	1.103	1.287	1.471	1.654	1.838	2.022	2.206
54.5	0.917	1.101	1.284	1.468	1.651	1.835	2.018	2.202
54.6	0.916	1.099	1.282	1.465	1.648	1.832	2.015	2.198
54.7	0.914	1.097	1.280	1.463	1.645	1.828	2.011	2.194
54.8	0.912	1.095	1.277	1.460	1.642	1.825	2.007	2.190
54.9	0.911	1.093	1.275	1.457	1.639	1.821	2.004	2.186

TIME	DISTANCE BETWEEN THE TRANSIT STATIONS, IN FEET.							
Sec's.	50.	60.	70.	80.	90.	100.	110.	120.
55.0	0.909	1.091	1.273	1.455	1.636	1.818	2.000	2.182
55.1	0.907	1.089	1.270	1.452	1.633	1.815	1.996	2.178
55.2	0.906	1.087	1.268	1.449	1.630	1.812	1.993	2.174
55.3	0.904	1.085	1.266	1.447	1.627	1.808	1.989	2.170
55.4	0.903	1.083	1.264	1.444	1.625	1.805	1.986	2.166
55.5	0.901	1.081	1.261	1.441	1.622	1.802	1.982	2.162
55.6	0.899	1.079	1.259	1.439	1.619	1.799	1.978	2.158
55.7	0.898	1.077	1.257	1.436	1.616	1.795	1.975	2.154
55.8	0.896	1.075	1.254	1.434	1.613	1.792	1.971	2.151
55.9	0.894	1.073	1.252	1.431	1.610	1.789	1.968	2.147
56.0	0.893	1.071	1.250	1.429	1.607	1.786	1.964	2.143
56.1	0.891	1.070	1.248	1.426	1.604	1.783	1.961	2.139
56.2	0.890	1.068	1.246	1.423	1.601	1.779	1.957	2.135
56.3	0.888	1.066	1.243	1.421	1.599	1.776	1.954	2.131
56.4	0.887	1.064	1.241	1.418	1.596	1.773	1.950	2.128
56.5	0.885	1.062	1.239	1.416	1.593	1.770	1.947	2.124
56.6	0.883	1.060	1.237	1.413	1.590	1.767	1.943	2.120
56.7	0.882	1.058	1.235	1.411	1.587	1.764	1.940	2.116
56.8	0.880	1.056	1.232	1.408	1.585	1.761	1.937	2.113
56.9	0.879	1.054	1.230	1.406	1.582	1.757	1.933	2.109
57.0	0.877	1.053	1.228	1.404	1.579	1.754	1.930	2.105
57.1	0.876	1.051	1.226	1.401	1.576	1.751	1.926	2.102
57.2	0.874	1.049	1.224	1.399	1.573	1.748	1.923	2.098
57.3	0.873	1.047	1.222	1.396	1.571	1.745	1.920	2.094
57.4	0.871	1.045	1.220	1.394	1.568	1.742	1.916	2.091
57.5	0.870	1.043	1.217	1.391	1.565	1.739	1.913	2.087
57.6	0.868	1.042	1.215	1.389	1.562	1.736	1.910	2.083
57.7	0.867	1.040	1.213	1.386	1.560	1.733	1.906	2.080
57.8	0.865	1.038	1.211	1.384	1.557	1.730	1.903	2.076
57.9	0.864	1.036	1.209	1.382	1.554	1.727	1.900	2.073
58.0	0.862	1.034	1.207	1.379	1.552	1.724	1.897	2.069
58.1	0.861	1.033	1.205	1.377	1.549	1.721	1.893	2.065
58.2	0.859	1.031	1.203	1.375	1.546	1.718	1.890	2.062
58.3	0.858	1.029	1.201	1.372	1.544	1.715	1.887	2.058
58.4	0.856	1.027	1.199	1.370	1.541	1.712	1.884	2.055
58.5	0.855	1.026	1.197	1.368	1.538	1.709	1.880	2.051
58.6	0.853	1.024	1.195	1.365	1.536	1.706	1.877	2.048
58.7	0.852	1.022	1.193	1.363	1.533	1.704	1.874	2.044
58.8	0.850	1.020	1.190	1.361	1.531	1.701	1.871	2.041
58.9	0.849	1.019	1.188	1.358	1.528	1.698	1.868	2.037
59.0	0.847	1.017	1.186	1.356	1.525	1.695	1.864	2.034
59.1	0.846	1.015	1.184	1.354	1.523	1.692	1.861	2.030
59.2	0.845	1.014	1.182	1.351	1.520	1.689	1.858	2.027
59.3	0.843	1.012	1.180	1.349	1.518	1.686	1.855	2.024
59.4	0.842	1.010	1.178	1.347	1.515	1.684	1.852	2.020
59.5	0.840	1.008	1.176	1.345	1.513	1.681	1.849	2.017
59.6	0.839	1.007	1.174	1.342	1.510	1.678	1.846	2.013
59.7	0.838	1.005	1.173	1.340	1.508	1.675	1.843	2.010
59.8	0.836	1.003	1.171	1.338	1.505	1.672	1.839	2.007
59.9	0.835	1.002	1.169	1.336	1.503	1.669	1.836	2.003

TABLE XXVIII—Continued.

TABLE OF VELOCITIES OF TUBES IN MEASURING FLUMES, IN FEET PER SECOND. THE TIME OCCUPIED IN PASSING FROM THE UPSTREAM TO THE DOWNSTREAM TRANSIT STATION, AND THE DISTANCE BETWEEN THEM, BEING GIVEN.

TIME.	DISTANCE BETWEEN THE TRANSIT STATIONS, IN FEET								TIME.	DISTANCE BETWEEN THE TRANSIT STATIONS, IN FEET.							
Sec's.	50.	60.	70.	80.	90.	100.	110.	120.	Sec's.	50.	60.	70.	80.	90.	100.	110.	120.
60.0	0.833	1.000	1.167	1.333	1.500	1.667	1.833	2.000	65.0	0.769	0.923	1.077	1.231	1.385	1.538	1.692	1.846
60.1	0.832	0.998	1.165	1.331	1.498	1.664	1.830	1.997	65.1	0.768	0.922	1.075	1.229	1.382	1.536	1.690	1.843
60.2	0.831	0.997	1.163	1.329	1.495	1.661	1.827	1.993	65.2	0.767	0.920	1.074	1.227	1.380	1.534	1.687	1.840
60.3	0.829	0.995	1.161	1.327	1.493	1.658	1.824	1.990	65.3	0.766	0.919	1.072	1.225	1.378	1.531	1.685	1.838
60.4	0.828	0.993	1.159	1.325	1.490	1.656	1.821	1.987	65.4	0.765	0.917	1.070	1.223	1.376	1.529	1.682	1.835
60.5	0.826	0.992	1.157	1.322	1.488	1.653	1.818	1.983	65.5	0.763	0.916	1.069	1.221	1.374	1.527	1.679	1.832
60.6	0.825	0.990	1.155	1.320	1.485	1.650	1.815	1.980	65.6	0.762	0.915	1.067	1.220	1.372	1.524	1.677	1.829
60.7	0.824	0.988	1.153	1.318	1.483	1.647	1.812	1.977	65.7	0.761	0.913	1.065	1.218	1.370	1.522	1.674	1.826
60.8	0.822	0.987	1.151	1.316	1.480	1.645	1.809	1.974	65.8	0.760	0.912	1.064	1.216	1.368	1.520	1.672	1.824
60.9	0.821	0.985	1.149	1.314	1.478	1.642	1.806	1.970	65.9	0.759	0.910	1.062	1.214	1.366	1.517	1.669	1.821
61.0	0.820	0.984	1.148	1.311	1.475	1.639	1.803	1.967	66.0	0.758	0.909	1.061	1.212	1.364	1.515	1.667	1.818
61.1	0.818	0.982	1.146	1.309	1.473	1.637	1.800	1.964	66.1	0.756	0.908	1.059	1.210	1.362	1.513	1.664	1.815
61.2	0.817	0.980	1.144	1.307	1.471	1.634	1.797	1.961	66.2	0.755	0.906	1.057	1.208	1.360	1.511	1.662	1.813
61.3	0.816	0.979	1.142	1.305	1.468	1.631	1.794	1.958	66.3	0.754	0.905	1.056	1.207	1.357	1.508	1.659	1.810
61.4	0.814	0.977	1.140	1.303	1.466	1.629	1.792	1.954	66.4	0.753	0.904	1.054	1.205	1.355	1.506	1.657	1.807
61.5	0.813	0.976	1.138	1.301	1.463	1.626	1.789	1.951	66.5	0.752	0.902	1.053	1.203	1.353	1.504	1.654	1.805
61.6	0.812	0.974	1.136	1.299	1.461	1.623	1.786	1.948	66.6	0.751	0.901	1.051	1.201	1.351	1.502	1.652	1.802
61.7	0.810	0.972	1.135	1.297	1.459	1.621	1.783	1.945	66.7	0.750	0.900	1.049	1.199	1.349	1.499	1.649	1.799
61.8	0.809	0.971	1.133	1.294	1.456	1.618	1.780	1.942	66.8	0.749	0.898	1.048	1.198	1.347	1.497	1.647	1.796
61.9	0.808	0.969	1.131	1.292	1.454	1.616	1.777	1.939	66.9	0.747	0.897	1.046	1.196	1.345	1.495	1.644	1.794
62.0	0.806	0.968	1.129	1.290	1.452	1.613	1.774	1.935	67.0	0.746	0.896	1.045	1.194	1.343	1.493	1.642	1.791
62.1	0.805	0.966	1.127	1.288	1.449	1.610	1.771	1.932	67.1	0.745	0.894	1.043	1.192	1.341	1.490	1.639	1.788
62.2	0.804	0.965	1.125	1.286	1.447	1.608	1.768	1.929	67.2	0.744	0.893	1.042	1.190	1.339	1.488	1.637	1.786
62.3	0.803	0.963	1.124	1.284	1.445	1.605	1.766	1.926	67.3	0.743	0.892	1.040	1.189	1.337	1.486	1.634	1.783
62.4	0.801	0.962	1.122	1.282	1.442	1.603	1.763	1.923	67.4	0.742	0.890	1.039	1.187	1.335	1.484	1.632	1.780
62.5	0.800	0.960	1.120	1.280	1.440	1.600	1.760	1.920	67.5	0.741	0.889	1.037	1.185	1.333	1.481	1.630	1.778
62.6	0.799	0.958	1.118	1.278	1.438	1.597	1.757	1.917	67.6	0.740	0.888	1.036	1.183	1.331	1.479	1.627	1.775
62.7	0.797	0.957	1.116	1.276	1.435	1.595	1.754	1.914	67.7	0.739	0.886	1.034	1.182	1.329	1.477	1.625	1.773
62.8	0.796	0.955	1.115	1.274	1.433	1.592	1.752	1.911	67.8	0.737	0.885	1.032	1.180	1.327	1.475	1.622	1.770
62.9	0.795	0.954	1.113	1.272	1.431	1.590	1.749	1.908	67.9	0.736	0.884	1.031	1.178	1.325	1.473	1.620	1.767
63.0	0.794	0.952	1.111	1.270	1.429	1.587	1.746	1.905	68.0	0.735	0.882	1.029	1.176	1.324	1.471	1.618	1.765
63.1	0.792	0.951	1.109	1.268	1.426	1.585	1.743	1.902	68.1	0.734	0.881	1.028	1.175	1.322	1.468	1.615	1.762
63.2	0.791	0.949	1.108	1.266	1.424	1.582	1.741	1.899	68.2	0.733	0.880	1.026	1.173	1.320	1.466	1.613	1.760
63.3	0.790	0.948	1.106	1.264	1.422	1.580	1.738	1.896	68.3	0.732	0.878	1.025	1.171	1.318	1.464	1.611	1.757
63.4	0.789	0.946	1.104	1.262	1.420	1.577	1.735	1.893	68.4	0.731	0.877	1.023	1.170	1.316	1.462	1.608	1.754
63.5	0.787	0.945	1.102	1.260	1.417	1.575	1.732	1.890	68.5	0.730	0.876	1.022	1.168	1.314	1.460	1.606	1.752
63.6	0.786	0.943	1.101	1.258	1.415	1.572	1.730	1.887	68.6	0.729	0.875	1.020	1.166	1.312	1.458	1.603	1.749
63.7	0.785	0.942	1.099	1.256	1.413	1.570	1.727	1.884	68.7	0.728	0.873	1.019	1.164	1.310	1.456	1.601	1.747
63.8	0.784	0.940	1.097	1.254	1.411	1.567	1.724	1.881	68.8	0.727	0.872	1.017	1.163	1.308	1.453	1.599	1.744
63.9	0.782	0.939	1.095	1.252	1.408	1.565	1.721	1.878	68.9	0.726	0.871	1.016	1.161	1.306	1.451	1.597	1.742
64.0	0.781	0.937	1.094	1.250	1.406	1.562	1.719	1.875	69.0	0.725	0.870	1.014	1.159	1.304	1.449	1.594	1.739
64.1	0.780	0.936	1.092	1.248	1.404	1.560	1.716	1.872	69.1	0.724	0.868	1.013	1.158	1.302	1.447	1.592	1.737
64.2	0.779	0.935	1.090	1.246	1.402	1.558	1.713	1.869	69.2	0.723	0.867	1.012	1.156	1.301	1.445	1.590	1.734
64.3	0.778	0.933	1.089	1.244	1.400	1.555	1.711	1.866	69.3	0.722	0.866	1.010	1.154	1.299	1.443	1.587	1.732
64.4	0.776	0.932	1.087	1.242	1.398	1.553	1.708	1.863	69.4	0.720	0.865	1.009	1.153	1.297	1.441	1.585	1.729
64.5	0.775	0.930	1.085	1.240	1.395	1.550	1.705	1.860	69.5	0.719	0.863	1.007	1.151	1.295	1.439	1.583	1.727
64.6	0.774	0.929	1.084	1.238	1.393	1.548	1.703	1.858	69.6	0.718	0.862	1.006	1.149	1.293	1.437	1.580	1.724
64.7	0.773	0.927	1.082	1.236	1.391	1.546	1.700	1.855	69.7	0.717	0.861	1.004	1.148	1.291	1.435	1.578	1.722
64.8	0.772	0.926	1.080	1.235	1.389	1.543	1.698	1.852	69.8	0.716	0.860	1.003	1.146	1.289	1.433	1.576	1.719
64.9	0.770	0.924	1.079	1.233	1.387	1.541	1.695	1.849	69.9	0.715	0.858	1.001	1.144	1.288	1.431	1.574	1.717

TABLE XXVIII — Continued.

TABLE OF VELOCITIES OF TUBES IN MEASURING FLUMES, IN FEET PER SECOND. THE TIME OCCUPIED IN PASSING FROM THE UPSTREAM TO THE DOWNSTREAM TRANSIT STATION, AND THE DISTANCE BETWEEN THEM, BEING GIVEN.

TIME	DISTANCE BETWEEN THE TRANSIT STATIONS, IN FEET.							
Sec's.	50.	60.	70.	80.	90.	100.	110.	120.
70.0	0.714	0.857	1.000	1.143	1.286	1.429	1.571	1.714
70.1	0.713	0.856	0.999	1.141	1.284	1.427	1.569	1.712
70.2	0.712	0.855	0.997	1.140	1.282	1.425	1.567	1.709
70.3	0.711	0.853	0.996	1.138	1.280	1.422	1.565	1.707
70.4	0.710	0.852	0.994	1.136	1.278	1.420	1.562	1.705
70.5	0.709	0.851	0.993	1.135	1.277	1.418	1.560	1.702
70.6	0.708	0.850	0.992	1.133	1.275	1.416	1.558	1.700
70.7	0.707	0.849	0.990	1.132	1.273	1.414	1.556	1.697
70.8	0.706	0.847	0.989	1.130	1.271	1.412	1.554	1.695
70.9	0.705	0.846	0.987	1.128	1.269	1.410	1.551	1.693
71.0	0.704	0.845	0.986	1.127	1.268	1.408	1.549	1.690
71.1	0.703	0.844	0.985	1.125	1.266	1.406	1.547	1.688
71.2	0.702	0.843	0.983	1.124	1.264	1.404	1.545	1.685
71.3	0.701	0.842	0.982	1.122	1.262	1.403	1.543	1.683
71.4	0.700	0.840	0.980	1.120	1.261	1.401	1.541	1.681
71.5	0.699	0.839	0.979	1.119	1.259	1.399	1.538	1.678
71.6	0.698	0.838	0.978	1.117	1.257	1.397	1.536	1.676
71.7	0.697	0.837	0.976	1.116	1.255	1.395	1.534	1.674
71.8	0.696	0.836	0.975	1.114	1.253	1.393	1.532	1.671
71.9	0.695	0.834	0.974	1.113	1.252	1.391	1.530	1.669
72.0	0.694	0.833	0.972	1.111	1.250	1.389	1.528	1.667
72.1	0.693	0.832	0.971	1.110	1.248	1.387	1.526	1.664
72.2	0.693	0.831	0.970	1.108	1.247	1.385	1.524	1.662
72.3	0.692	0.830	0.968	1.107	1.245	1.383	1.521	1.660
72.4	0.691	0.829	0.967	1.105	1.243	1.381	1.519	1.657
72.5	0.690	0.828	0.966	1.103	1.241	1.379	1.517	1.655
72.6	0.689	0.826	0.964	1.102	1.240	1.377	1.515	1.653
72.7	0.688	0.825	0.963	1.100	1.238	1.376	1.513	1.651
72.8	0.687	0.824	0.962	1.099	1.236	1.374	1.511	1.648
72.9	0.686	0.823	0.960	1.097	1.235	1.372	1.509	1.646
73.0	0.685	0.822	0.959	1.096	1.233	1.370	1.507	1.644
73.1	0.684	0.821	0.958	1.094	1.231	1.368	1.505	1.642
73.2	0.683	0.820	0.956	1.093	1.230	1.366	1.503	1.639
73.3	0.682	0.819	0.955	1.091	1.228	1.364	1.501	1.637
73.4	0.681	0.817	0.954	1.090	1.226	1.362	1.499	1.635
73.5	0.680	0.816	0.952	1.088	1.224	1.361	1.497	1.633
73.6	0.679	0.815	0.951	1.087	1.223	1.359	1.495	1.630
73.7	0.678	0.814	0.950	1.085	1.221	1.357	1.493	1.628
73.8	0.678	0.813	0.949	1.084	1.220	1.355	1.491	1.626
73.9	0.677	0.812	0.947	1.083	1.218	1.353	1.488	1.624
74.0	0.676	0.811	0.946	1.081	1.216	1.351	1.486	1.622
74.1	0.675	0.810	0.945	1.080	1.215	1.350	1.484	1.619
74.2	0.674	0.809	0.943	1.078	1.213	1.348	1.482	1.617
74.3	0.673	0.808	0.942	1.077	1.211	1.346	1.480	1.615
74.4	0.672	0.806	0.941	1.075	1.210	1.344	1.478	1.613
74.5	0.671	0.805	0.940	1.074	1.208	1.342	1.477	1.611
74.6	0.670	0.804	0.938	1.072	1.206	1.340	1.475	1.609
74.7	0.669	0.803	0.937	1.071	1.205	1.339	1.473	1.606
74.8	0.668	0.802	0.936	1.070	1.203	1.337	1.471	1.604
74.9	0.668	0.801	0.935	1.068	1.202	1.335	1.469	1.602
75.0	0.667	0.800	0.933	1.067	1.200	1.333	1.467	1.600
75.1	0.666	0.799	0.932	1.065	1.198	1.332	1.465	1.598
75.2	0.665	0.798	0.931	1.064	1.197	1.330	1.463	1.596
75.3	0.664	0.797	0.930	1.062	1.195	1.328	1.461	1.594
75.4	0.663	0.796	0.928	1.061	1.194	1.326	1.459	1.592
75.5	0.662	0.795	0.927	1.060	1.192	1.325	1.457	1.589
75.6	0.661	0.794	0.926	1.058	1.190	1.323	1.455	1.587
75.7	0.661	0.793	0.925	1.057	1.189	1.321	1.453	1.585
75.8	0.660	0.792	0.923	1.055	1.187	1.319	1.451	1.583
75.9	0.659	0.791	0.922	1.054	1.186	1.318	1.449	1.581
76.0	0.658	0.789	0.921	1.053	1.184	1.316	1.447	1.579
76.1	0.657	0.788	0.920	1.051	1.183	1.314	1.445	1.577
76.2	0.656	0.787	0.919	1.050	1.181	1.312	1.444	1.575
76.3	0.655	0.786	0.917	1.048	1.180	1.311	1.442	1.573
76.4	0.654	0.785	0.916	1.047	1.178	1.309	1.440	1.571
76.5	0.654	0.784	0.915	1.046	1.176	1.307	1.438	1.569
76.6	0.653	0.783	0.914	1.044	1.175	1.305	1.436	1.567
76.7	0.652	0.782	0.913	1.043	1.173	1.304	1.434	1.565
76.8	0.651	0.781	0.911	1.042	1.172	1.302	1.432	1.562
76.9	0.650	0.780	0.910	1.040	1.170	1.300	1.430	1.560
77.0	0.649	0.779	0.909	1.039	1.169	1.299	1.429	1.558
77.1	0.649	0.778	0.908	1.038	1.167	1.297	1.427	1.556
77.2	0.648	0.777	0.907	1.036	1.166	1.295	1.425	1.554
77.3	0.647	0.776	0.906	1.035	1.164	1.294	1.423	1.552
77.4	0.646	0.775	0.904	1.034	1.163	1.292	1.421	1.550
77.5	0.645	0.774	0.903	1.032	1.161	1.290	1.419	1.548
77.6	0.644	0.773	0.902	1.031	1.160	1.289	1.418	1.546
77.7	0.644	0.772	0.901	1.030	1.158	1.287	1.416	1.544
77.8	0.643	0.771	0.900	1.028	1.157	1.285	1.414	1.542
77.9	0.642	0.770	0.899	1.027	1.155	1.284	1.412	1.540
78.0	0.641	0.769	0.897	1.026	1.154	1.282	1.410	1.538
78.1	0.640	0.768	0.896	1.024	1.152	1.280	1.408	1.536
78.2	0.639	0.767	0.895	1.023	1.151	1.279	1.407	1.535
78.3	0.639	0.766	0.894	1.022	1.149	1.277	1.405	1.533
78.4	0.638	0.765	0.893	1.020	1.148	1.276	1.403	1.531
78.5	0.637	0.764	0.892	1.019	1.146	1.274	1.401	1.529
78.6	0.636	0.763	0.891	1.018	1.145	1.272	1.399	1.527
78.7	0.635	0.762	0.889	1.017	1.144	1.271	1.398	1.525
78.8	0.635	0.761	0.888	1.015	1.142	1.269	1.396	1.523
78.9	0.634	0.760	0.887	1.014	1.141	1.267	1.394	1.521
79.0	0.633	0.759	0.886	1.013	1.139	1.266	1.392	1.519
79.1	0.632	0.759	0.885	1.011	1.138	1.264	1.391	1.517
79.2	0.631	0.758	0.884	1.010	1.136	1.263	1.389	1.515
79.3	0.631	0.757	0.883	1.009	1.135	1.261	1.387	1.513
79.4	0.630	0.756	0.882	1.008	1.134	1.259	1.385	1.511
79.5	0.629	0.755	0.881	1.006	1.132	1.258	1.384	1.509
79.6	0.628	0.754	0.879	1.005	1.131	1.256	1.382	1.508
79.7	0.627	0.753	0.878	1.004	1.129	1.255	1.380	1.506
79.8	0.627	0.752	0.877	1.003	1.128	1.253	1.378	1.504
79.9	0.626	0.751	0.876	1.001	1.126	1.252	1.377	1.502

TABLE XXVIII — Continued.

TABLE OF VELOCITIES OF TUBES IN MEASURING FLUMES, IN FEET PER SECOND. THE TIME OCCUPIED IN PASSING FROM THE UPSTREAM TO THE DOWNSTREAM TRANSIT STATION, AND THE DISTANCE BETWEEN THEM, BEING GIVEN.

TIME.	DISTANCE BETWEEN THE TRANSIT STATIONS, IN FEET.								TIME.	DISTANCE BETWEEN THE TRANSIT STATIONS, IN FEET.							
Sec's.	50.	60.	70.	80.	90.	100.	110.	120.	Sec's.	50.	60.	70.	80.	90.	100.	110.	120.
80.0	0.625	0.750	0.875	1.000	1.125	1.250	1.375	1.500	85.0	0.588	0.706	0.824	0.941	1.059	1.176	1.294	1.412
80.1	0.624	0.749	0.874	0.999	1.124	1.248	1.373	1.498	85.1	0.588	0.705	0.823	0.940	1.058	1.175	1.293	1.410
80.2	0.623	0.748	0.873	0.998	1.122	1.247	1.372	1.496	85.2	0.587	0.704	0.822	0.939	1.056	1.174	1.291	1.408
80.3	0.623	0.747	0.872	0.996	1.121	1.245	1.370	1.494	85.3	0.586	0.703	0.821	0.938	1.055	1.172	1.290	1.407
80.4	0.622	0.746	0.871	0.995	1.119	1.244	1.368	1.493	85.4	0.585	0.703	0.820	0.937	1.054	1.171	1.288	1.405
80.5	0.621	0.745	0.870	0.994	1.118	1.242	1.366	1.491	85.5	0.585	0.702	0.819	0.936	1.053	1.170	1.287	1.404
80.6	0.620	0.744	0.868	0.993	1.117	1.241	1.365	1.489	85.6	0.584	0.701	0.818	0.935	1.051	1.168	1.285	1.402
80.7	0.620	0.743	0.867	0.991	1.115	1.239	1.363	1.487	85.7	0.583	0.700	0.817	0.933	1.050	1.167	1.284	1.400
80.8	0.619	0.743	0.866	0.990	1.114	1.238	1.361	1.485	85.8	0.583	0.699	0.816	0.932	1.049	1.166	1.282	1.399
80.9	0.618	0.742	0.865	0.989	1.112	1.236	1.360	1.483	85.9	0.582	0.698	0.815	0.931	1.048	1.164	1.281	1.397
81.0	0.617	0.741	0.864	0.988	1.111	1.235	1.358	1.481	86.0	0.581	0.698	0.814	0.930	1.047	1.163	1.279	1.395
81.1	0.617	0.740	0.863	0.986	1.110	1.233	1.356	1.480	86.1	0.581	0.697	0.813	0.929	1.045	1.161	1.278	1.394
81.2	0.616	0.739	0.862	0.985	1.108	1.232	1.355	1.478	86.2	0.580	0.696	0.812	0.928	1.044	1.160	1.276	1.392
81.3	0.615	0.738	0.861	0.984	1.107	1.230	1.353	1.476	86.3	0.579	0.695	0.811	0.927	1.043	1.159	1.275	1.390
81.4	0.614	0.737	0.860	0.983	1.106	1.229	1.351	1.474	86.4	0.579	0.694	0.810	0.926	1.042	1.157	1.273	1.389
81.5	0.613	0.736	0.859	0.982	1.104	1.227	1.350	1.472	86.5	0.578	0.694	0.809	0.925	1.040	1.156	1.272	1.387
81.6	0.613	0.735	0.858	0.980	1.103	1.225	1.348	1.471	86.6	0.577	0.693	0.808	0.924	1.039	1.155	1.270	1.386
81.7	0.612	0.734	0.857	0.979	1.102	1.224	1.346	1.469	86.7	0.577	0.692	0.807	0.923	1.038	1.153	1.269	1.384
81.8	0.611	0.733	0.856	0.978	1.100	1.222	1.345	1.467	86.8	0.576	0.691	0.806	0.922	1.037	1.152	1.267	1.382
81.9	0.611	0.733	0.855	0.977	1.099	1.221	1.343	1.465	86.9	0.575	0.690	0.806	0.921	1.036	1.151	1.266	1.381
82.0	0.610	0.732	0.854	0.976	1.098	1.220	1.341	1.463	87.0	0.575	0.690	0.805	0.920	1.034	1.149	1.264	1.379
82.1	0.609	0.731	0.853	0.974	1.096	1.218	1.340	1.462	87.1	0.574	0.689	0.804	0.918	1.033	1.148	1.263	1.378
82.2	0.608	0.730	0.852	0.973	1.095	1.217	1.338	1.460	87.2	0.573	0.688	0.803	0.917	1.032	1.147	1.261	1.376
82.3	0.608	0.729	0.851	0.972	1.094	1.215	1.337	1.458	87.3	0.573	0.687	0.802	0.916	1.031	1.145	1.260	1.375
82.4	0.607	0.728	0.850	0.971	1.092	1.214	1.335	1.456	87.4	0.572	0.686	0.801	0.915	1.030	1.144	1.259	1.373
82.5	0.606	0.727	0.848	0.970	1.091	1.212	1.333	1.455	87.5	0.571	0.686	0.800	0.914	1.029	1.143	1.257	1.371
82.6	0.605	0.726	0.847	0.969	1.090	1.211	1.332	1.453	87.6	0.571	0.685	0.799	0.913	1.027	1.142	1.256	1.370
82.7	0.605	0.726	0.846	0.967	1.088	1.209	1.330	1.451	87.7	0.570	0.684	0.798	0.912	1.026	1.140	1.254	1.368
82.8	0.604	0.725	0.845	0.966	1.087	1.208	1.329	1.449	87.8	0.569	0.683	0.797	0.911	1.025	1.139	1.253	1.367
82.9	0.603	0.724	0.844	0.965	1.086	1.206	1.327	1.448	87.9	0.569	0.683	0.796	0.910	1.024	1.138	1.251	1.365
83.0	0.602	0.723	0.843	0.964	1.084	1.205	1.325	1.446	88.0	0.568	0.682	0.795	0.909	1.023	1.136	1.250	1.364
83.1	0.602	0.722	0.842	0.963	1.083	1.203	1.324	1.444	88.1	0.568	0.681	0.795	0.908	1.022	1.135	1.249	1.362
83.2	0.601	0.721	0.841	0.962	1.082	1.202	1.322	1.442	88.2	0.567	0.680	0.794	0.907	1.020	1.134	1.247	1.361
83.3	0.600	0.720	0.840	0.960	1.080	1.200	1.321	1.441	88.3	0.566	0.680	0.793	0.906	1.019	1.133	1.246	1.359
83.4	0.600	0.719	0.839	0.959	1.079	1.199	1.319	1.439	88.4	0.566	0.679	0.792	0.905	1.018	1.131	1.244	1.357
83.5	0.599	0.719	0.838	0.958	1.078	1.198	1.317	1.437	88.5	0.565	0.678	0.791	0.904	1.017	1.130	1.243	1.356
83.6	0.598	0.718	0.837	0.957	1.077	1.196	1.316	1.435	88.6	0.564	0.677	0.790	0.903	1.016	1.129	1.242	1.354
83.7	0.597	0.717	0.836	0.956	1.075	1.195	1.314	1.434	88.7	0.564	0.676	0.789	0.902	1.015	1.127	1.240	1.353
83.8	0.597	0.716	0.835	0.955	1.074	1.193	1.313	1.432	88.8	0.563	0.676	0.788	0.901	1.014	1.126	1.239	1.351
83.9	0.596	0.715	0.834	0.954	1.073	1.192	1.311	1.430	88.9	0.562	0.675	0.787	0.900	1.012	1.125	1.237	1.350
84.0	0.595	0.714	0.833	0.952	1.071	1.190	1.310	1.429	89.0	0.562	0.674	0.787	0.899	1.011	1.124	1.236	1.348
84.1	0.595	0.713	0.832	0.951	1.070	1.189	1.308	1.427	89.1	0.561	0.673	0.786	0.898	1.010	1.122	1.235	1.347
84.2	0.594	0.713	0.831	0.950	1.069	1.188	1.306	1.425	89.2	0.561	0.673	0.785	0.897	1.009	1.121	1.233	1.345
84.3	0.593	0.712	0.830	0.949	1.068	1.186	1.305	1.423	89.3	0.560	0.672	0.784	0.896	1.008	1.120	1.232	1.344
84.4	0.592	0.711	0.829	0.948	1.066	1.185	1.303	1.422	89.4	0.559	0.671	0.783	0.895	1.007	1.119	1.230	1.342
84.5	0.592	0.710	0.828	0.947	1.065	1.183	1.302	1.420	89.5	0.559	0.670	0.782	0.894	1.006	1.117	1.229	1.341
84.6	0.591	0.709	0.827	0.946	1.064	1.182	1.300	1.418	89.6	0.558	0.67[illegible]	0.781	0.893	1.004	1.116	1.228	1.339
84.7	0.590	0.708	0.826	0.945	1.063	1.181	1.299	1.417	89.7	0.557	0.66[illegible]	0.780	0.892	1.003	1.115	1.226	1.338
84.8	0.590	0.708	0.825	0.943	1.061	1.179	1.297	1.415	89.8	0.557	0.66[illegible]	0.780	0.891	1.002	1.114	1.225	1.336
84.9	0.589	0.707	0.824	0.942	1.060	1.178	1.296	1.413	89.9	0.536	0.667	0.779	0.890	1.001	1.112	1.224	1.335

TABLE XXVIII—Continued.

TABLE OF VELOCITIES OF TUBES IN MEASURING FLUMES, IN FEET PER SECOND. THE TIME OCCUPIED IN PASSING FROM THE UPSTREAM TO THE DOWNSTREAM TRANSIT STATION, AND THE DISTANCE BETWEEN THEM, BEING GIVEN.

TIME	DISTANCE BETWEEN THE TRANSIT STATIONS, IN FEET.							
Sec's.	50.	60.	70.	80.	90.	100.	110.	120.
90.0	0.556	0.667	0.778	0.889	1.000	1.111	1.222	1.333
90.1	0.555	0.666	0.777	0.888	0.999	1.110	1.221	1.332
90.2	0.554	0.665	0.776	0.887	0.998	1.109	1.220	1.330
90.3	0.554	0.664	0.775	0.886	0.997	1.107	1.218	1.329
90.4	0.553	0.664	0.774	0.885	0.996	1.106	1.217	1.327
90.5	0.552	0.663	0.773	0.884	0.994	1.105	1.215	1.326
90.6	0.552	0.662	0.773	0.883	0.993	1.104	1.214	1.325
90.7	0.551	0.662	0.772	0.882	0.992	1.103	1.213	1.323
90.8	0.551	0.661	0.771	0.881	0.991	1.101	1.211	1.322
90.9	0.550	0.660	0.770	0.880	0.990	1.100	1.210	1.320
91.0	0.549	0.659	0.769	0.879	0.989	1.099	1.209	1.319
91.1	0.549	0.659	0.768	0.878	0.988	1.098	1.207	1.317
91.2	0.548	0.658	0.768	0.877	0.987	1.096	1.206	1.316
91.3	0.548	0.657	0.767	0.876	0.986	1.095	1.205	1.314
91.4	0.547	0.656	0.766	0.875	0.985	1.094	1.204	1.313
91.5	0.546	0.656	0.765	0.874	0.984	1.093	1.202	1.311
91.6	0.546	0.655	0.764	0.873	0.983	1.092	1.201	1.310
91.7	0.545	0.654	0.763	0.872	0.981	1.091	1.200	1.309
91.8	0.545	0.654	0.763	0.871	0.980	1.089	1.198	1.307
91.9	0.544	0.653	0.762	0.871	0.979	1.088	1.197	1.306
92.0	0.543	0.652	0.761	0.870	0.978	1.087	1.196	1.304
92.1	0.543	0.651	0.760	0.869	0.977	1.086	1.194	1.303
92.2	0.542	0.651	0.759	0.868	0.976	1.085	1.193	1.302
92.3	0.542	0.650	0.758	0.867	0.975	1.083	1.192	1.300
92.4	0.541	0.649	0.758	0.866	0.974	1.082	1.190	1.299
92.5	0.541	0.649	0.757	0.865	0.973	1.081	1.189	1.297
92.6	0.540	0.648	0.756	0.864	0.972	1.080	1.188	1.296
92.7	0.539	0.647	0.755	0.863	0.971	1.079	1.187	1.294
92.8	0.539	0.647	0.754	0.862	0.970	1.078	1.185	1.293
92.9	0.538	0.646	0.753	0.861	0.969	1.076	1.184	1.292
93.0	0.538	0.645	0.753	0.860	0.968	1.075	1.183	1.290
93.1	0.537	0.644	0.752	0.859	0.967	1.074	1.182	1.289
93.2	0.536	0.644	0.751	0.858	0.966	1.073	1.180	1.288
93.3	0.536	0.643	0.750	0.857	0.965	1.072	1.179	1.286
93.4	0.535	0.642	0.749	0.857	0.964	1.071	1.178	1.285
93.5	0.535	0.642	0.749	0.856	0.963	1.070	1.176	1.283
93.6	0.534	0.641	0.748	0.855	0.962	1.068	1.175	1.282
93.7	0.534	0.640	0.747	0.854	0.961	1.067	1.174	1.281
93.8	0.533	0.640	0.746	0.853	0.959	1.066	1.173	1.279
93.9	0.532	0.639	0.745	0.852	0.958	1.065	1.171	1.278
94.0	0.532	0.638	0.745	0.851	0.957	1.064	1.170	1.277
94.1	0.531	0.638	0.744	0.850	0.956	1.063	1.169	1.275
94.2	0.531	0.637	0.743	0.849	0.955	1.062	1.168	1.274
94.3	0.530	0.636	0.742	0.848	0.954	1.060	1.166	1.273
94.4	0.530	0.636	0.742	0.847	0.953	1.059	1.165	1.271
94.5	0.529	0.635	0.741	0.847	0.952	1.058	1.164	1.270
94.6	0.529	0.634	0.740	0.846	0.951	1.057	1.163	1.268
94.7	0.528	0.634	0.739	0.845	0.950	1.056	1.162	1.267
94.8	0.527	0.633	0.738	0.844	0.949	1.055	1.160	1.266
94.9	0.527	0.632	0.738	0.843	0.948	1.054	1.159	1.264
95.0	0.526	0.632	0.737	0.842	0.947	1.053	1.158	1.263
95.1	0.526	0.631	0.736	0.841	0.946	1.052	1.157	1.262
95.2	0.525	0.630	0.735	0.840	0.945	1.050	1.155	1.261
95.3	0.525	0.630	0.735	0.839	0.944	1.049	1.154	1.259
95.4	0.524	0.629	0.734	0.839	0.943	1.048	1.153	1.258
95.5	0.524	0.628	0.733	0.838	0.942	1.047	1.152	1.257
95.6	0.523	0.628	0.732	0.837	0.941	1.046	1.151	1.255
95.7	0.522	0.627	0.731	0.836	0.940	1.045	1.149	1.254
95.8	0.522	0.626	0.731	0.835	0.939	1.044	1.148	1.253
95.9	0.521	0.626	0.730	0.834	0.938	1.043	1.147	1.251
96.0	0.521	0.625	0.729	0.833	0.937	1.042	1.146	1.250
96.1	0.520	0.624	0.728	0.832	0.937	1.041	1.145	1.249
96.2	0.520	0.624	0.728	0.832	0.936	1.040	1.143	1.247
96.3	0.519	0.623	0.727	0.831	0.935	1.038	1.142	1.246
96.4	0.519	0.622	0.726	0.830	0.934	1.037	1.141	1.245
96.5	0.518	0.622	0.725	0.829	0.933	1.036	1.140	1.244
96.6	0.518	0.621	0.725	0.828	0.932	1.035	1.139	1.242
96.7	0.517	0.620	0.724	0.827	0.931	1.034	1.138	1.241
96.8	0.517	0.620	0.723	0.826	0.930	1.033	1.136	1.240
96.9	0.516	0.619	0.722	0.826	0.929	1.032	1.135	1.238
97.0	0.515	0.619	0.722	0.825	0.928	1.031	1.134	1.237
97.1	0.515	0.618	0.721	0.824	0.927	1.030	1.133	1.236
97.2	0.514	0.617	0.720	0.823	0.926	1.029	1.132	1.235
97.3	0.514	0.617	0.719	0.822	0.925	1.028	1.131	1.233
97.4	0.513	0.616	0.719	0.821	0.924	1.027	1.129	1.232
97.5	0.513	0.615	0.718	0.821	0.923	1.026	1.128	1.231
97.6	0.512	0.615	0.717	0.820	0.922	1.025	1.127	1.230
97.7	0.512	0.614	0.716	0.819	0.921	1.024	1.126	1.228
97.8	0.511	0.613	0.716	0.818	0.920	1.022	1.125	1.227
97.9	0.511	0.613	0.715	0.817	0.919	1.021	1.124	1.226
98.0	0.510	0.612	0.714	0.816	0.918	1.020	1.122	1.224
98.1	0.510	0.612	0.714	0.815	0.917	1.019	1.121	1.223
98.2	0.509	0.611	0.713	0.815	0.916	1.018	1.120	1.222
98.3	0.509	0.610	0.712	0.814	0.916	1.017	1.119	1.221
98.4	0.508	0.610	0.711	0.813	0.915	1.016	1.118	1.220
98.5	0.508	0.609	0.711	0.812	0.914	1.015	1.117	1.218
98.6	0.507	0.609	0.710	0.811	0.913	1.014	1.116	1.217
98.7	0.507	0.608	0.709	0.811	0.912	1.013	1.114	1.216
98.8	0.506	0.607	0.709	0.810	0.911	1.012	1.113	1.215
98.9	0.506	0.607	0.708	0.809	0.910	1.011	1.112	1.213
99.0	0.505	0.606	0.707	0.808	0.909	1.010	1.111	1.212
99.1	0.505	0.605	0.706	0.807	0.908	1.009	1.110	1.211
99.2	0.504	0.605	0.706	0.806	0.907	1.008	1.109	1.210
99.3	0.504	0.604	0.705	0.806	0.906	1.007	1.108	1.208
99.4	0.503	0.604	0.704	0.805	0.905	1.006	1.107	1.207
99.5	0.503	0.603	0.704	0.804	0.905	1.005	1.106	1.206
99.6	0.502	0.602	0.703	0.803	0.904	1.004	1.104	1.205
99.7	0.502	0.602	0.702	0.802	0.903	1.003	1.103	1.204
99.8	0.501	0.601	0.701	0.802	0.902	1.002	1.102	1.202
99.9	0.501	0.601	0.701	0.801	0.901	1.001	1.101	1.201
100.0	0.500	0.600	0.700	0.800	0.900	1.000	1.100	1.200

TABLE XXIX.

VALUES OF THE COEFFICIENT $\left(1 - 0.116\,(\sqrt{D} - 0.1)\right)$.

D	Value of the Coefficient.	Logarithm of the Coefficient.	*D*	Value of the Coefficient.	Logarithm of the Coefficient.
0,000	1,01160	0,0050088	0,050	0,98566	$\bar{1}$.9937271
0,001	1,00798	0,0034304	0,051	0,98540	$\bar{1}$.9936126
0,002	1,00641	0,0027749	0,052	0,98515	$\bar{1}$.9935024
0,003	1,00525	0,0022741	0,053	0,98489	$\bar{1}$.9933877
0,004	1,00426	0,0018462	0,054	0,98464	$\bar{1}$.9932775
0,005	1,00340	0,0014741	0,055	0,98440	$\bar{1}$.9931716
0,006	1,00261	0,0011320	0,056	0,98415	$\bar{1}$.9930613
0,007	1,00189	0,0008200	0,057	0,98391	$\bar{1}$.9929554
0,008	1,00122	0,0005295	0,058	0,98366	$\bar{1}$.9928450
0,009	1,00060	0,0002605	0,059	0,98342	$\bar{1}$.9927390
0,010	1,00000	0,0000000	0,060	0,98319	$\bar{1}$.9926375
0,011	0,99943	$\bar{1}$.9997524	0,061	0,98295	$\bar{1}$.9925314
0,012	0,99889	$\bar{1}$.9995177	0,062	0,98272	$\bar{1}$.9924298
0,013	0,99837	$\bar{1}$.9992915	0,063	0,98248	$\bar{1}$.9923237
0,014	0,99787	$\bar{1}$.9990740	0,064	0,98225	$\bar{1}$.9922220
0,015	0,99733	$\bar{1}$.9988650	0,065	0,98203	$\bar{1}$.9921248
0,016	0,99693	$\bar{1}$.9986647	0,066	0,98180	$\bar{1}$.9920230
0,017	0,99648	$\bar{1}$.9984686	0,067	0,98157	$\bar{1}$.9919213
0,018	0,99604	$\bar{1}$.9982768	0,068	0,98135	$\bar{1}$.9918239
0,019	0,99561	$\bar{1}$.9980893	0,069	0,98113	$\bar{1}$.9917266
0,020	0,99520	$\bar{1}$.9979104	0,070	0,98091	$\bar{1}$.9916292
0,021	0,99479	$\bar{1}$.9977314	0,071	0,98069	$\bar{1}$.9915317
0,022	0,99439	$\bar{1}$.9975567	0,072	0,98047	$\bar{1}$.9914343
0,023	0,99401	$\bar{1}$.9973908	0,073	0,98026	$\bar{1}$.9913413
0,024	0,99363	$\bar{1}$.9972247	0,074	0,98004	$\bar{1}$.9912438
0,025	0,99326	$\bar{1}$.9970629	0,075	0,97983	$\bar{1}$.9911507
0,026	0,99290	$\bar{1}$.9969055	0,076	0,97962	$\bar{1}$.9910576
0,027	0,99254	$\bar{1}$.9967480	0,077	0,97941	$\bar{1}$.9909645
0,028	0,99219	$\bar{1}$.9965948	0,078	0,97920	$\bar{1}$.9908714
0,029	0,99185	$\bar{1}$.9964460	0,079	0,97900	$\bar{1}$.9907827
0,030	0,99151	$\bar{1}$.9962971	0,080	0,97879	$\bar{1}$.9906895
0,031	0,99118	$\bar{1}$.9961525	0,081	0,97859	$\bar{1}$.9906008
0,032	0,99085	$\bar{1}$.9960079	0,082	0,97838	$\bar{1}$.9905076
0,033	0,99053	$\bar{1}$.9958676	0,083	0,97818	$\bar{1}$.9904188
0,034	0,99021	$\bar{1}$.9957273	0,084	0,97798	$\bar{1}$.9903300
0,035	0,98990	$\bar{1}$.9955913	0,085	0,97778	$\bar{1}$.9902411
0,036	0,98959	$\bar{1}$.9954553	0,086	0,97758	$\bar{1}$.9901523
0,037	0,98929	$\bar{1}$.9953236	0,087	0,97738	$\bar{1}$.9900634
0,038	0,98899	$\bar{1}$.9951919	0,088	0,97719	$\bar{1}$.9899790
0,039	0,98869	$\bar{1}$.9950601	0,089	0,97699	$\bar{1}$.9898901
0,040	0,98840	$\bar{1}$.9949327	0,090	0,97680	$\bar{1}$.9898057
0,041	0,98811	$\bar{1}$.9948053	0,091	0,97661	$\bar{1}$.9897212
0,042	0,98783	$\bar{1}$.9946822	0,092	0,97641	$\bar{1}$.9896322
0,043	0,98755	$\bar{1}$.9945591	0,093	0,97622	$\bar{1}$.9895477
0,044	0,98727	$\bar{1}$.9944359	0,094	0,97604	$\bar{1}$.9894676
0,045	0,98699	$\bar{1}$.9943128	0,095	0,97583	$\bar{1}$.9893831
0,046	0,98672	$\bar{1}$.9941939	0,096	0,97566	$\bar{1}$.9892985
0,047	0,98645	$\bar{1}$.9940751	0,097	0,97547	$\bar{1}$.9892159
0,048	0,98619	$\bar{1}$.9939606	0,098	0,97529	$\bar{1}$.9891338
0,049	0,98592	$\bar{1}$.9938417	0,099	0,97510	$\bar{1}$.9890492
0,050	0,98566	$\bar{1}$.9937271	0,100	0,97492	$\bar{1}$.9889690

TABLE XXX.

VELOCITIES, IN FEET PER SECOND, DUE TO HEADS FROM 0 TO 4.99 FEET.

Head.	0	1	2	3	4	5	6	7	8	9
0.0	0.000	0.802	1.134	1.389	1.604	1.793	1.965	2.122	2.268	2.406
.1	2.536	2.660	2.778	2.892	3.001	3.106	3.208	3.307	3.403	3.496
.2	3.587	3.675	3.762	3.846	3.929	4.010	4.090	4.167	4.244	4.319
.3	4.393	4.465	4.537	4.607	4.677	4.745	4.812	4.878	4.944	5.009
.4	5.072	5.135	5.198	5.259	5.320	5.380	5.440	5.498	5.557	5.614
.5	5.671	5.728	5.783	5.839	5.894	5.948	6.002	6.055	6.108	6.160
.6	6.212	6.264	6.315	6.366	6.416	6.466	6.516	6.565	6.614	6.662
.7	6.710	6.758	6.805	6.852	6.899	6.946	6.992	7.038	7.083	7.129
.8	7.173	7.218	7.263	7.307	7.351	7.394	7.438	7.481	7.524	7.566
.9	7.609	7.651	7.693	7.734	7.776	7.817	7.858	7.899	7.940	7.980
1.0	8.020	8.060	8.100	8.140	8.179	8.218	8.257	8.296	8.335	8.373
.1	8.412	8.450	8.488	8.526	8.563	8.601	8.638	8.675	8.712	8.749
.2	8.786	8.822	8.859	8.895	8.931	8.967	9.003	9.038	9.074	9.109
.3	9.144	9.180	9.214	9.249	9.284	9.319	9.353	9.387	9.422	9.456
.4	9.490	9.523	9.557	9.591	9.624	9.658	9.691	9.724	9.757	9.790
.5	9.823	9.855	9.888	9.920	9.953	9.985	10.017	10.049	10.081	10.113
.6	10.145	10.176	10.208	10.240	10.271	10.302	10.333	10.364	10.395	10.426
.7	10.457	10.488	10.518	10.549	10.579	10.610	10.640	10.670	10.700	10.730
.8	10.760	10.790	10.820	10.850	10.879	10.909	10.938	10.967	10.997	11.026
9	11.055	11.084	11.113	11.142	11.171	11.200	11.228	11.257	11.285	11.314
2.0	11.342	11.371	11.399	11.427	11.455	11.483	11.511	11.539	11.567	11.595
.1	11.622	11.650	11.678	11.705	11.733	11.760	11.787	11.814	11.842	11.869
.2	11.896	11.923	11.950	11.977	12.004	12.030	12.057	12.084	12.110	12.137
.3	12.163	12.190	12.216	12.242	12.269	12.295	12.321	12.347	12.373	12.399
.4	12.425	12.451	12.477	12.502	12.528	12.554	12.579	12.605	12.630	12.656
.5	12.681	12.706	12.732	12.757	12.782	12.807	12.832	12.857	12.882	12.907
.6	12.932	12.957	12.982	13.007	13.031	13.056	13.081	13.105	13.130	13.154
.7	13.179	13.203	13.227	13.252	13.276	13.300	13.324	13.348	13.372	13.396
.8	13.420	13.444	13.468	13.492	13.516	13.540	13.563	13.587	13.611	13.634
.9	13.658	13.681	13.705	13.728	13.752	13.775	13.798	13.822	13.845	13.868
3.0	13.891	13.915	13.938	13.961	13.984	14.007	14.030	14.053	14.075	14.098
.1	14.121	14.144	14.166	14.189	14.212	14.234	14.257	14.280	14.302	14.325
.2	14.347	14.369	14.392	14.414	14.436	14.459	14.481	14.503	14.525	14.547
.3	14.569	14.591	14.613	14.635	14.657	14.679	14.701	14.723	14.745	14.767
.4	14.789	14.810	14.832	14.854	14.875	14.897	14.918	14.940	14.961	14.983
.5	15.004	15.026	15.047	15.069	15.090	15.111	15.132	15.154	15.175	15.196
.6	15.217	15.238	15.259	15.281	15.302	15.323	15.344	15.364	15.385	15.406
.7	15.427	15.448	15.469	15.490	15.510	15.531	15.552	15.572	15.593	15.614
.8	15.634	15.655	15.675	15.696	15.716	15.737	15.757	15.778	15.798	15.818
.9	15.839	15.859	15.879	15.899	15.920	15.940	15.960	15.980	16.000	16.020
4.0	16.040	16.060	16.080	16.100	16.120	16.140	16.160	16.180	16.200	16.220
.1	16.240	16.259	16.279	16.299	16.319	16.338	16.358	16.378	16.397	16.417
.2	16.437	16.456	16.476	16.495	16.515	16.534	16.554	16.573	16.592	16.612
.3	16.631	16.650	16.670	16.689	16.708	16.727	16.747	16.766	16.785	16.804
.4	16.823	16.842	16.862	16.881	16.900	16.919	16.938	16.957	16.976	16.994
.5	17.013	17.032	17.051	17.070	17.089	17.108	17.126	17.145	17.164	17.183
.6	17.201	17.220	17.239	17.257	17.276	17.295	17.313	17.332	17.350	17.369
.7	17.387	17.406	17.424	17.443	17.461	17.480	17.498	17.516	17.535	17.553
.8	17.571	17.590	17.608	17.626	17.644	17.663	17.681	17.699	17.717	17.735
.9	17.753	17.772	17.790	17.808	17.826	17.844	17.862	17.880	17.898	17.916

TABLE XXX — Continued.

VELOCITIES, IN FEET PER SECOND, DUE TO HEADS FROM 5 TO 9.99 FEET.

Head.	0	1	2	3	4	5	6	7	8	9
5.0	17.934	17.952	17.970	17.987	18.005	18.023	18.041	18.059	18.077	18.094
.1	18.112	18.130	18.148	18.165	18.183	18.201	18.218	18.236	18.254	18.271
.2	18.289	18.306	18.324	18.342	18.359	18.377	18.394	18.412	18.429	18.446
.3	18.464	18.481	18.499	18.516	18.533	18.551	18.568	18.585	18.603	18.620
.4	18.637	18.655	18.672	18.689	18.706	18.723	18.741	18.758	18.775	18.792
.5	18.809	18.826	18.843	18.860	18.877	18.894	18.911	18.928	18.945	18.962
.6	18.979	18.996	19.013	19.030	19.047	19.064	19.081	19.098	19.114	19.131
.7	19.148	19.165	19.182	19.198	19.215	19.232	19.248	19.265	19.282	19.299
.8	19.315	19.332	19.348	19.365	19.382	19.398	19.415	19.431	19.448	19.464
.9	19.481	19.497	19.514	19.530	19.547	19.563	19.580	19.596	19.613	19.629
6.0	19.645	19.662	19.678	19.694	19.711	19.727	19.743	19.760	19.776	19.792
.1	19.808	19.825	19.841	19.857	19.873	19.889	19.906	19.922	19.938	19.954
.2	19.970	19.986	20.002	20.018	20.034	20.050	20.067	20.083	20.099	20.115
.3	20.131	20.147	20.162	20.178	20.194	20.210	20.226	20.242	20.258	20.274
.4	20.290	20.306	20.321	20.337	20.353	20.369	20.385	20.400	20.416	20.432
.5	20.448	20.463	20.479	20.495	20.510	20.526	20.542	20.557	20.573	20.589
.6	20.604	20.620	20.635	20.651	20.667	20.682	20.698	20.713	20.729	20.744
.7	20.760	20.775	20.791	20.806	20.822	20.837	20.853	20.868	20.883	20.899
.8	20.914	20.929	20.945	20.960	20.976	20.991	21.006	21.021	21.037	21.052
.9	21.067	21.083	21.098	21.113	21.128	21.144	21.159	21.174	21.189	21.204
7.0	21.219	21.235	21.250	21.265	21.280	21.295	21.310	21.325	21.340	21.355
.1	21.370	21.386	21.401	21.416	21.431	21.446	21.461	21.476	21.491	21.506
.2	21.520	21.535	21.550	21.565	21.580	21.595	21.610	21.625	21.640	21.655
.3	21.669	21.684	21.699	21.714	21.729	21.743	21.758	21.773	21.788	21.803
.4	21.817	21.832	21.847	21.861	21.876	21.891	21.906	21.920	21.935	21.950
.5	21.964	21.979	21.993	22.008	22.023	22.037	22.052	22.066	22.081	22.096
.6	22.110	22.125	22.139	22.154	22.168	22.183	22.197	22.212	22.226	22.241
.7	22.255	22.270	22.284	22.298	22.313	22.327	22.342	22.356	22.370	22.385
.8	22.399	22.414	22.428	22.442	22.457	22.471	22.485	22.499	22.514	22.528
.9	22.542	22.557	22.571	22.585	22.599	22.614	22.628	22.642	22.656	22.670
8.0	22.685	22.699	22.713	22.727	22.741	22.755	22.769	22.784	22.798	22.812
.1	22.826	22.840	22.854	22.868	22.882	22.896	22.910	22.924	22.938	22.952
.2	22.966	22.980	22.994	23.008	23.022	23.036	23.050	23.064	23.078	23.092
.3	23.106	23.120	23.134	23.148	23.162	23.175	23.189	23.203	23.217	23.231
.4	23.245	23.259	23.272	23.286	23.300	23.314	23.328	23.341	23.355	23.369
.5	23.383	23.396	23.410	23.424	23.438	23.451	23.465	23.479	23.492	23.506
.6	23.520	23.534	23.547	23.561	23.574	23.588	23.602	23.615	23.629	23.643
.7	23.656	23.670	23.683	23.697	23.711	23.724	23.738	23.751	23.765	23.778
.8	23.792	23.805	23.819	23.832	23.846	23.859	23.873	23.886	23.900	23.913
.9	23.927	23.940	23.953	23.967	23.980	23.994	24.007	24.020	24.034	24.047
9.0	24.061	24.074	24.087	24.101	24.114	24.127	24.141	24.154	24.167	24.181
.1	24.194	24.207	24.220	24.234	24.247	24.260	24.274	24.287	24.300	24.313
.2	24.326	24.340	24.353	24.366	24.379	24.392	24.406	24.419	24.432	24.445
.3	24.458	24.471	24.485	24.498	24.511	24.524	24.537	24.550	24.563	24.576
.4	24.589	24.603	24.616	24.629	24.642	24.655	24.668	24.681	24.694	24.707
.5	24.720	24.733	24.746	24.759	24.772	24.785	24.798	24.811	24.824	24.837
.6	24.850	24.863	24.876	24.888	24.901	24.914	24.927	24.940	24.953	24.966
.7	24.979	24.992	25.005	25.017	25.030	25.043	25.056	25.069	25.082	25.094
.8	25.107	25.120	25.133	25.146	25.158	25.171	25.184	25.197	25.209	25.222
.9	25.235	25.248	25.260	25.273	25.286	25.299	25.311	25.324	25.337	25.349

TABLE XXX—Continued.

VELOCITIES, IN FEET PER SECOND, DUE TO HEADS FROM 10 TO 14.99 FEET.

Head.	0	1	2	3	4	5	6	7	8	9
10.0	25.362	25.375	25.387	25.400	25.413	25.425	25.438	25.451	25.463	25.476
.1	25.489	25.501	25.514	25.526	25.539	25.552	25.564	25.577	25.589	25.602
.2	25.614	25.627	25.640	25.652	25.665	25.677	25.690	25.702	25.715	25.728
.3	25.740	25.752	25.765	25.777	25.790	25.802	25.815	25.827	25.839	25.852
.4	25.864	25.877	25.889	25.902	25.914	25.926	25.939	25.951	25.964	25.976
.5	25.988	26.001	26.013	26.026	26.038	26.050	26.063	26.075	26.087	26.099
.6	26.112	26.124	26.136	26.149	26.161	26.173	26.186	26.198	26.210	26.222
.7	26.235	26.247	26.259	26.272	26.284	26.296	26.308	26.320	26.333	26.345
.8	26.357	26.369	26.381	26.394	26.406	26.418	26.430	26.442	26.454	26.467
.9	26.479	26.491	26.503	26.515	26.527	26.540	26.552	26.564	26.576	26.588
11.0	26.600	26.612	26.624	26.636	26.648	26.660	26.672	26.684	26.697	26.709
.1	26.721	26.733	26.745	26.757	26.769	26.781	26.793	26.805	26.817	26.829
.2	26.841	26.853	26.865	26.877	26.889	26.901	26.913	26.924	26.936	26.948
.3	26.960	26.972	26.984	26.936	27.008	27.020	27.032	27.044	27.056	27.067
.4	27.079	27.091	27.103	27.115	27.127	27.139	27.150	27.162	27.174	27.186
.5	27.198	27.210	27.221	27.233	27.245	27.257	27.269	27.280	27.292	27.304
.6	27.316	27.328	27.339	27.351	27.363	27.375	27.386	27.398	27.410	27.422
.7	27.433	27.445	27.457	27.468	27.480	27.492	27.504	27.515	27.527	27.539
.8	27.550	27.562	27.574	27.585	27.597	27.609	27.620	27.632	27.644	27.655
9	27.667	27.678	27.690	27.702	27.713	27.725	27.736	27.748	27.760	27.771
12.0	27.783	27.794	27.806	27.817	27.829	27.841	27.852	27.864	27.875	27.887
.1	27.898	27.910	27.921	27.933	27.944	27.956	27.967	27.979	27.990	28.002
.2	28.013	28.025	28.036	28.048	28.059	28.071	28.082	28.094	28.105	28.117
.3	28.128	28.139	28.151	28.162	28.174	28.185	28.196	28.208	28.219	28.231
.4	28.242	28.253	28.265	28.276	28.288	28.299	28.310	28.322	28.333	28.344
.5	28.356	28.367	28.378	28.390	28.401	28.412	28.424	28.435	28.446	28.458
.6	28.469	28.480	28.491	28.503	28.514	28.525	28.537	28.548	28.559	28.570
.7	28.582	28.593	28.604	28.615	28.627	28.638	28.649	28.660	28.672	28.683
.8	28.694	28.705	28.716	28.727	28.739	28.750	28.761	28.772	28.783	28.795
.9	28.806	28.817	28.828	28.839	28.850	28.862	28.873	28.884	28.895	28.906
13.0	28.917	28.928	28.939	28.951	28.962	28.973	28.984	28.995	29.006	29.017
.1	29.028	29.039	29.050	29.061	29.073	29.084	29.095	29.106	29.117	29.128
.2	29.139	29.150	29.161	29.172	29.183	29.194	29.205	29.216	29.227	29.238
.3	29.249	29.260	29.271	29.282	29.293	29.304	29.315	29.326	29.337	29.348
.4	29.359	29.370	29.381	29.392	29.403	29.413	29.424	29.435	29.446	29.457
.5	29.468	29.479	29.490	29.501	29.512	29.523	29.533	29.514	29.555	29.566
.6	29.577	29.588	29.599	29.610	29.620	29.631	29.642	29.653	29.664	29.675
.7	29.686	29.696	29.707	29.718	29.729	29.740	29.751	29.761	29.772	29.783
.8	29.794	29.805	29.815	29.826	29.837	29.848	29.858	29.869	29.880	29.891
.9	29.901	29.912	29.923	29.934	29.944	29.955	29.966	29.977	29.987	29.998
14.0	30.009	30.020	30.030	30.041	30.052	30.062	30.073	30.084	30.094	30.105
.1	30.116	30.126	30.137	30.148	30.159	30.169	30.180	30.190	30.201	30.212
.2	30.222	30.233	30.244	30.254	30.265	30.276	30.286	30.297	30.307	30.318
.3	30.329	30.339	30.350	30.360	30.371	30.382	30.392	30.403	30.413	30.424
.4	30.435	30.445	30.456	30.466	30.477	30.487	30.498	30.508	30.519	30.529
.5	30.540	30.551	30.561	30.572	30.582	30.593	30.603	30.614	30.624	30.635
.6	30.645	30.656	30.666	30.677	30.687	30.698	30.708	30.719	30.729	30.739
.7	30.750	30.760	30.771	30.781	30.792	30.802	30.813	30.823	30.833	30.844
.8	30.854	30.865	30.875	30.886	30.896	30.906	30.917	30.927	30.938	30.948
.9	30.958	30.969	30.979	30.990	31.000	31.010	31.021	31.031	31.041	31.052

TABLE XXX — CONTINUED.

VELOCITIES, IN FEET PER SECOND, DUE TO HEADS FROM 15 TO 19.99 FEET.

Head.	0	1	2	3	4	5	6	7	8	9
15.0	31.062	31.072	31.083	31.093	31.103	31.114	31.124	31.134	31.145	31.155
.1	31.165	31.176	31.186	31.196	31.207	31.217	31.227	31.238	31.248	31.258
.2	31.268	31.279	31.289	31.299	31.310	31.320	31.330	31.340	31.351	31.361
.3	31.371	31.381	31.392	31.402	31.412	31.422	31.433	31.443	31.453	31.463
.4	31.474	31.484	31.494	31.504	31.514	31.525	31.535	31.545	31.555	31.565
.5	31.576	31.586	31.596	31.606	31.616	31.626	31.637	31.647	31.657	31.667
.6	31.677	31.687	31.698	31.708	31.718	31.728	31.738	31.748	31.758	31.768
.7	31.779	31.789	31.799	31.809	31.819	31.829	31.839	31.849	31.859	31.870
.8	31.880	31.830	31.900	31.910	31.920	31.930	31.940	31.950	31.960	31.970
.9	31.980	31.990	32.000	32.011	32.021	32.031	32.041	32.051	32.061	32.071
16.0	32.081	32.091	32.101	32.111	32.121	32.131	32.141	32.151	32.161	32.171
.1	32.181	32.191	32.201	32.211	32.221	32.231	32.241	32.251	32.261	32 271
.2	32.281	32.291	32.301	32.311	32.321	32.330	32.340	32.350	32.360	32.370
.3	32.380	32.390	32.400	32.410	32.420	32.430	32.440	32.450	32.460	32.470
.4	32.480	32.489	32.499	32.509	32.519	32.529	32.539	32.549	32.559	32.569
.5	32.579	32.588	32.598	32.608	32.618	32.628	32.637	32.647	32.657	32.667
.6	32.677	32.687	32.696	32.706	32.716	32.726	32 736	32.746	32.755	32.765
.7	32.775	32.785	32.795	32.804	32.814	32.824	32.834	32.844	32.854	32.863
.8	32.873	32.883	32.893	32.903	32.912	32.922	32.932	32.941	32.951	32.961
.9	32.971	32.980	32.990	33.000	33.010	33.019	33.029	33.039	33.049	33.058
17.0	33.068	33.078	33.088	33.097	33.107	33.117	33.126	33.136	33.146	33.156
.1	33.165	33.175	33.185	33.194	33.204	33.214	33.223	33.233	33.243	33.252
.2	33.262	33.272	33.281	33.291	33.301	33.310	33.320	33.330	33.339	33.349
.3	33.359	33.368	33.378	33.388	33.397	33.407	33.416	33.426	33.436	33.445
.4	33.455	33.465	33.474	33.484	33.493	33.503	33.513	33.522	33.532	33.541
.5	33.551	33.560	33.570	33.580	33.589	33.599	33.608	33.618	33.628	33.637
.6	33.647	33.656	33.666	33.675	33.685	33.694	33.704	33.713	33.723	33.733
.7	33.742	33.752	33.761	33.771	33.780	33.790	33.799	33.809	33.818	33.828
.8	33.837	33.847	33.856	33.866	33.875	33.885	33.894	33.904	33.913	33.923
.9	33.932	33.942	33.951	33.961	33.970	33.980	33.989	33.998	34.008	34.017
18.0	34.027	34.036	34.046	34.055	34.065	34.074	34.083	34.093	34.102	34.112
.1	34.121	34.131	34.140	34.149	34.159	34.168	34.178	34.187	34.197	34.206
.2	34.215	34.225	34.234	34.244	34.253	34.262	34 272	34.281	34.290	34.300
.3	34.309	34.319	34.328	34.337	34.347	34.356	34.365	34.375	34.384	34.393
.4	34.403	34.412	34.422	34.431	34.440	34.450	34.459	34.468	34.478	34.487
.5	34.496	34.505	34.515	34.524	34.533	34.543	34.552	34.561	34.574	34.580
.6	34.589	34.599	34.608	34.617	34.626	34.636	34.645	34.654	34.664	34.673
.7	34.682	34.691	34.701	34.710	34.719	34.728	34.738	34.747	34.756	34.766
.8	34.775	34.784	34.793	34.802	34.812	34.821	34.830	34.839	34.849	34.858
.9	34.867	34.876	34.886	34.895	34.904	34.913	34.922	34.932	34.941	34.950
19.0	34.959	34.968	34.978	34.987	34.996	35.005	35.014	35.024	35.033	35.042
.1	35.051	35.060	35.069	35.079	35.088	35.097	35.106	35.115	35.124	35.134
.2	35.143	35.152	35.161	35.170	35.179	35.188	35.198	35.207	35.216	35.225
.3	35.234	35.243	35.252	35.262	35.271	35.280	35.289	35.298	35.307	35.316
.4	35.325	35.334	35.344	35.353	35.362	35.371	35.380	35.389	35.398	35.407
.5	35.416	35.425	35.434	35.443	35.453	35.462	35.471	35.480	35.489	35.498
.6	35.507	35.516	35.525	35.534	35.543	35.552	35.561	35.570	35.579	35.588
.7	35.597	35.606	35.615	35.624	35.634	35.643	35.652	35.661	35.670	35.679
.8	35.688	35.697	35.706	35.715	35.724	35.733	35.742	35.751	35.760	35.769
.9	35.778	35.787	35.796	35.805	35.814	35.823	35.832	35.841	35.849	35.858

TABLE XXX—Continued.

VELOCITIES, IN FEET PER SECOND, DUE TO HEADS FROM 20 TO 24.99 FEET.

Head.	0	1	2	3	4	5	6	7	8	9
20.0	35.867	35.876	35.885	35.894	35.903	35.912	35.921	35.930	35.939	35.948
.1	35.957	35.966	35.975	35.984	35.993	36.002	36.011	36.020	36.028	36.037
.2	36.046	36.055	36.064	36.073	36.082	36.091	36.100	36.109	36.118	36.127
.3	36.135	36.144	36.153	36.162	36.171	36.180	36.189	36.198	36.207	36.215
.4	36.224	36.233	36.242	36.251	36.260	36.269	36.278	36.286	36.295	36.304
.5	36.313	36.322	36.331	36.340	36.348	36.357	36.366	36.375	36.384	36.393
.6	36.401	36.410	36.419	36.428	36.437	36.446	36.454	36.463	36.472	36.481
.7	36.490	36.499	36.507	36.516	36.525	36.534	36.543	36.551	36.560	36.569
.8	36.578	36.587	36.595	36.604	36.613	36.622	36.630	36.639	36.648	36.657
.9	36.666	36.674	36.683	36.632	36.701	36.709	36.718	36.727	36.736	36.744
21.0	36.753	36.762	36.771	36.779	36.788	36.797	36.806	36.814	36.823	36.832
.1	36.841	36.849	36.858	36.867	36.875	36.884	36.893	36.902	36.910	36.919
.2	36.928	36.936	36.945	36.954	36.963	36.971	36.980	36.989	36.997	37.006
.3	37.015	37.023	37.032	37.041	37.049	37.058	37.067	37.076	37.084	37.093
.4	37.102	37.110	37.119	37.128	37.136	37.145	37.154	37.162	37.171	37.179
.5	37.188	37.197	37.205	37.214	37.223	37.231	37.240	37.249	37.257	37.266
.6	37.275	37.283	37.292	37.300	37.309	37.318	37.326	37.335	37.343	37.352
.7	37.361	37.369	37.378	37.387	37.395	37.404	37.412	37.421	37.430	37.438
.8	37.447	37.455	37.464	37.472	37.481	37.490	37.498	37.506	37.515	37.524
9	37.532	37.541	37.550	37.558	37.567	37.575	37.584	37.592	37.601	37.610
22.0	37.618	37.627	37.635	37.644	37.652	37.661	37.669	37.678	37.686	37.695
.1	37.703	37.712	37.721	37.729	37.738	37.746	37.755	37.763	37.772	37.780
.2	37.789	37.797	37.806	37.814	37.823	37.832	37.840	37.848	37.857	37.865
.3	37.874	37.882	37.891	37.899	37.908	37.916	37.925	37.933	37.942	37.950
.4	37.959	37.967	37.975	37.984	37.992	38.001	38.009	38.018	38.026	38.035
.5	38.043	38.052	38.060	38.068	38.077	38.085	38.094	38.102	38.111	38.119
.6	38.128	38.136	38.144	38.153	38.161	38.170	38.178	38.187	38.195	38.203
.7	38.212	38.220	38.229	38.237	38.246	38.254	38.262	38.271	38.279	38.288
.8	38.296	38.304	38.313	38.321	38.330	38.338	38.346	38.355	38.363	38.371
.9	38.380	38.388	38.397	38.405	38.413	38.422	38.430	38.438	38.447	38.455
23.0	38.464	38.472	38.480	38.489	38.497	38.505	38.514	38.522	38.530	38.539
.1	38.547	38.555	38.564	38.572	38.580	38.589	38.597	38.605	38.614	38.622
.2	38.630	38.638	38.647	38.655	38.664	38.672	38.680	38.689	38.697	38.705
.3	38.714	38.722	38.730	38.738	38.747	38.755	38.763	38.772	38.780	38.788
.4	38.797	38.805	38.813	38.821	38.830	38.838	38.846	38.855	38.863	38.871
.5	38.879	38.888	38.896	38.904	38.912	38.921	38.929	38.937	38.945	38.954
.6	38.962	38.970	38.978	38.987	38.995	39.003	39.011	39.020	39.028	39.036
.7	39.044	39.053	39.061	39.069	39.077	39.086	39.094	39.102	39.110	39.119
.8	39.127	39.135	39.143	39.151	39.160	39.168	39.176	39.184	39.192	39.201
.9	39.209	39.217	39.225	39.233	39.242	39.250	39.258	39.266	39.274	39.283
24.0	39.291	39.299	39.307	39.315	39.324	39.332	39.340	39.348	39.356	39.364
.1	39.373	39.381	39.389	39.397	39.405	39.413	39.422	39.430	39.438	39.446
.2	39.454	39.462	39.470	39.479	39.487	39.495	39.503	39.511	39.519	39.527
.3	39.536	39.544	39.552	39.560	39.568	39.576	39.584	39.592	39.601	39.609
.4	39.617	39.625	39.633	39.641	39.649	39.657	39.666	39.674	39.682	39.690
.5	39.698	39.706	39.714	39.722	39.730	39.738	39.747	39.755	39.763	39.771
.6	39.779	39.787	39.795	39.803	39.811	39.819	39.827	39.835	39.844	39.852
.7	39.860	39.868	39.876	39.884	39.892	39.900	39.908	39.916	39.924	39.932
.8	39.940	39.948	39.956	39.964	39.972	39.981	39.989	39.997	40.005	40.013
.9	40.021	40.029	40.037	40.045	40.053	40.061	40.069	40.077	40.085	40.093

TABLE XXX — Continued.

VELOCITIES, IN FEET PER SECOND, DUE TO HEADS FROM 25 TO 29.99 FEET.

Head.	0	1	2	3	4	5	6	7	8	9
25.0	40.101	40.109	40.117	40.125	40.133	40.141	40.149	40.157	40.165	40.173
.1	40.181	40.189	40.197	40.205	40.213	40.221	40.229	40.237	40.245	40.253
.2	40.261	40.269	40.277	40.285	40.293	40.301	40.309	40.317	40.325	40.333
.3	40.341	40.349	40.357	40.365	40.373	40.381	40.389	40.397	40.405	40.413
.4	40.421	40.428	40.436	40.444	40.452	40.460	40.468	40.476	40.484	40.492
.5	40.500	40.508	40.516	40.524	40.532	40.540	40.548	40.556	40.563	40.571
.6	40.579	40.587	40.595	40.603	40.611	40.619	40.627	40.635	40.643	40.651
.7	40.659	40.666	40.674	40.682	40.690	40.698	40.706	40.714	40.722	40.730
.8	40.738	40.745	40.753	40.761	40.769	40.777	40.785	40.793	40.801	40.809
.9	40.816	40.824	40.832	40.840	40.848	40.856	40.864	40.872	40.879	40.887
26.0	40.895	40.903	40.911	40.919	40.927	40.934	40.942	40.950	40.958	40.966
.1	40.974	40.982	40.989	40.997	41.005	41.013	41.021	41.029	41.036	41.044
.2	41.052	41.060	41.068	41.076	41.083	41.091	41.099	41.107	41.115	41.123
.3	41.130	41.138	41.146	41.154	41.162	41.169	41.177	41.185	41.193	41.201
.4	41.209	41.216	41.224	41.232	41.240	41.248	41.255	41.263	41.271	41.279
.5	41.287	41.294	41.302	41.310	41.318	41.325	41.333	41.341	41.349	41.357
.6	41.364	41.372	41.380	41.388	41.395	41.403	41.411	41.419	41.426	41.434
.7	41.442	41.450	41.458	41.465	41.473	41.481	41.489	41.496	41.504	41.512
.8	41.520	41.527	41.535	41.543	41.551	41.558	41.566	41.574	41.581	41.589
.9	41.597	41.605	41.612	41.620	41.628	41.636	41.643	41.651	41.659	41.666
27.0	41.674	41.682	41.690	41.697	41.705	41.713	41.720	41.728	41.736	41.744
.1	41.751	41.759	41.767	41.774	41.782	41.790	41.797	41.805	41.813	41.821
.2	41.828	41.836	41.844	41.851	41.859	41.867	41.874	41.882	41.890	41.897
.3	41.905	41.913	41.920	41.928	41.936	41.943	41.951	41.959	41.967	41.974
.4	41.982	41.989	41.997	42.005	42.012	42.020	42.028	42.035	42.043	42.051
.5	42.058	42.066	42.074	42.081	42.089	42.096	42.104	42.112	42.119	42.127
.6	42.135	42.142	42.150	42.158	42.165	42.173	42.180	42.188	42.196	42.203
.7	42.211	42.219	42.226	42.234	42.241	42.249	42.257	42.264	42.272	42.279
.8	42.287	42.295	42.302	42.310	42.317	42.325	42.333	42.340	42.348	42.355
.9	42.363	42.371	42.378	42.386	42.393	42.401	42.409	42.416	42.424	42.431
28.0	42.439	42.446	42.454	42.462	42.469	42.477	42.484	42.492	42.499	42.507
.1	42.515	42.522	42.530	42.537	42.545	42.552	42.560	42.568	42.575	42.583
.2	42.590	42.598	42.605	42.613	42.620	42.628	42.635	42.643	42.651	42.658
.3	42.666	42.673	42.681	42.688	42.696	42.703	42.711	42.718	42.726	42.733
.4	42.741	42.748	42.756	42.764	42.771	42.779	42.786	42.794	42.801	42.809
.5	42.816	42.824	42.831	42.839	42.846	42.854	42.861	42.869	42.876	42.884
.6	42.891	42.899	42.906	42.914	42.921	42.929	42.936	42.914	42.951	42.959
.7	42.966	42.974	42.981	42.989	42.996	43.004	43.011	43.018	43.026	43.033
.8	43.041	43.048	43.056	43.063	43.071	43.078	43.086	43.093	43.101	43.108
.9	43.116	43.123	43.130	43.138	43.145	43.153	43.160	43.168	43.175	43.183
29.0	43.190	43.198	43.205	43.212	43.220	43.227	43.235	43.243	43.250	43.257
.1	43.264	43.272	43.279	43.287	43.294	43.302	43.309	43.316	43.324	43.331
.2	43.339	43.346	43.354	43.361	43.368	43.376	43.383	43.391	43.398	43.405
.3	43.413	43.420	43.428	43.435	43.443	43.450	43.457	43.465	43.472	43.480
.4	43.487	43.494	43.502	43.509	43.517	43.524	43.531	43.539	43.546	43.553
.5	43.561	43.568	43.576	43.583	43.590	43.598	43.605	43.612	43.620	43.627
.6	43.635	43.642	43.649	43.657	43.664	43.671	43.679	43.686	43.694	43.701
.7	43.708	43.716	43.723	43.730	43.738	43.745	43.752	43.760	43.767	43.774
.8	43.782	43.789	43.796	43.804	43.811	43.818	43.826	43.833	43.840	43.848
.9	43.855	43.862	43.870	43.877	43.884	43.892	43.899	43.906	43.914	43.921

TABLE XXX—Continued.

VELOCITIES, IN FEET PER SECOND, DUE TO HEADS FROM 30 TO 34.99 FEET.

Head.	0	1	2	3	4	5	6	7	8	9
30,0	43.928	43.936	43.943	43.950	43.958	43.965	43.972	43.980	43.987	43.994
.1	44.002	44.009	44.016	44.024	44.031	44.038	44.045	44.053	44.060	44.067
.2	44.075	44 082	44.089	44.097	44.104	44.111	44.118	44.126	44.133	44.140
.3	44.148	44.155	44.162	44.169	44.177	44.184	44.191	44.198	44.206	44.213
.4	44.220	44.228	44.235	44.242	44.249	44.257	44.264	44.271	44.278	44.286
.5	44.293	44.300	44.308	44.315	44.322	44.329	44.337	44.344	44.351	44.358
.6	44.366	44.373	44.380	44.387	44.395	44.402	44.409	44.416	44.423	44.431
.7	44.438	44.445	44.452	44.460	44.467	44.474	44.481	44.489	44.496	44.503
.8	44.510	44.518	44.525	44.532	44.539	44.546	44.554	44.561	44.568	44.575
.9	44.582	44.590	44.597	44.604	44.611	44 619	44.626	44.633	44.640	44.647
31.0	44.655	44.662	44.669	44.676	44.683	41.691	44.698	44.705	44.712	44.719
.1	44.727	44.734	44.741	44.748	44.755	44.762	44.770	44.777	44 784	44.791
.2	44.798	44.806	44.813	44.820	44.827	41.834	44.841	44.849	44.856	44.863
.3	44.870	44.877	44.884	44.892	44.899	44.906	44.913	44.920	44.927	44.935
.4	44.942	44.949	44.956	44.963	44.970	44.978	44.985	44.992	44.999	45.006
.5	45.013	45.020	45.028	45.035	45.042	45.049	45.056	45.063	45.070	45.078
.6	45.085	45.092	45.099	45.106	45.113	45.120	45.127	45.135	45.142	45.149
.7	45.156	45.163	45.170	45.177	45.184	45.192	45.199	45.206	45.213	45.220
.8	45.227	45.234	45.241	45.248	45.256	45.263	45.270	45.277	45.284	45.291
.9	45.298	45.305	45.312	45.319	45.327	45.334	45.341	45.348	45.355	45.362
32.0	45.369	45.376	45.383	45.390	45.397	45.405	45.412	45.419	45.426	45.433
.1	45.440	45.447	45.454	45.461	45.468	45.475	45.482	45.489	45.497	45.504
.2	45.511	45.518	45.525	45.532	45.539	45.546	45.553	45.560	45.567	45.574
.3	45.581	45.588	45.595	45.602	45.609	45.617	45.624	45.631	45.638	45.645
.4	45.652	45.659	45.666	45.673	45.680	45.687	45.694	45.701	45.708	45.715
.5	45.722	45.729	45.736	45.743	45.750	45.757	45.764	45.771	45.778	45.785
.6	45.792	45.799	45.807	45.814	45.821	45.828	45.835	45.842	45.849	45.856
.7	45.863	45.870	45.877	45.884	45.891	45.898	45.905	45.912	45.919	45.926
.8	45.933	45.940	45.947	45.954	45.961	45.968	45.975	45.982	45.989	45.996
.9	46.003	46.010	46.017	46.024	46.031	46.038	46.045	46.052	46.059	46.066
33 0	46.073	46.080	46.086	46.093	46.100	46.107	46.114	46.121	46.128	46 135
.1	46.142	46.149	46.156	46.163	46.170	46.177	46.184	46.191	46.198	46.205
.2	46.212	46.219	46.226	46.233	46.240	46.247	46.254	46.261	46.268	46.275
.3	46.281	46.288	46.295	46.302	46.309	46.316	46.323	46.330	46.337	46.344
.4	46.351	46.358	46.365	46.372	46.379	46.386	46.393	46.399	46.406	46.413
.5	46.420	46.427	46.434	46.441	46.448	46.455	46.462	46.469	46.476	46.483
.6	46.489	46.496	46.503	46.510	46.517	46.524	46.531	46.538	46.545	46.552
.7	46.559	46.566	46.572	46.579	46.586	46.593	46.600	46.607	46.614	46.621
.8	46.628	46.635	46.642	46.648	46.655	46.662	46.669	46.676	46.683	46.690
.9	46.697	46.703	46.710	46.717	46.724	46.731	46.739	46.745	46.752	46 759
34.0	46.765	46.772	46.779	46.786	46.793	46.800	46.807	46.814	46.820	46.827
.1	46.834	46.841	46.848	46.855	46.862	46.868	46.875	46.882	46.889	46.896
.2	46.903	46.910	46.916	46.923	46.930	46.937	46.944	46.951	46.958	46.964
.3	46.971	46.978	46.985	46.992	46.999	47.005	47.012	47.019	47.026	47.033
.4	47.040	47.047	47.053	47.060	47.067	47.074	47.081	47.088	47.094	47.101
.5	47.108	47.115	47.122	47.128	47.135	47.142	47.149	47.156	47.163	47.169
.6	47.176	47.183	47.190	47.197	47.203	47.210	47.217	47.224	47.231	47.238
.7	47.244	47.251	47.258	47.265	47.272	47.278	47.285	47.292	47.299	47.306
.8	47.312	47.319	47.326	47.333	47.340	47.346	47.353	47.360	47.367	47.374
.9	47.380	47.387	47.394	47.401	47.407	47.414	47.421	47.428	47.435	47.441

TABLE XXX—Continued.

VELOCITIES, IN FEET PER SECOND, DUE TO HEADS FROM 35 TO 39.99 FEET.

Head.	0	1	2	3	4	5	6	7	8	9
35.0	47.448	47.455	47.462	47.469	47.475	47.482	47.489	47.496	47.502	47.509
.1	47.516	47.523	47.529	47.536	47.543	47.550	47.556	47.563	47.570	47.577
.2	47.584	47.590	47.597	47.604	47.611	47.617	47.624	47.631	47.638	47.644
.3	47.651	47.658	47.665	47.671	47.678	47.685	47.692	47.698	47.705	47.712
.4	47.719	47.725	47.732	47.739	47.745	47.752	47.759	47.766	47.772	47.779
.5	47.786	47.793	47.799	47.806	47.813	47.819	47.826	47.833	47.840	47.846
.6	47.853	47.860	47.867	47.873	47.880	47.887	47.893	47.900	47.907	47.914
.7	47.920	47.927	47.934	47.940	47.947	47.954	47.961	47.967	47.974	47.981
.8	47.987	47.994	48.001	48.007	48.014	48.021	48.028	48.034	48.041	48.048
.9	48.054	48.061	48.068	48.074	48.081	48.088	48.094	48.101	48.108	48.115
36.0	48.121	48.128	48.134	48.141	48.148	48.155	48.161	48.168	48.175	48.181
.1	48.188	48.195	48.201	48.208	48.215	48.221	48.228	48.235	48.241	48.248
.2	48.255	48.261	48.268	48.275	48.281	48.288	48.295	48.302	48.308	48 315
.3	48.321	48.328	48.335	48.341	48.348	48.355	48.361	48.368	48.375	48.381
.4	48.388	48.394	48.401	48.408	48.414	48.421	48.428	48.434	48.441	48.448
.5	48.454	48.461	48.467	48.474	48.481	48.487	48.494	48.501	48.507	48.514
.6	48.521	48.527	48.534	48.540	48.547	48.554	48.560	48.567	48.574	48.580
.7	48.587	48.593	48.600	48.607	48.613	48.620	48.626	48.633	48.640	48.646
.8	48.653	48.660	48.666	48.673	48.679	48.686	48.693	48.699	48.706	48.712
.9	48.719	48.726	48.732	48.739	48.745	48.752	48.759	48.765	48.771	48.778
37.0	48.785	48.792	48.798	48.805	48.811	48.818	48.824	48.831	48.838	48.844
.1	48.851	48.857	48.864	48.871	48.877	48.884	48.890	48.897	48.903	48.910
.2	48.917	48.923	48.930	48.936	48.943	48.950	48.956	48.963	48.969	48.976
.3	48.982	48.989	48.995	49.002	49.009	49.015	49.022	49.028	49.035	49.041
.4	49.048	49.055	49.061	49.068	49.074	49.081	49.087	49.094	49.100	49.107
.5	49.113	49.120	49.127	49.133	49.140	49.146	49.153	49.159	49.166	49.172
.6	49.179	49.185	49.192	49.199	49.205	49.212	49.218	49.225	49.231	49.238
.7	49.244	49.251	49.257	49.264	49.270	49.277	49.283	49.290	49.297	49.303
.8	49.310	49.316	49.323	49.329	49.336	49.312	49.349	49.355	49.362	49.368
.9	49.375	49.381	49.388	49.394	49.401	49.407	49.414	49.420	49.427	49.433
38.0	49.440	49.446	49.453	49.459	49.466	49.472	49.479	49.485	49.492	49.498
.1	49.505	49.511	49.518	49.524	49.531	49.537	49.544	49.550	49.557	49.563
.2	49.570	49.576	49.583	49.589	49.596	49.602	49.609	49.615	49.622	49.628
.3	49.635	49.641	49.648	49.654	49.661	49.667	49.673	49.680	49.686	49.693
.4	49.699	49.706	49.712	49.719	49.725	49.732	49.738	49.745	49.751	49.758
.5	49.764	49.770	49.777	49.783	49.790	49.796	49.803	49.809	49.816	49.822
.6	49.829	49.835	49.842	49.848	49.854	49.861	49.867	49.874	49.880	49.887
.7	49.893	49.900	49.906	49.912	49.919	49.925	49.932	49.938	49.945	49.951
.8	49.958	49.964	49.970	49.977	49.983	49.990	49.996	50.003	50.009	50.015
.9	50.022	50.028	50.035	50.041	50.048	50.054	50.060	50.967	50.073	50.080
39.0	50.086	50.093	50.099	50.105	50.112	50.118	50.125	50.131	50.137	50.144
.1	50.150	50.157	50.163	50.170	50.176	50.182	50.189	50.195	50.202	50.208
.2	50.214	50.221	50.227	50.234	50.240	50.246	50.253	50.259	50.266	50.272
.3	50.278	50.285	50.291	50.298	50.304	50.310	50.317	50.323	50.330	50.336
.4	50.342	50.349	50.355	50.362	50.368	50.374	50.381	50.387	50.393	50.400
.5	50.406	50.413	50.419	50.425	50.432	50.438	50.444	50.451	50.457	50.464
.6	50.470	50.476	50.483	50.489	50.495	50.502	50.508	50.515	50.521	50.527
.7	50.534	50.540	50.546	50.553	50.559	50.565	50.572	50.578	50.585	50.591
.8	50.597	50.604	50.610	50.616	50.623	50.629	50.635	50.642	50.648	50.654
.9	50.661	50.667	50.673	50.680	50.686	50.692	50.699	50.705	50.712	50.718

TABLE XXX—Continued.

VELOCITIES, IN FEET PER SECOND, DUE TO HEADS FROM 40 TO 44.99 FEET.

Head.	0	1	2	3	4	5	6	7	8	9
40.0	50.724	50.731	50.737	50.743	50.750	50.756	50.762	50.769	50.775	50.781
.1	50.788	50.794	50.800	50.807	50.813	50.819	50.826	50.832	50.838	50.845
.2	50.851	50.857	50.863	50.870	50.876	50.882	50.889	50.895	50.901	50.908
.3	50.914	50.920	50.927	50.933	50.939	50.946	50.952	50.958	50.965	50.971
.4	50.977	50.983	50.990	50.996	51.002	51.009	51.015	51.021	51.028	51.034
.5	51.040	51.047	51.053	51.059	51.065	51.072	51.078	51.084	51.091	51.097
.6	51.103	51.110	51.116	51.122	51.128	51.135	51.141	51.147	51.154	51.160
.7	51.166	51.172	51.179	51.185	51.191	51.198	51.204	51.210	51.216	51.223
.8	51.229	51.235	51.241	51.248	51.254	51.260	51.267	51.273	51.279	51.285
.9	51.292	51.298	51.304	51.310	51.317	51.323	51.329	51.336	51.342	51.318
41.0	51.354	51.361	51.367	51.373	51.379	51.386	51.392	51.398	51.404	51.411
.1	51.417	51.423	51.429	51.436	51.442	51.448	51.454	51.461	51.467	51.473
.2	51.479	51.486	51.492	51.498	51.504	51.511	51.517	51.523	51.529	51.536
.3	51.542	51.548	51.554	51.561	51.567	51.573	51.579	51.586	51.592	51.598
.4	51.604	51.610	51.617	51.623	51.629	51.635	51.642	51.648	51.654	51.660
.5	51.667	51.673	51.679	51.685	51.691	51.698	51.704	51.710	51.716	51.723
.6	51.729	51.735	51.741	51.747	51.754	51.760	51.766	51.772	51.778	51.785
.7	51.791	51.797	51.803	51.809	51.816	51.822	51.828	51.834	51.841	51.847
.8	51.853	51.859	51.865	51.872	51.878	51.884	51.890	51.896	51.903	51.903
.9	51.915	51.921	51.927	51.934	51.940	51.946	51.952	51.958	51.965	51.971
42.0	51.977	51.983	51.989	51.995	52.002	52.008	52.014	52.020	52.026	52.032
.1	52.039	52.045	52.051	52.057	52.063	52.070	52.076	52.082	52.088	52.094
.2	52.100	52.107	52.113	52.119	52.125	52.131	52.137	52.144	52.150	52.156
.3	52.162	52.168	52.174	52.181	52.187	52.193	52.199	52.205	52.211	52.218
.4	52.224	52.230	52.236	52.242	52.248	52.255	52.261	52.267	52.273	52.279
.5	52.285	52.291	52.298	52.304	52.310	52.316	52.322	52.328	52.334	52.341
.6	52.347	52.353	52.359	52.365	52.371	52.377	52.384	52.390	52.396	52.402
.7	52.408	52.414	52.420	52.427	52.433	52.439	52.445	52.451	52.457	52.463
.8	52.470	52.476	52.482	52.488	52.494	52.500	52.506	52.512	52.519	52.525
.9	52.531	52.537	52.543	52.549	52.555	52.561	52.567	52.574	52.580	52.586
43.0	52.592	52.598	52.604	52.610	52.616	52.623	52.629	52.635	52.641	52.647
.1	52.653	52.659	52.665	52.671	52.678	52.684	52.690	52.696	52.702	52.708
.2	52.714	52.720	52.726	52.732	52.738	52.745	52.751	52.757	52.763	52.769
.3	52.775	52.781	52.787	52.793	52.799	52.806	52.812	52.818	52.824	52.830
.4	52.836	52.842	52.848	52.854	52.869	52.866	52.873	52.879	52.885	52.891
.5	52.897	52.903	52.909	52.915	52.921	52.927	52.933	52.939	52.945	52.952
.6	52.958	52.964	52.970	52.976	52.982	52.988	52.994	53.000	53.006	53.012
.7	53.018	53.024	53.030	53.037	53.043	53.049	53.055	53.061	53.067	53.073
.8	53.079	53.085	53.091	53.097	53.103	53.109	53.115	53.121	53.127	53.133
.9	53.139	53.146	53.152	53.158	53.164	53.170	53.176	53.182	53.188	53.194
44.0	53.200	53.206	53.212	53.218	53.224	53.230	53.236	53.242	53.248	53.254
.1	53.260	53.266	53.272	53.279	53.285	53.291	53.297	53.303	53.309	53.315
.2	53.321	53.327	53.333	53.339	53.345	53.351	53.357	53.363	53.369	53.375
.3	53.381	53.387	53.393	53.399	53.405	53.411	53.417	53.423	53.429	53.435
.4	53.441	53.447	53.453	53.459	53.465	53.471	53.477	53.483	53.489	53.495
.5	53.501	53.507	53.513	53.519	53.525	53.531	53.537	53.543	53.549	53.555
.6	53.561	53.567	53.573	53.579	53.586	53.592	53.598	53.604	53.610	53.616
.7	53.621	53.627	53.633	53.639	53.645	53.651	53.657	53.663	53.669	53.675
.8	53.681	53.687	53.693	53.699	53.705	53.711	53.717	53.723	53.729	53.735
.9	53.741	53.747	53.753	53.759	53.765	53.771	53.777	53.783	53.789	53.795

TABLE XXX — Continued.

VELOCITIES, IN FEET PER SECOND, DUE TO HEADS FROM 45 TO 49.99 FEET.

Head.	0	1	2	3	4	5	6	7	8	9
45,0	53.801	53.807	53.813	53.819	53.825	53.831	53.837	53.843	53.849	53.855
.1	53.861	53.867	53.873	53.879	53.885	53.891	53.897	53.903	53.909	53.915
.2	53.921	53.927	53.932	53.938	53.944	53.950	53.956	53.962	53.968	53.974
.3	53.980	53.986	53.992	53.998	54.004	54.010	54.016	54.022	54.028	54.034
.4	54.040	54.046	54.052	54.058	54.064	54.069	54.075	54.081	54.087	54.093
.5	54.039	54.105	54.111	54.117	54.123	54.129	54.135	54.141	54.147	54.153
.6	54.159	54.165	54.170	54.176	54.182	54.188	54.194	54.200	54.206	54.212
.7	54.218	54.224	54.230	54.236	54.242	54.248	54.254	54.259	54.265	54.271
.8	54.277	54.283	54.289	54.295	54.301	54.307	54.313	54.319	54.325	54.331
.9	54.336	54.342	54.348	54.354	54.360	54.366	54.372	54.378	54.384	54.390
46.0	54.396	54.402	54.407	54.413	54.419	54.425	54.431	54.437	54.443	54.449
.1	54.455	54.461	54.467	54.472	54.478	54.484	54.490	54.496	54.502	54.508
.2	54.514	54.520	54.526	54.531	54.537	54.543	54.549	54.555	54.561	54.567
.3	54.573	54.579	54.585	54.590	54.596	54.602	54.608	54.614	54.620	54.626
.4	54.632	54.638	54.643	54.649	54.655	54.661	54.667	54.673	54.679	54.685
.5	54.690	54.696	54.702	54.708	54.714	54.720	54.726	54.732	54.737	54.743
.6	54.749	54.755	54.761	54.767	54.773	54.779	54.784	54.790	54.796	54.802
.7	54.808	54.814	54.820	54.826	54.831	54.837	54.843	54.849	54.855	54.861
.8	54.867	54.872	54.878	54.884	54.890	54.896	54.902	54.908	54.913	54.919
.9	54.925	54.931	54.937	54.943	54.949	54.954	54.960	54.966	54.972	54.978
47.0	54.984	54.990	54.995	55.001	55.007	55.013	55.019	55.025	55.030	55.036
.1	55.042	55.048	55.054	55.060	55.066	55.071	55.077	55.083	55.089	55.095
.2	55.401	55.106	55.112	55.118	55.124	55.130	55.436	55.141	55.147	55.153
.3	55.159	55.165	55.171	55.176	55.182	55.188	55.194	55.200	55.206	55.211
.4	55.217	55.223	55.229	55.235	55.240	55.246	55.252	55.258	55.264	55.270
.5	55.275	55.281	55.287	55.293	55.299	55.304	55.310	55.316	55.322	55.328
.6	55.334	55.339	55.345	55.351	55.357	55.363	55.368	55.374	55.380	55.386
.7	55.392	55.397	55.403	55.409	55.415	55.421	55.426	55.432	55.438	55.444
.8	55.450	55.455	55.461	55.467	55.473	55.479	55.484	55.490	55.496	55.502
.9	55.508	55.513	55.519	55.525	55.531	55.537	55.542	55.548	55.554	55.560
48.0	55.566	55.571	55.577	55.583	55.589	55.595	55.600	55.606	55.612	55.618
.1	55.623	55.629	55.635	55.641	55.647	55.652	55.658	55.664	55.670	55.675
.2	55.681	55.687	55.693	55.699	55.704	55.710	55.716	55.722	55.727	55.733
.3	55.739	55.745	55.750	55.756	55.762	55.768	55.774	55.779	55.785	55.791
.4	55.797	55.802	55.808	55.814	55.820	55.825	55.831	55.837	55.843	55.818
.5	55.854	55.860	55.866	55.872	55.877	55.883	55.889	55.895	55.900	55.906
.6	55.912	55.918	55.923	55.929	55.935	55.941	55.946	55.952	55.958	55.964
.7	55.969	55.975	55.981	55.987	55.992	55.998	56.004	56.009	56.015	56.021
.8	56.027	56.032	56.038	56.044	56.050	56.055	56.061	56.067	56.073	56.078
.9	56.084	56.090	56.096	56.101	56.107	56.113	56.118	56.124	56.130	56.136
49.0	56.141	56.147	56.153	56.159	56.164	56.170	56.176	56.181	56.187	56.193
.1	56.199	56.204	56.210	56.216	56.222	56.227	56.233	56.239	56.244	56.250
.2	56.256	56.262	56.267	56.273	56.279	56.284	56.290	56.296	56.302	56.307
.3	56.313	56.319	56.324	56.330	56.336	56.342	56.347	56.353	56.359	56.364
.4	56.370	56.376	56.381	56.387	56.393	56.399	56.404	56.410	56.416	58.421
.5	56.427	56.433	56.439	56.444	56.450	56.456	56.461	56.467	56.473	56.478
.6	56.484	56.490	56.495	56.501	56.507	56.518	56.518	56.524	56.530	56.535
.7	56.541	56.547	56.552	56.558	56.564	56.569	56.575	56.581	56.586	56.592
.8	56.598	56.604	56.609	56.615	56.621	56.626	56.632	56.638	56.643	56.649
.9	56.655	56.660	56.666	56.672	56.677	56.683	56.689	56.694	56.700	56.706

LIBRARY
OF THE
UNIVERSITY
OF
CALIFORNIA

LIBRARY
OF THE
UNIVERSITY
OF
CALIFORNIA

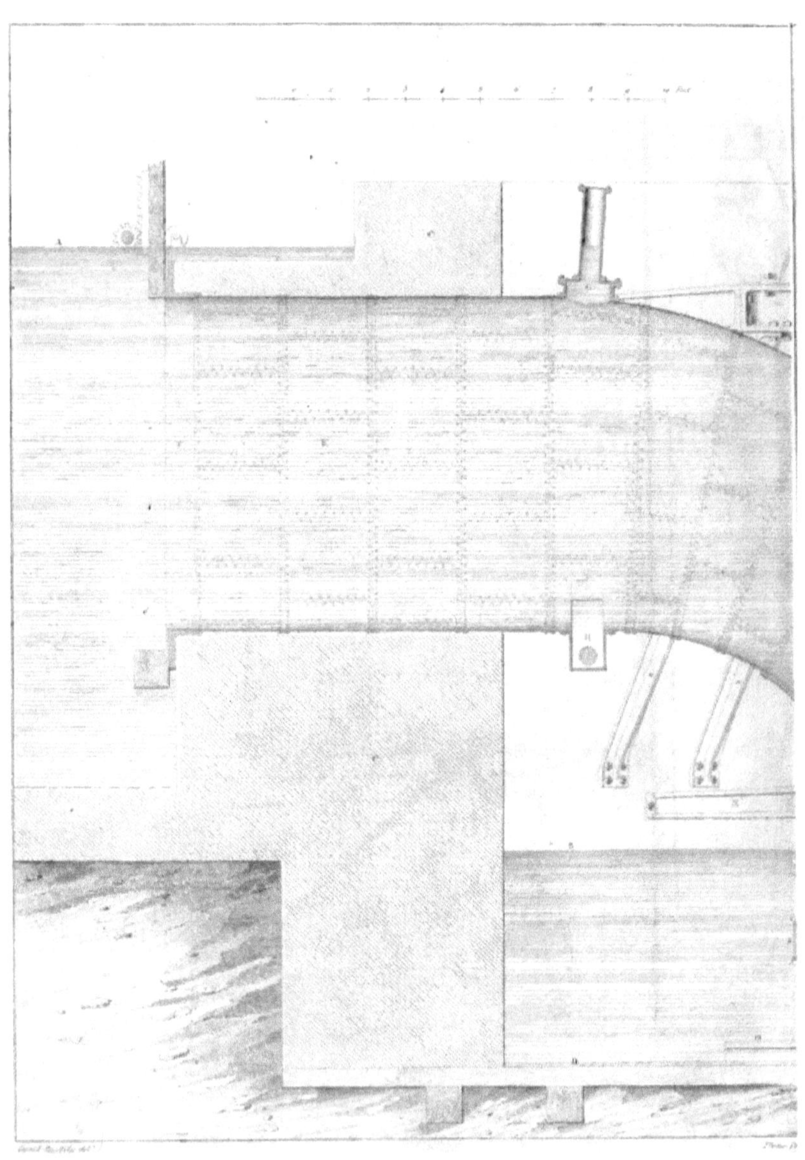

LIBRARY
OF THE
UNIVERSITY
OF
CALIFORNIA

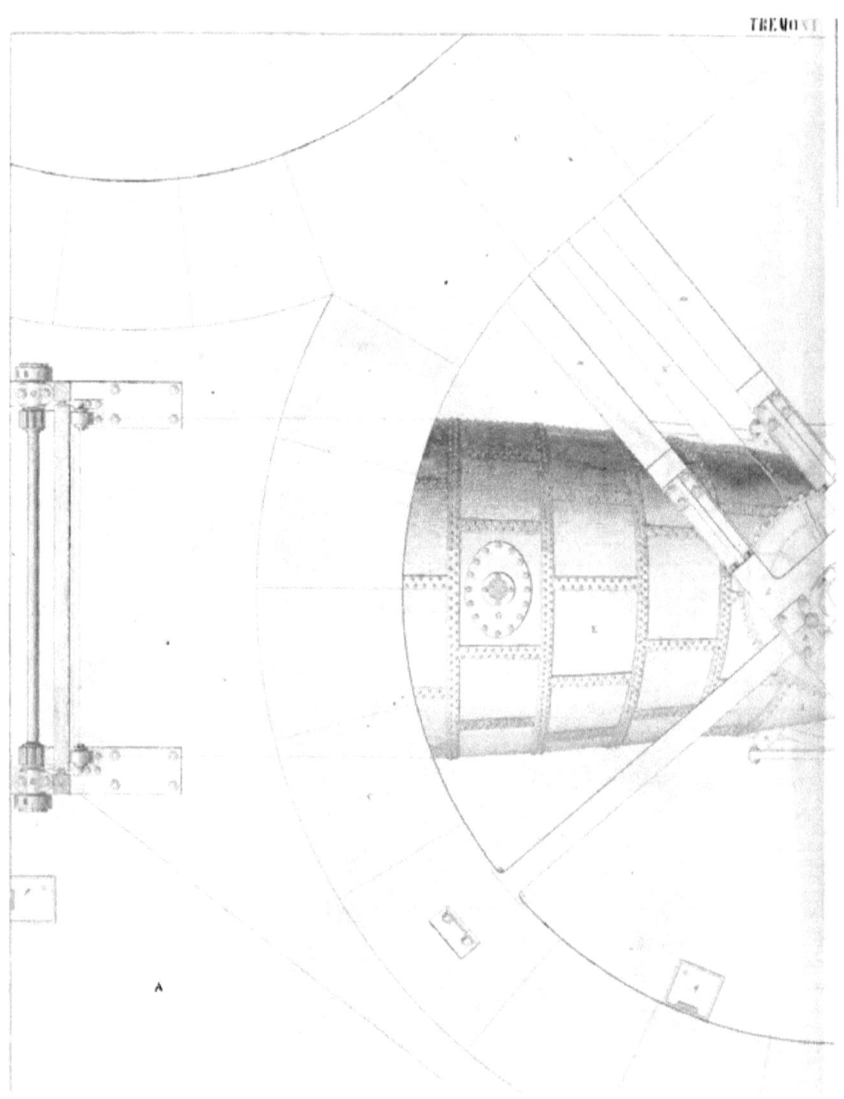
TREMONT
A

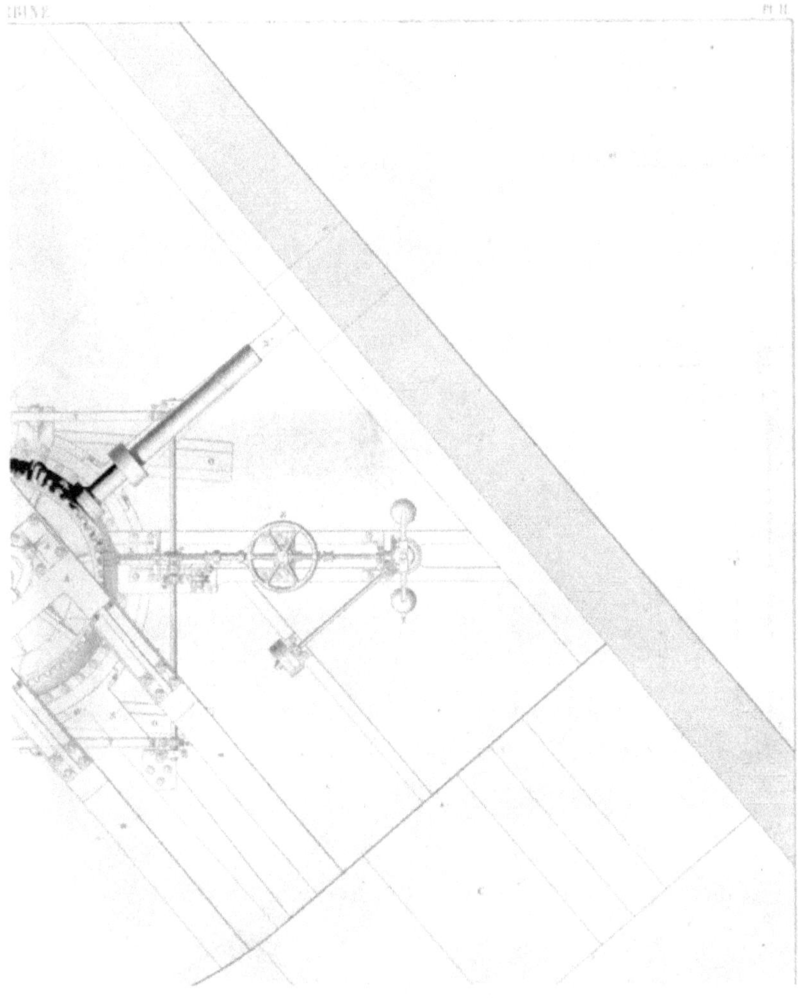

LIBRARY
OF THE
UNIVERSITY
OF
CALIFORNIA

Scale for Fig. 1 & 3
Scale for Fig. 2
Fig. 2
Fig. 3

LIBRARY
OF THE
UNIVERSITY
OF
CALIFORNIA

APPARATUS USED IN THE EXPE

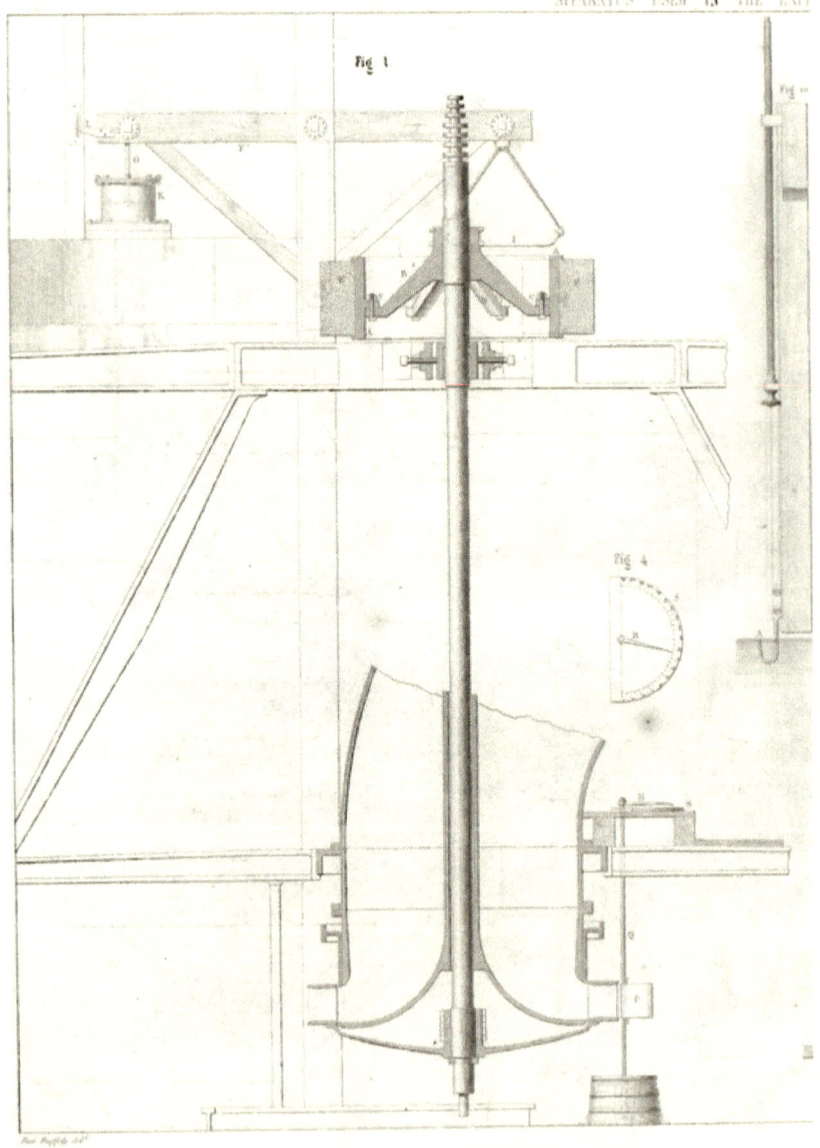

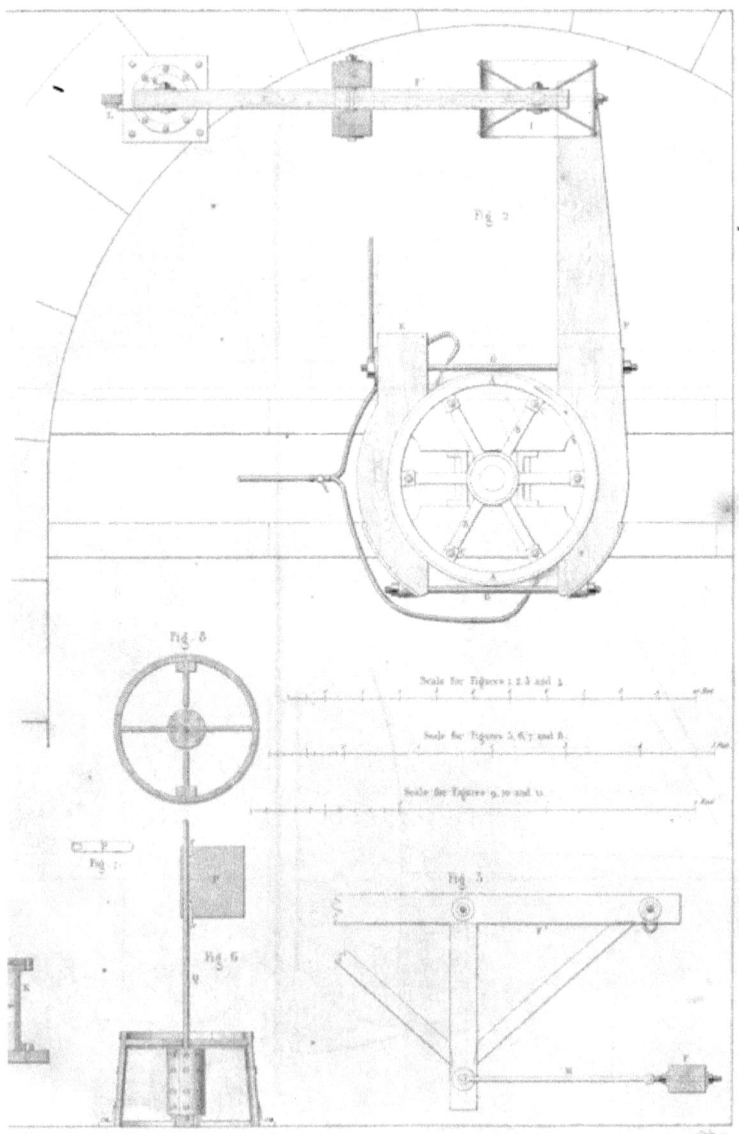
Fig. 2
Fig. 8
Scale for Figures 1, 2, 3 and 4
Scale for Figures 5, 6, 7 and 8.
Scale for Figures 9, 10 and 11
Fig. 7
Fig. 6
Fig. 3

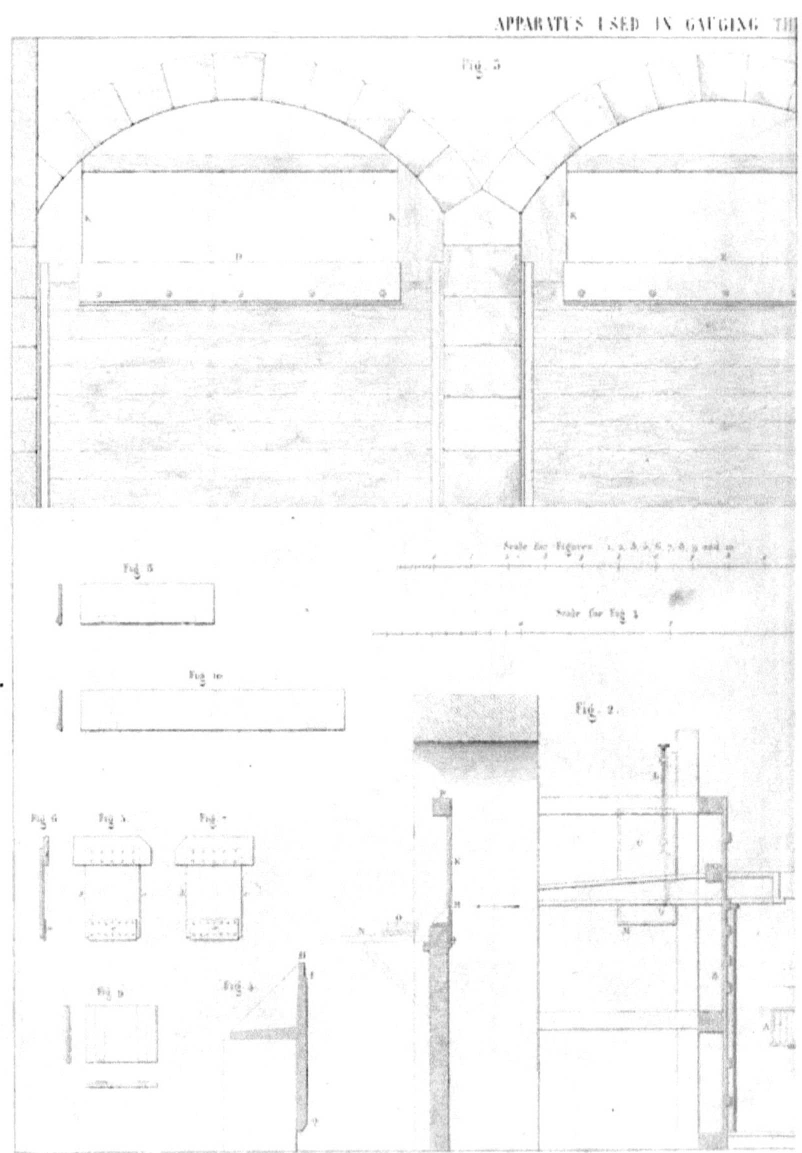
Fig. 3
Scale for Figures 1, 2, 3, 5, 6, 7, 8, 9 and 10
Scale for Fig. 4
Fig. 8
Fig. 10
Fig. 2.
Fig. 6
Fig. 5.
Fig. 7
Fig. 9
Fig. 4.

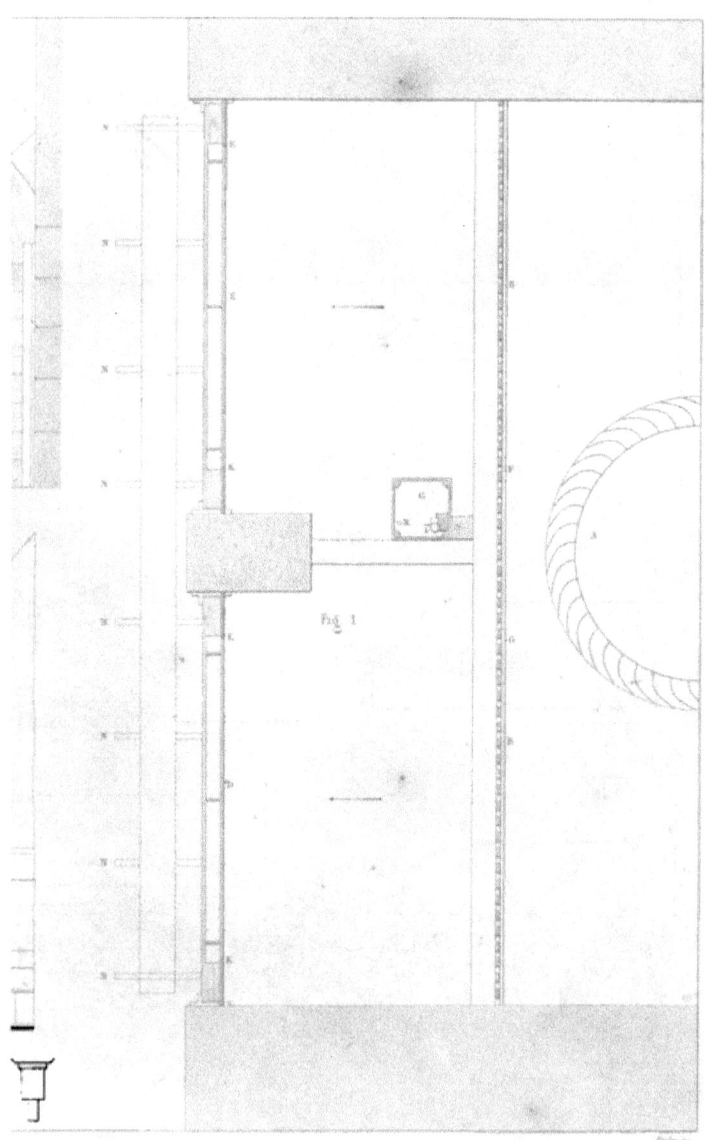

LIBRARY
OF THE
UNIVERSITY
OF
CALIFORNIA

EXPERIMENTS ON

Fig. 3
Fig. 2

UNIVERSITY
OF

APPARATUS USED IN THE EXPERIMENTS OF

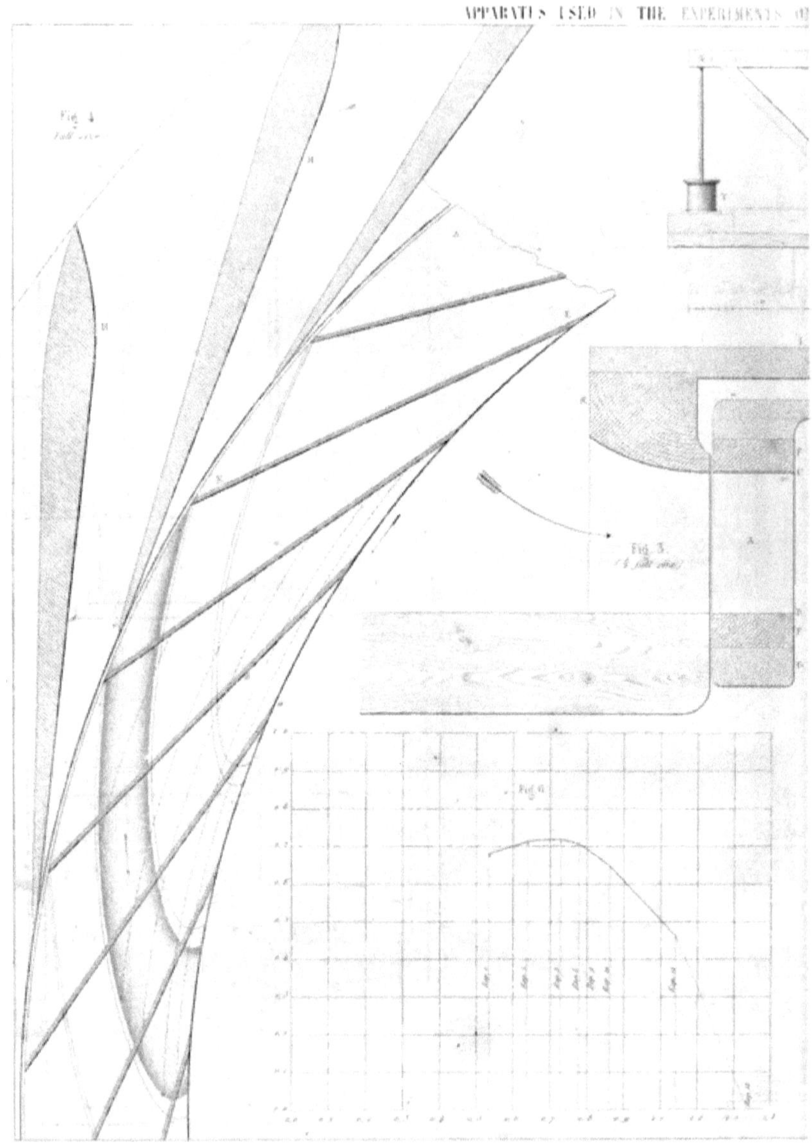

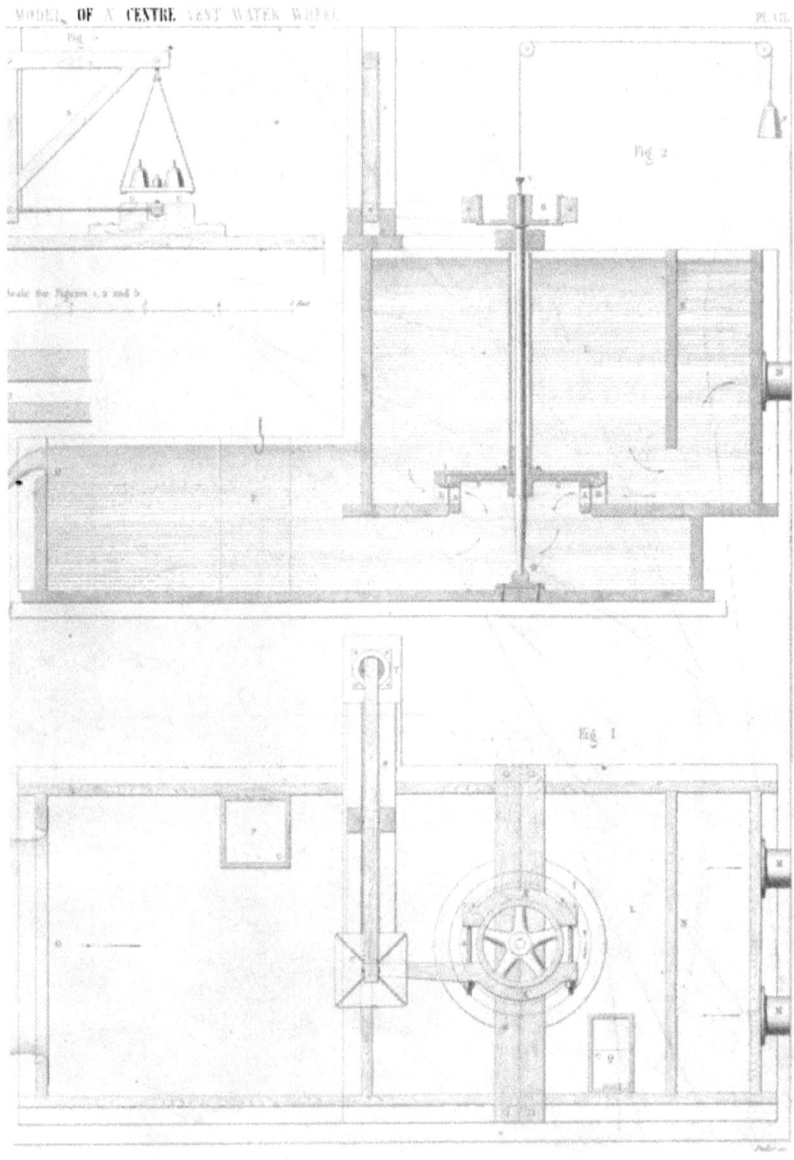
MODEL OF A CENTRE VENT WATER WHEEL
PL. II.
Fig. 2
Scale for Figures 1, 2 and 3
Fig. 1

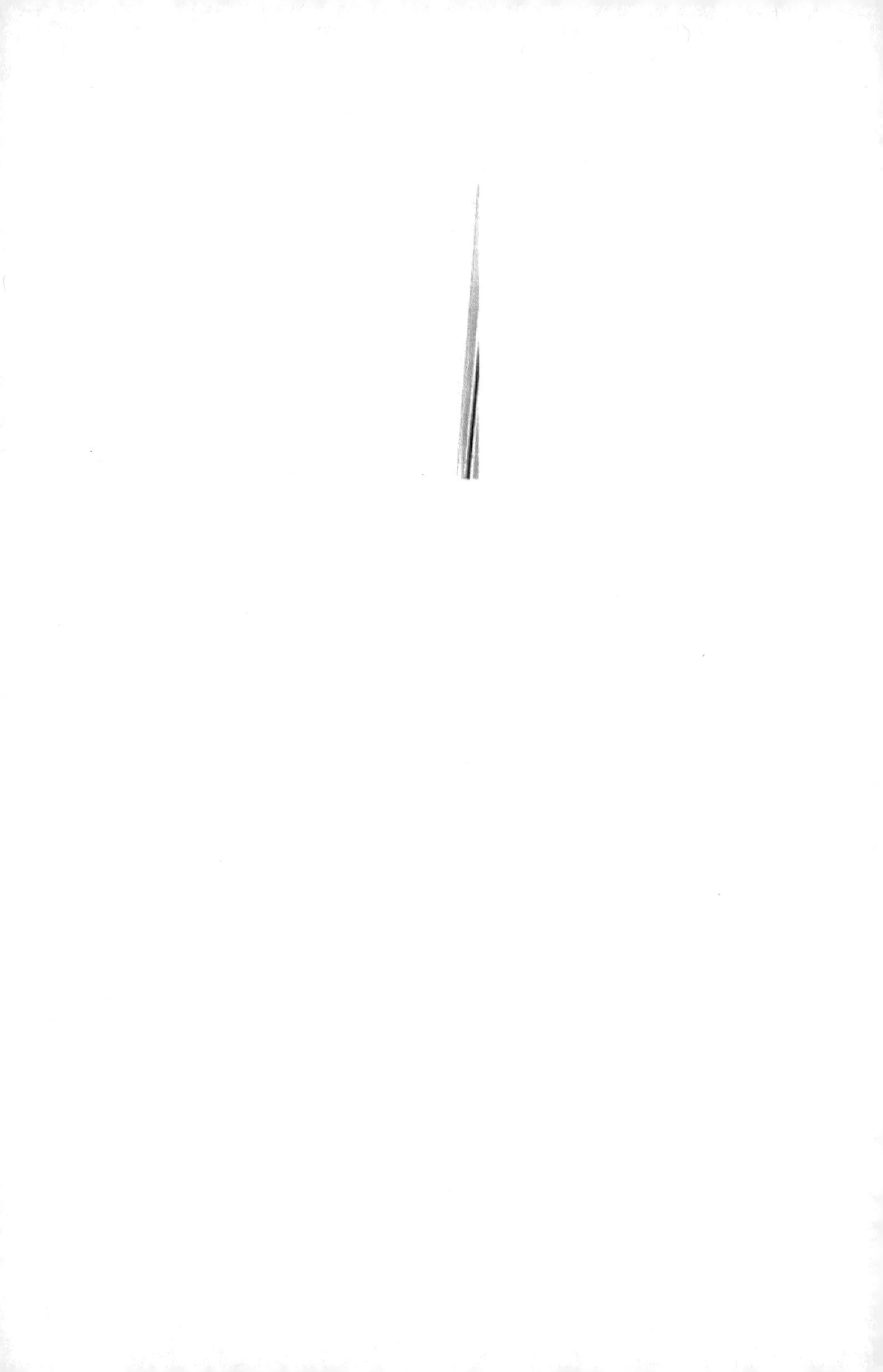

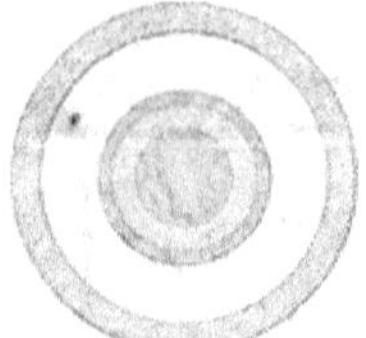

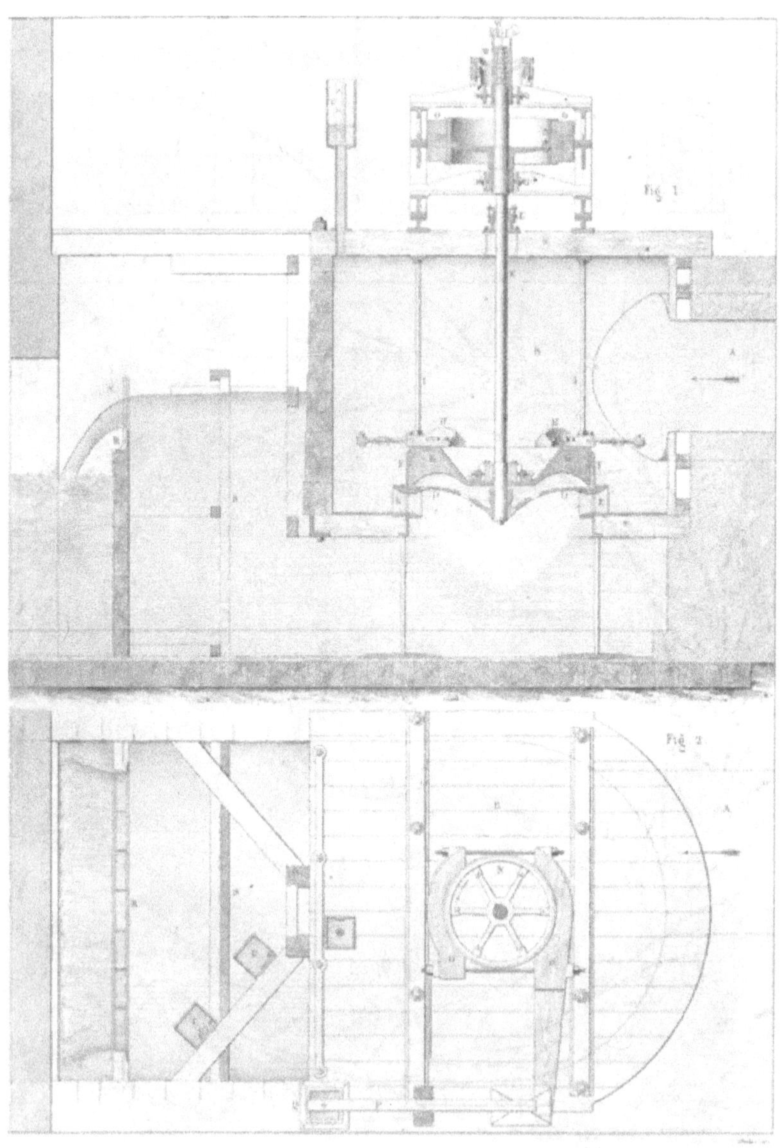
Fig. 1
Fig. 2

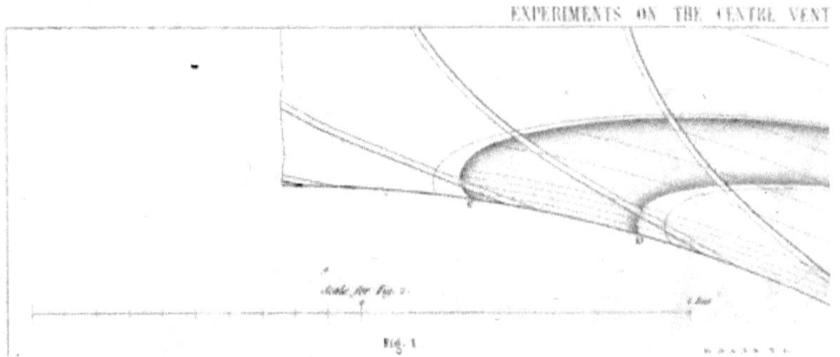

Fig. 1.

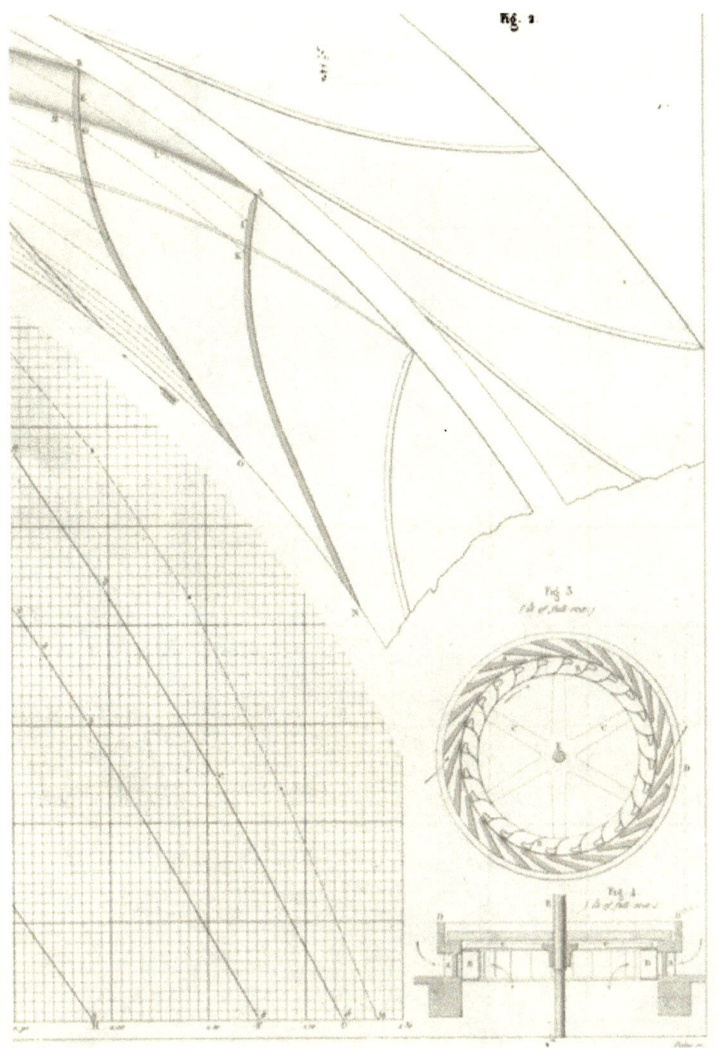
Fig. 2
Fig. 3
Fig. 4

LIBRARY
OF THE
UNIVERSITY
OF
CALIFORNIA

Fig. 5.
Fig. 6.
Fig. 9.
Fig. 10.
Fig. 13.
Fig. 14.

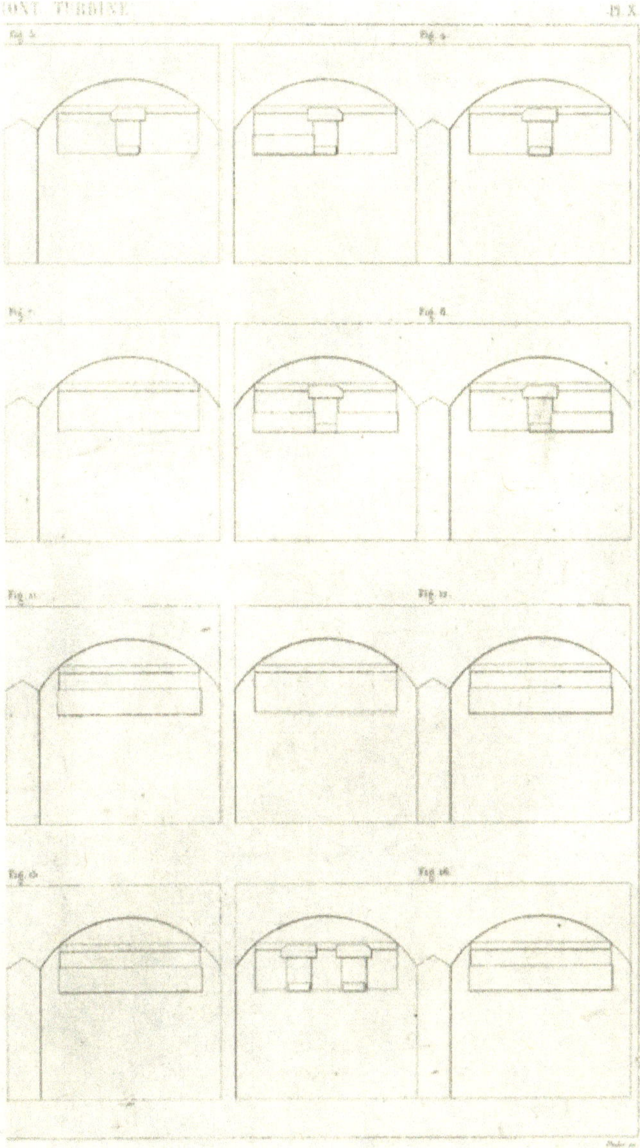

LIBRARY
OF THE
UNIVERSITY
OF
CALIFORNIA

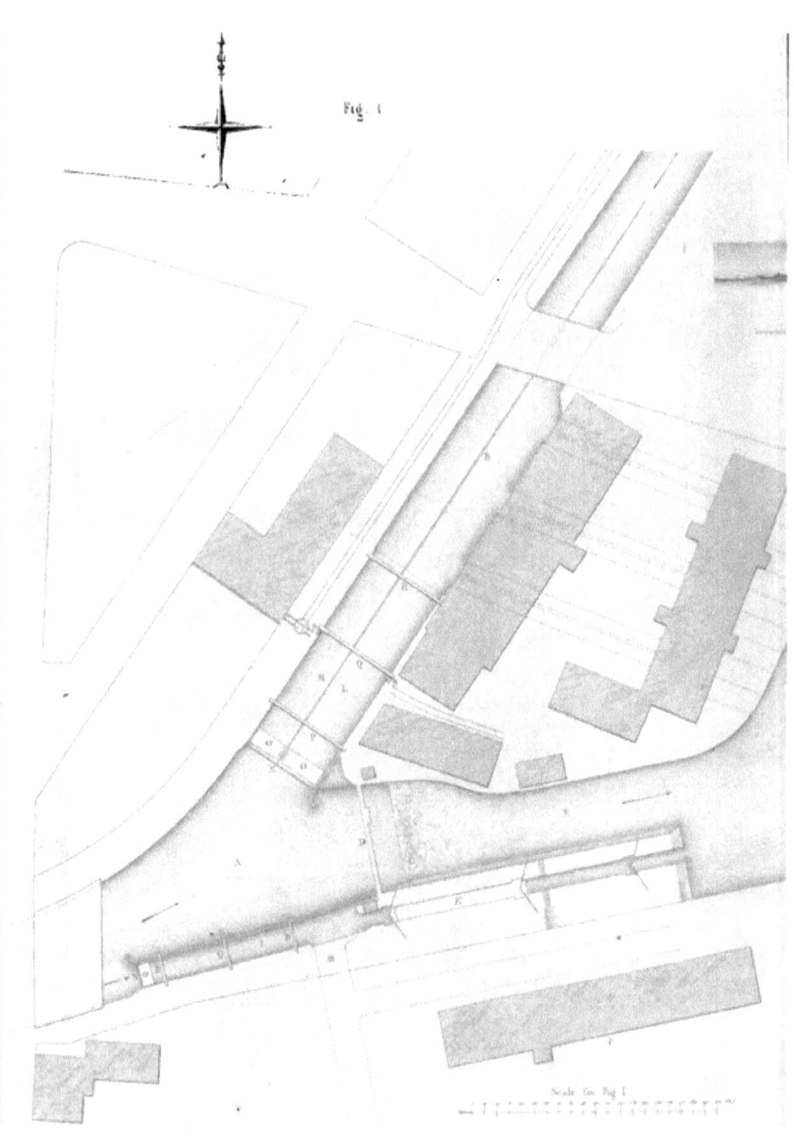
Fig. 1
Scale for Fig. 1

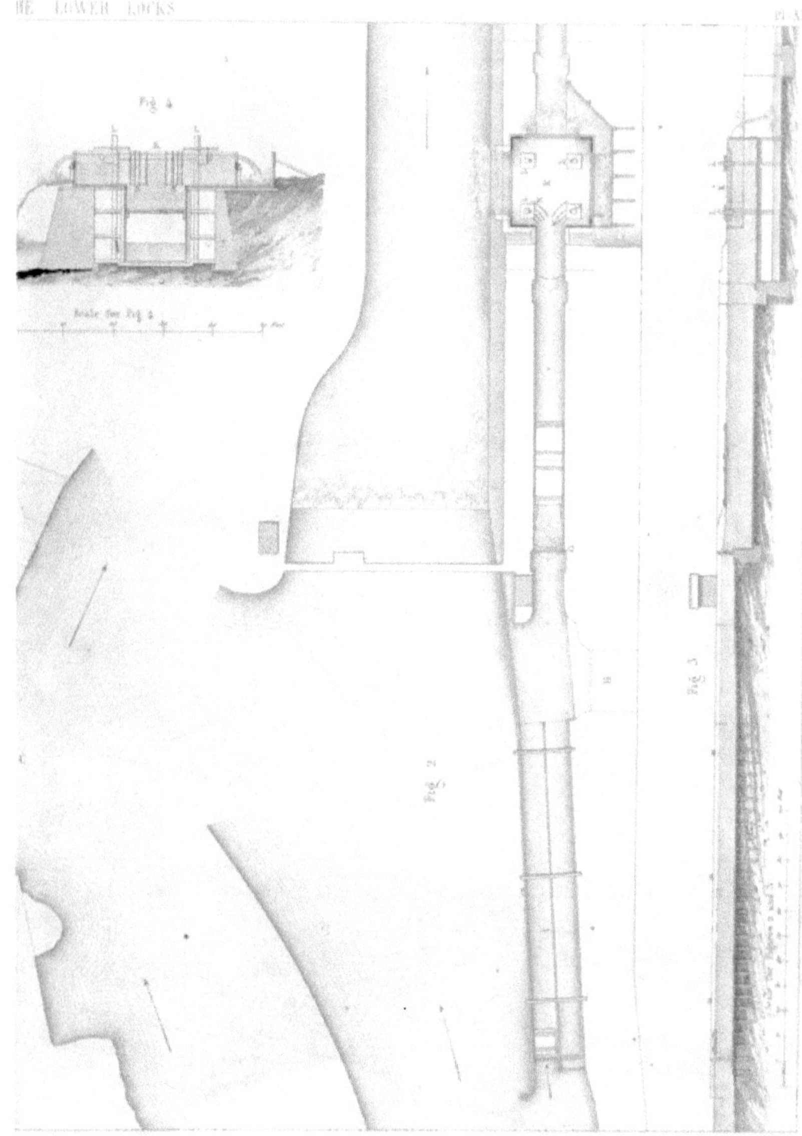
Fig. 4

LIBRARY
OF THE
UNIVERSITY
OF
CALIFORNIA

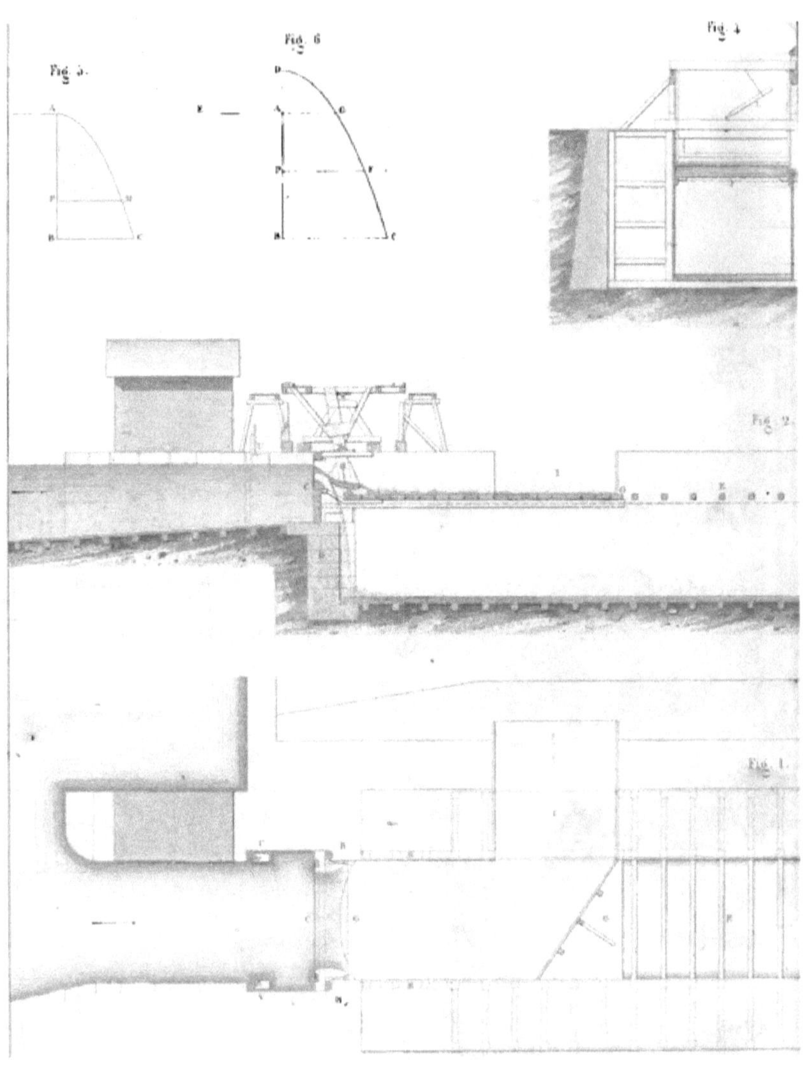

Fig. 5.
Fig. 6
Fig. 4
Fig. 2.
Fig. 1.

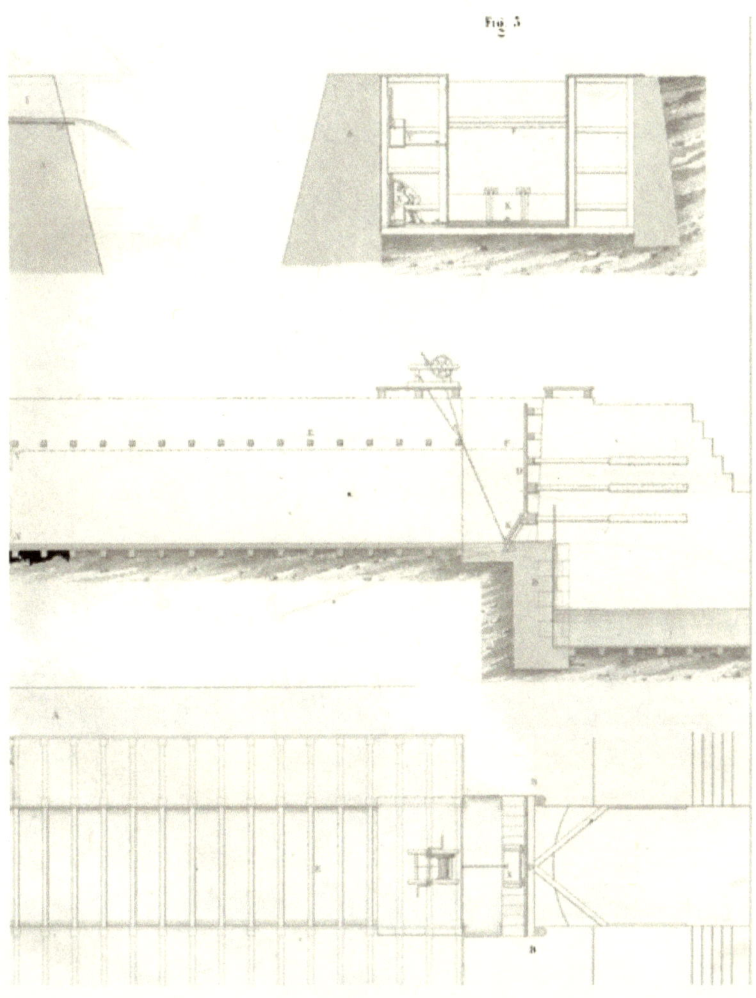

LIBRARY
OF THE
UNIVERSITY
OF
CALIFORNIA

LIBRARY
OF THE
UNIVERSITY
OF
CALIFORNIA

Fig. 1

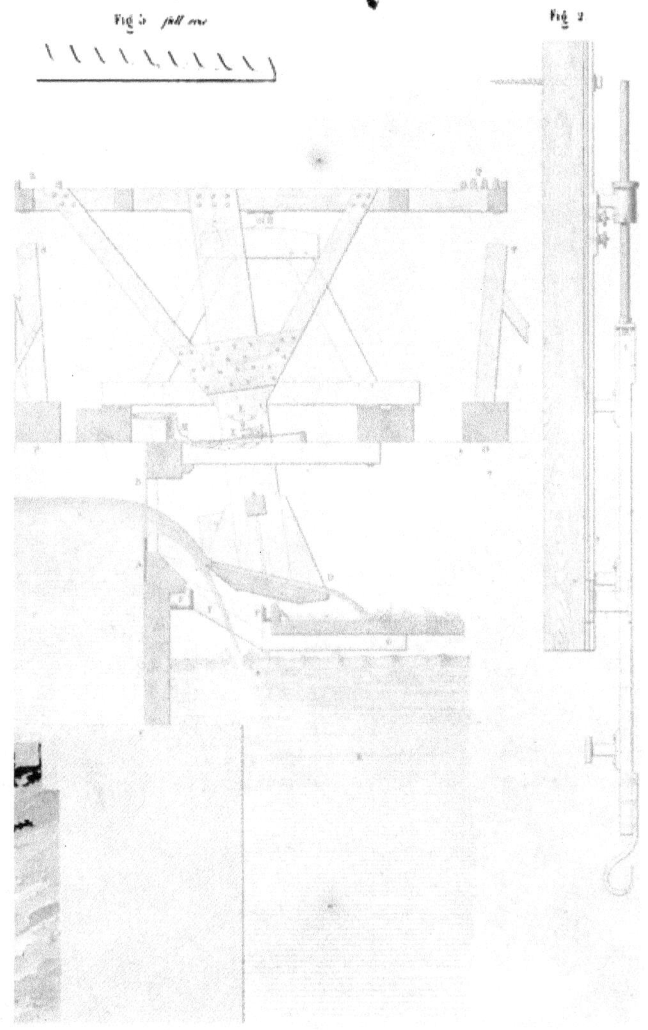

LIBRARY
OF THE
UNIVERSITY
OF
CALIFORNIA

LIBRARY
OF THE
UNIVERSITY
OF
CALIFORNIA

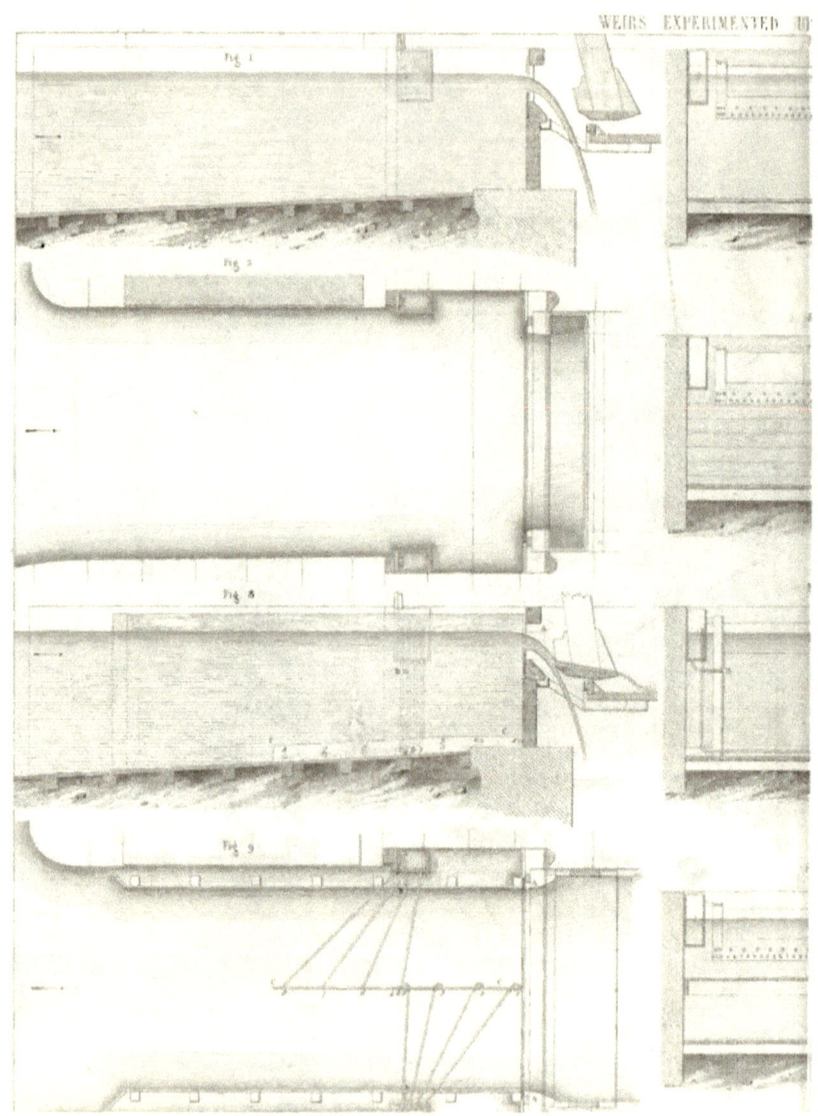
WEIRS EXPERIMENTED
Fig 1
Fig 2
Fig 8
Fig 9

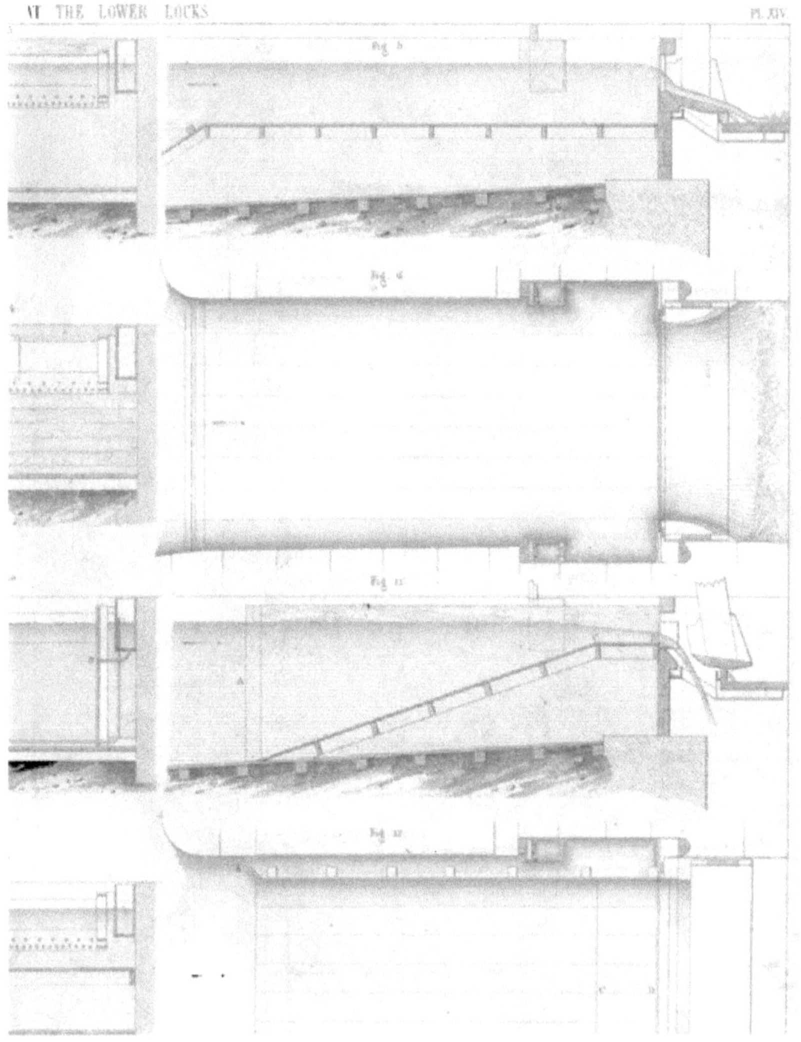

LIBRARY
OF THE
UNIVERSITY
OF
CALIFORNIA

Fig. 4.
Fig. 2
Fig. 5.
Fig. 3
Fig. 1
Scale for Figs. 2 and 4.

Fig. 8.

LIBRARY
OF THE
UNIVERSITY
OF
CALIFORNIA

EXPERIMENTS ON THE FLOW

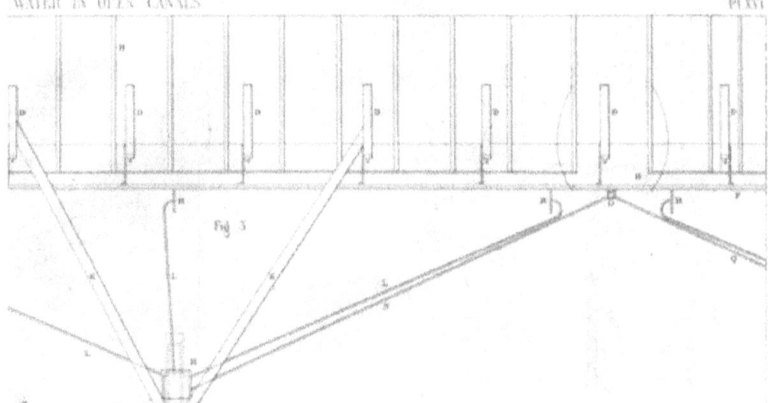
Fig 3

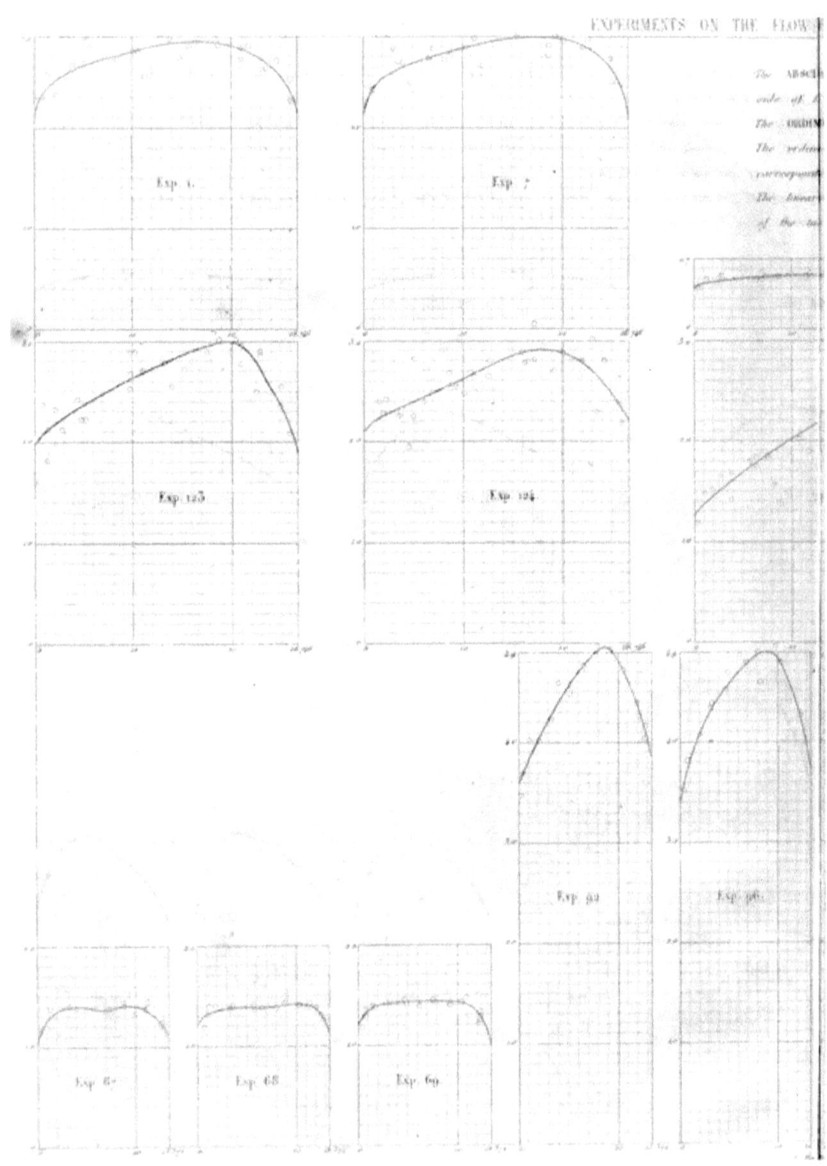
EXPERIMENTS ON THE FLOW
Exp. 1.
Exp. 7
Exp. 123
Exp. 124
Exp. 92
Exp. 96.
Exp. 68
Exp. 69

hed ○ represent the velocities of the several tubes in feet per second.
curved lines, represent the mean velocities of the tubes in feet per second in
of the flume.
ne on this plate, is one fourth of that of the originals from which the mean velocities

Exp. 107.

Exp. 108.

Exp. 139.

Exp. 140.

Exp. 109.

Exp. 110.

Exp. 111.

Exp. 112.

LIBRARY
OF THE
UNIVERSITY
OF
CALIFORNIA

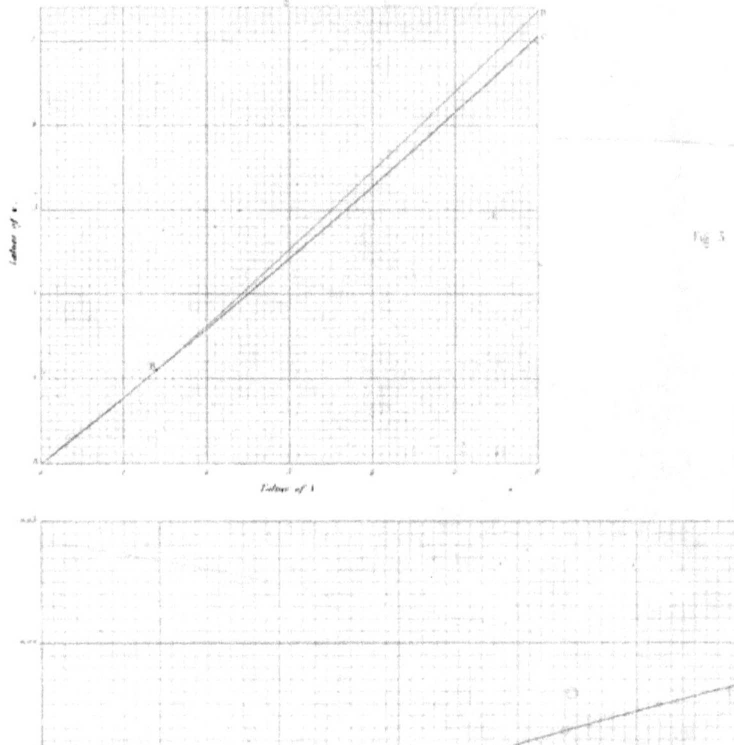

LIBRARY
OF THE
UNIVERSITY
OF
CALIFORNIA

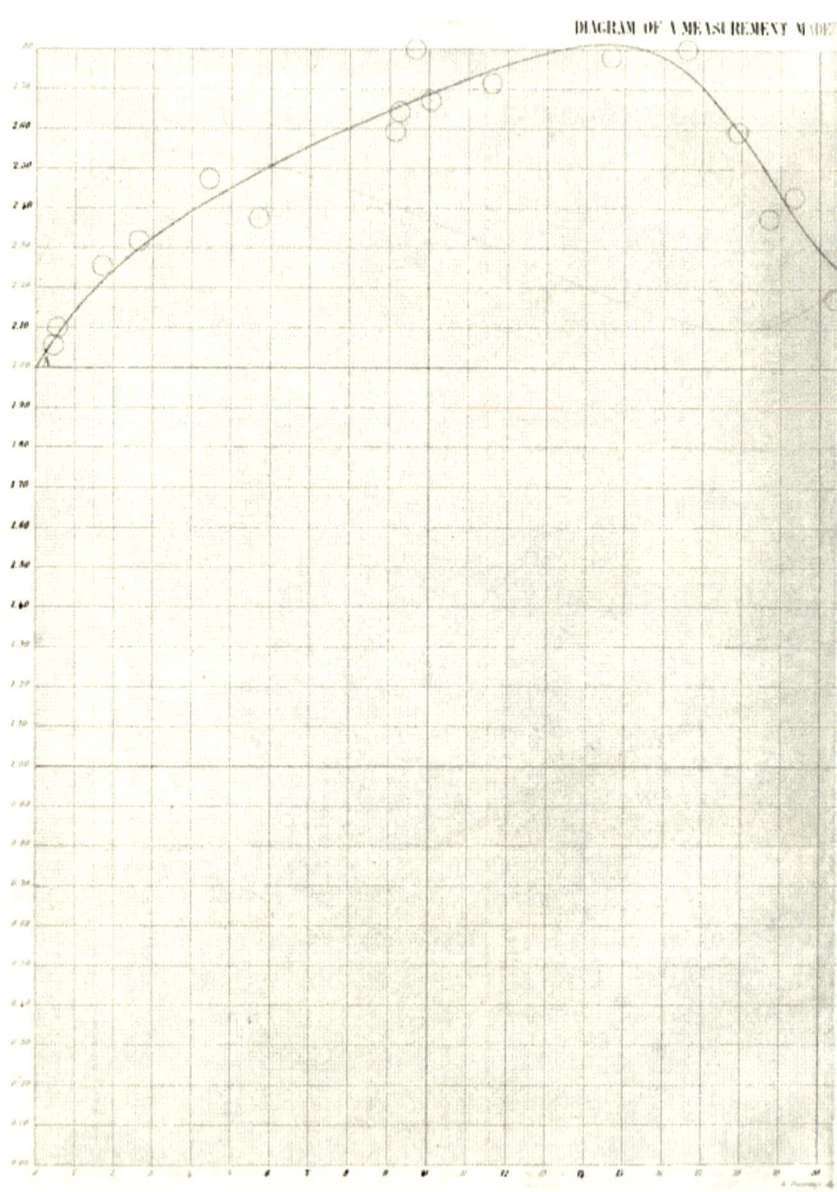
DIAGRAM OF A MEASUREMENT MADE

EXPERIMENTS ON THE FLOW OF WATER THRO

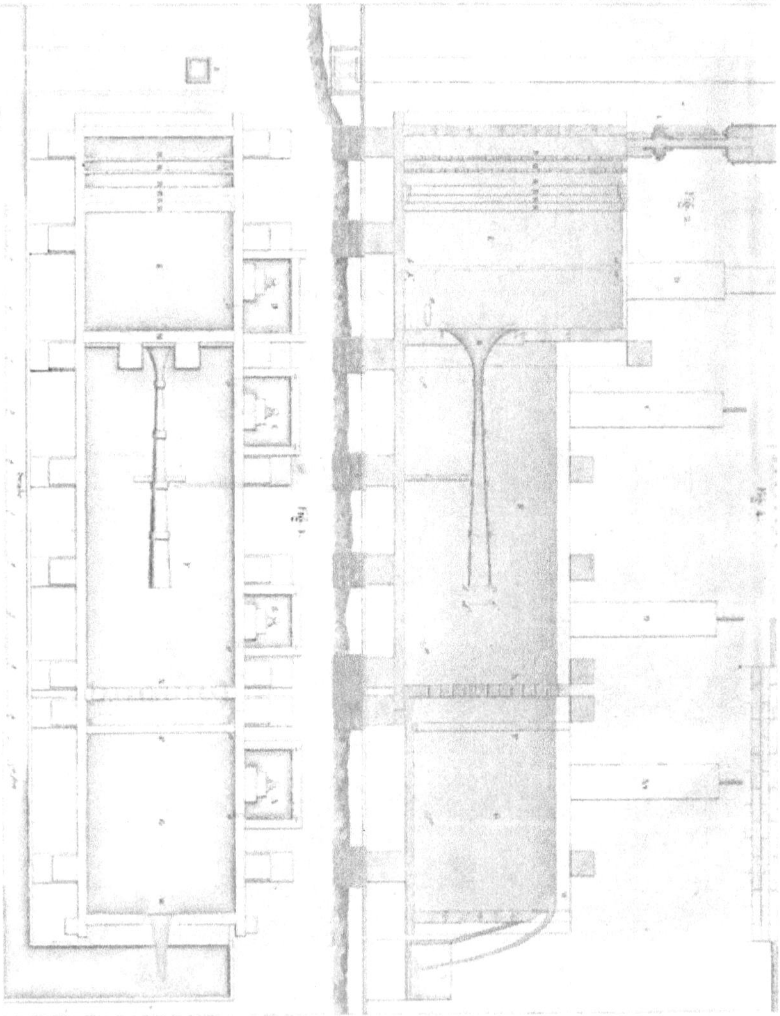

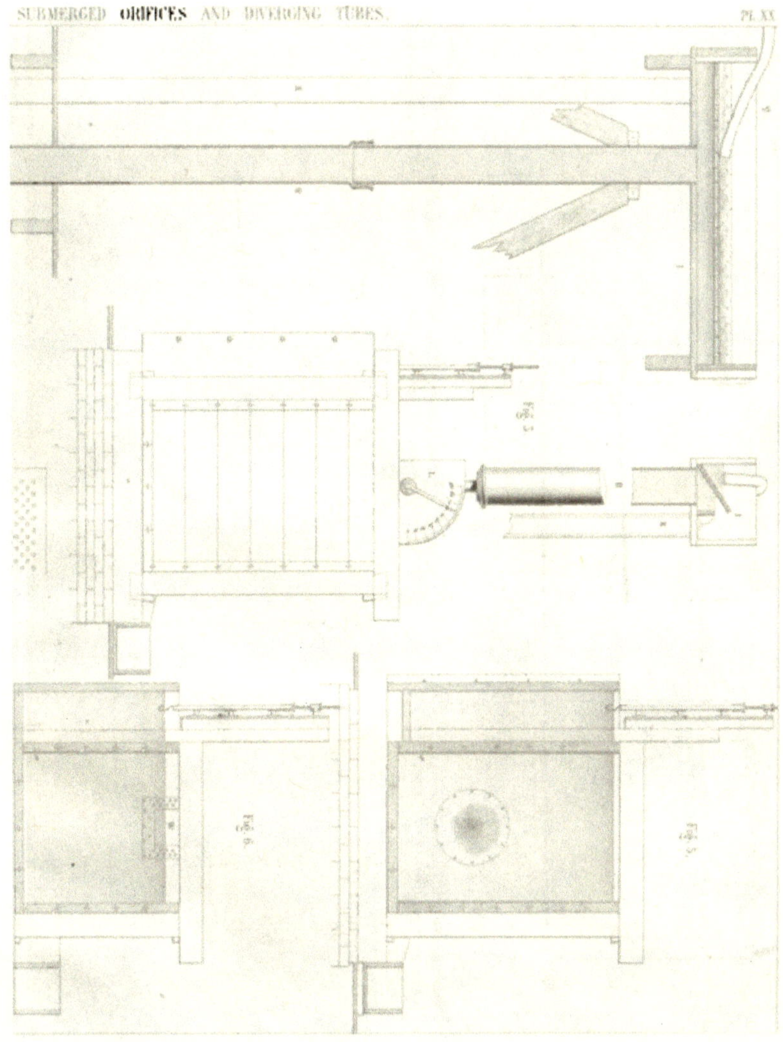
Fig. 3.
Fig. 6.
Fig. 5.

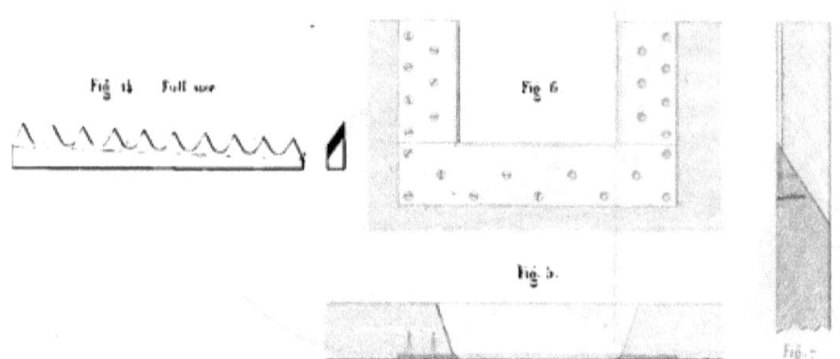
Fig 14 Full size
Fig 6
Fig. 5.
Fig. 7

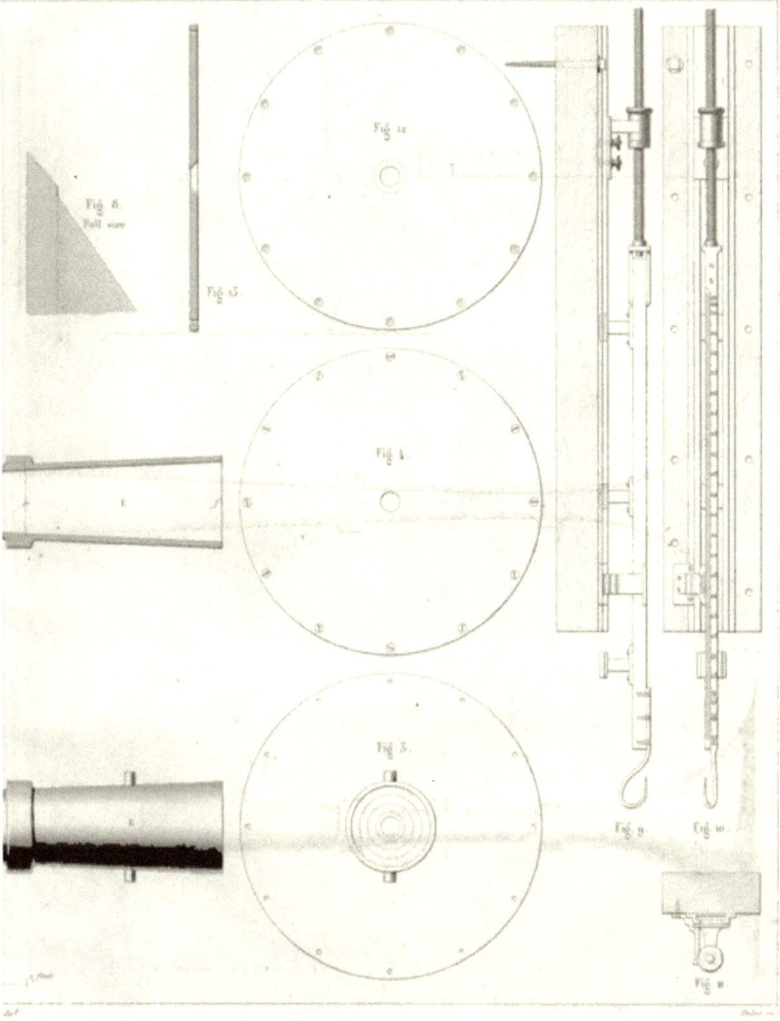

Fig. 12.
Fig. 8.
Full size
Fig. 13.
Fig. 4.
Fig. 5.
Fig. 9.
Fig. 10.
Fig. 11.

LIBRARY
OF THE
UNIVERSITY

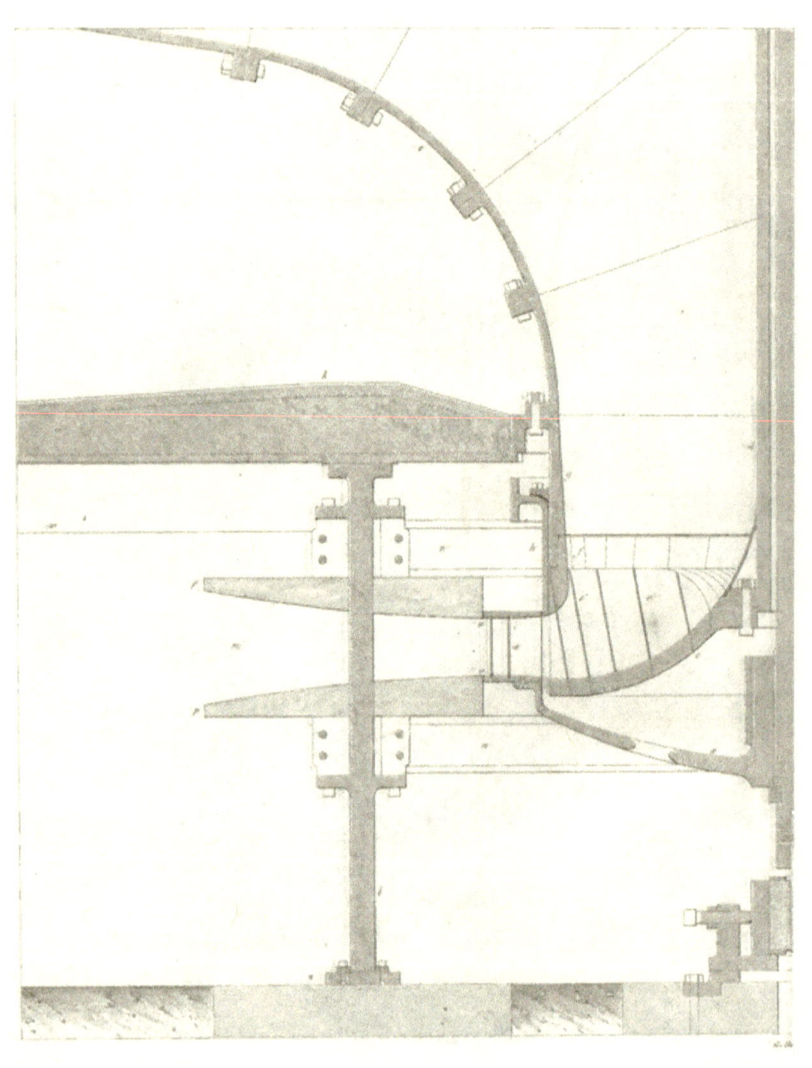

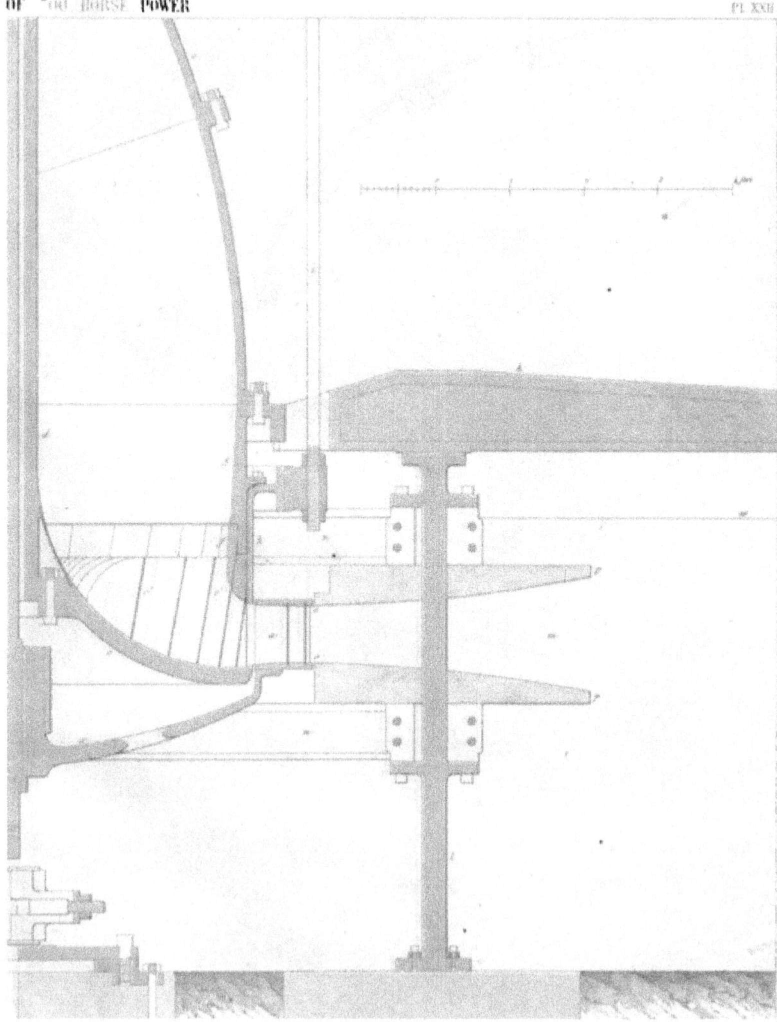

LIBRARY
OF THE
UNIVERSITY
OF
CALIFORNIA

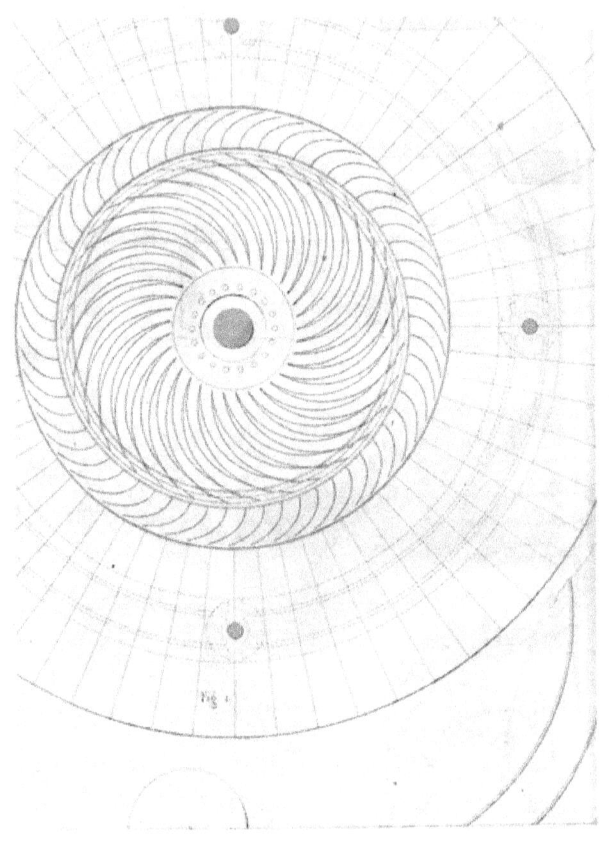
Fig. 1.

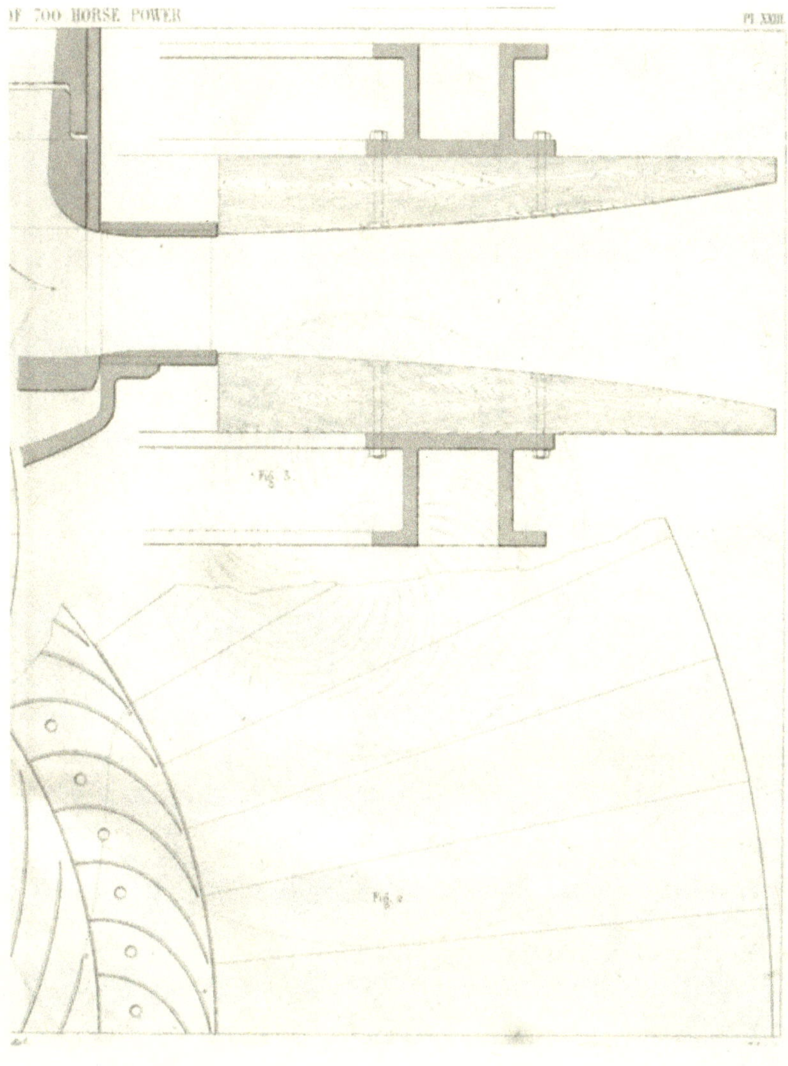
Fig. 3.
Fig. 2

www.ingramcontent.com/pod-product-compliance
Lightning Source LLC
Chambersburg PA
CBHW030259240426
43673CB00040B/1001

* 9 7 8 3 7 4 4 7 4 3 5 5 6 *